Strained-Layer Superlattices:
Physics

SEMICONDUCTORS
AND SEMIMETALS
Volume 32

Semiconductors and Semimetals

A Treatise

Edited by R. K. Willardson
ENIMONT AMERICA INC.
PHOENIX, ARIZONA

Albert C. Beer
BATTELLE COLUMBUS LABORATORIES
COLUMBUS, OHIO

Strained-Layer Superlattices: Physics

SEMICONDUCTORS
AND SEMIMETALS

Volume 32

Volume Editor

THOMAS P. PEARSALL

DEPARTMENT OF ELECTRICAL ENGINEERING
UNIVERSITY OF WASHINGTON
SEATTLE, WASHINGTON

ACADEMIC PRESS, INC.
Harcourt Brace Jovanovich, Publishers
Boston San Diego New York
London Sydney Tokyo Toronto

THIS BOOK IS PRINTED ON ACID-FREE PAPER. ∞

COPYRIGHT © 1990 BY AT&T BELL LABORATORIES, INC.
ALL RIGHTS RESERVED.
NO PART OF THIS PUBLICATION MAY BE REPRODUCED OR
TRANSMITTED IN ANY FORM OR BY ANY MEANS, ELECTRONIC
OR MECHANICAL, INCLUDING PHOTOCOPY, RECORDING, OR
ANY INFORMATION STORAGE AND RETRIEVAL SYSTEM, WITHOUT
PERMISSION IN WRITING FROM THE PUBLISHER.

ACADEMIC PRESS, INC.
1250 Sixth Avenue, San Diego, CA 92101

United Kingdom Edition published by
ACADEMIC PRESS LIMITED
24–28 Oval Road, London NW1 7DX

The Library of Congress has cataloged this serial publication as follows:

Semiconductors and semimetals.—Vol. 1-—New York: Academic Press, 1966-

v.: ill.; 24 cm.

Irregular.
Each vol. has also a distinctive title.
Edited by R. K. Willardson and Albert C. Beer.
ISSN 0080-8784 = Semiconductors and semimetals

1. Semiconductors—Collected works. 2. Semimetals—Collected works.
I. Willardson, Robert K.   II. Beer, Albert C.
QC610.9.S48   621.3815′2—dc19   85-642319

ISBN 0-12-752132-1

Printed in the United States of America
90 91 92 93   9 8 7 6 5 4 3 2 1

Contents

LIST OF CONTRIBUTORS	vii
PREFACE	ix

Chapter 1 Strained-Layer Superlattices
T. P. Pearsall

I. Introduction	1
II. Some Principles of Strained-Layer Epitaxy	3
III. Applications of Strained-Layer Epitaxy	7
IV. Strained-Layer Superlattices—Opportunities for Materials Science	10
References	14

Chapter 2 Effects of Homogeneous Strain on the Electronic and Vibrational Levels in Semiconductors
Fred H. Pollak

I. Introduction	17
II. Effects of Homogeneous Deformation on Electronic Energy Levels	18
III. Effects of External Stress on Quantum States	40
IV. Influence of Homogeneous Deformation on $q \approx 0$ Optical Phonons	43
V. Summary	48
Appendices	49
References	52

Chapter 3 Optical Studies of Strained III-V Heterolayers
J. Y. Marzin, J. M. Gérard, P. Voisin, and J. A. Brum

I. Introduction	56
II. Structural Aspects	57
III. Effects of Strain on the Band Structures	64
IV. Moderately Strained Systems	82
V. Large-Strain Systems	101
VI. Conclusion	114
Acknowledgment	115
References	115

Chapter 4 Structurally Induced States from Strain and Confinement
R. People and S. A. Jackson

I. Introduction	119
II. Optical and Electronic Properties of Lattice Mismatch-Induced Strained Layers	134
III. Zone-Folding Effects in Ultrashort-Period Strained-Layer Superlattices	164
IV. Other Effects in Strained-Layer Heterostructures	169
V. Summary	171
Acknowledgments	171
References	171

Chapter 5 Microscopic Phenomena in Ordered Superlattices
M. Jaros

I. Introduction	175
II. Phenomenology of the Breakdown of the Effective-Mass Approximation in Semiconductor Superlattices	179
III. Microscopic Theory of Ordered Superlattices	189
IV. Quantitative Assessment of Microscopic Phenomena	203
Acknowledgments	254
References	254

INDEX	259
CONTENTS OF PREVIOUS VOLUMES	267

List of Contributors

Numbers in parenthesis indicate the pages on which the authors' contributions begin.

J. A. BRUM, *IBM Thomas J. Watson Research Center, P.O. Box 218, Yorktown Heights, New York 10598* (55)

J. M. GÉRARD, *Centre National d'Etudes des Télécommunications, 196 avenue H. Ravera, F-92220 Bagneux, France* (55)

M. JAROS, *Physics Department, The University, Newcastle upon Tyne, United Kingdom* (175)

S. A. JACKSON, *AT&T Bell Laboratories, Murray Hill, New Jersey 07974* (119)

J. Y. MARZIN, *Centre National d'Etudes des Télécommunications, 196 avenue H. Ravera, F-92220 Bagneux, France* (55)

T. P. PEARSALL, *Department of Electrical Engineering, University of Washington, Seattle, Washington 98195* (1)

R. PEOPLE, *AT&T Bell Laboratories, Murray Hill, New Jersey 07974* (119)

F. H. POLLAK, *Department of Physics, Brooklyn College of the City University of New York, Brooklyn, New York 11210* (17)

P. VOISIN, *Departement de Physique Ecole Normale Supérieure, 24 rue Lhomond, F-75005 Paris, France* (55)

Preface

During the last decade, it has been acknowledged that the technology of silicon integrated circuits is approaching fundamental limits set by the atomic nature of matter. It is no longer possible to count on a doubling of chip capacity every few years. At the same time, the telecommunications and recording industries have driven the development of an economical and reliable optoelectronics technology. Optoelectronics offers *new functionality* to conventional silicon-based circuits. Development and integration of this new functionality is essential to the continued expansion of the information processing capacity of integrated circuits. Yet a vast gulf continues to separate the technologies for optoelectronics and silicon-based devices because they are based on dissimilar materials. These two volumes (Volumes 32 and 33) on strained-layer superlattices are dedicated to the idea that this gulf will be short-lived.

In 1982 Gordon Osbourn of Sandia Laboratories made the link between strained-layer structures and the need for new functionality in integrated circuit design.[1] Osbourn considered the conventional wisdom that all strain in semiconducting devices was bad and stood it on its head by proposing that the strain associated with the heteroepitaxy of dissimilar materials may itself offer *new functionality*, whose advantages may far outweigh the disadvantages of the presence of strain. In 1986, Temkin et al.[2] tested a device that illustrates the possibilities opened up by this breakthrough in thinking: a silicon-based photodiode with an absorption edge, strain-shifted to the 1.3–1.5 μm window for optical fiber communications.

The physics and technology of semiconductor strained-layer superlattices are surveyed in this two-volume set. Of course, the field of activity is wide and growing. The contents of this set should not be viewed as a review, but rather as a milestone in research and development that will play an important part in the evolution of semiconductor device technology.

Thomas P. Pearsall
March 28, 1990

[1] Osbourn, G. C. (1982), *J. Appl. Phys.* **53**, 1586.
[2] Temkin, H., Pearsall, T. P., Bean, J. C., Logan, R. A., and Luryi, S. (1986), *Appl. Phys. Lett.* **48**, 330.

CHAPTER 1

Strained-Layer Superlattices

T. P. Pearsall

DEPARTMENT OF ELECTRICAL ENGINEERING
UNIVERSITY OF WASHINGTON
SEATTLE, WASHINGTON

I.	INTRODUCTION	1
II.	SOME PRINCIPLES OF STRAINED-LAYER EPITAXY	3
	A. *Critical Layer Thickness*	3
	B. *Planar Nucleation and Growth in Strained-Layer Epitaxy*	4
	C. *Achieving Uniform Compositional Growth—Clustering*	5
III.	APPLICATIONS OF STRAINED-LAYER EPITAXY	7
	A. *Strain as a Tool*	7
	B. *Strained-Layer Devices*	8
IV.	STRAINED-LAYER SUPERLATTICES—OPPORTUNITIES FOR MATERIALS SCIENCE	10
	REFERENCES	14

I. Introduction

Strained-layer epitaxy is a stepchild of the semiconductor laser. Following the demonstration of infrared lasers from lead salt compounds at low temperatures (Butler *et al.*, 1964) and the continuous (CW) operation at room temperature of GaAs heterojunction lasers in 1969 (Casey and Panish, 1978), there has been a steadily growing interest in materials systems involving heterostructures between different semiconductors. Many of these materials systems have the common feature that the difference in lattice parameter between the heterostructure constituents is small, often much less than 0.1% in order to minimize the strain and dislocations associated with lattice mismatch (Pearsall *et al.*, 1976a). Liquid-phase epitaxy (LPE) was used in these initial experiments, and it is easy to demonstrate that the nucleation of a film on a substrate of different composition is difficult, if not impossible, when the lattice parameter mismatch exceeds 1%. Experimental attempts to get around this problem by growing compositionally graded buffer layers have been only partially successful. This kind of grading does permit the epitaxial growth of materials with a substantial difference in lattice parameter from that of the substrate. The mismatch strain is partially relieved by the formation of a network of misfit dislocations, rendering the material unsuitable for many device applications. A result typical of this effort is shown in Fig. 1 (Pearsall et al., 1976b).

The surface of the mesa in Fig. 1 consists of a 2-μm-thick layer of GaAs$_{0.88}$Sb$_{0.12}$ grown on a graded buffer region, which is in turn grown on a GaAs substrate. The lattice mismatch between the uppermost epitaxial layer and the substrate is 8×10^{-3}. The presence of a substantial network of misfit dislocations is evident in the crosshatched pattern on the surface of the mesa, with a dislocation density of approximately 10^7cm^{-2}. It can also be seen in this figure that this concentration of defects is not present in the substrate whose surface can be seen around the periphery of the mesa. Misfit dislocations are created in the structure shown in Fig. 1 when the layer

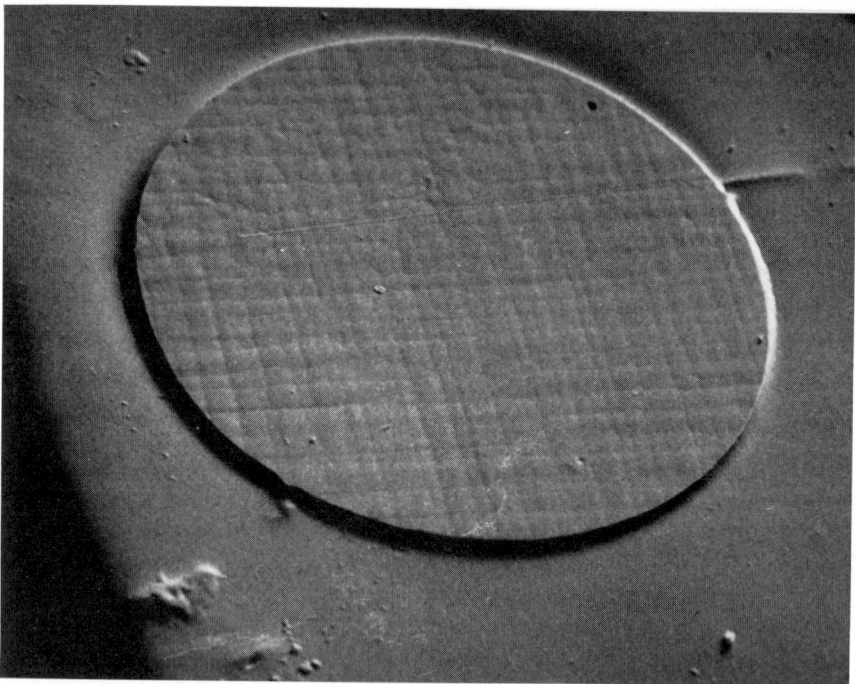

FIG. 1. Growth of lattice-mismatched GaAs$_{0.88}$Sb$_{0.12}$ on (001)GaAs by liquid-phase epitaxy. This layer is 2 μm thick, well over the critical thickness limit, and some of the lattice mismatch strain has been relieved by the formation of misfit dislocations. These dislocations form a network easily visible on the surface of the epitaxial layer.

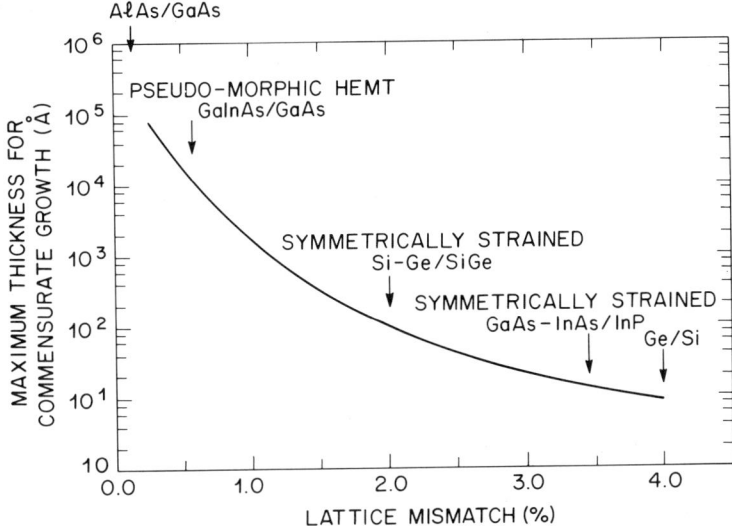

FIG. 2. Critical layer thickness for some representative heterostructures. The development of vapor-phase epitaxy techniques such as metal-organic chemical-vapor deposition (MO-CVD) and molecular-beam epitaxy (MBE) have greatly expanded the range of possibilities for strained-layer growth from those accessible by liquid-phase epitaxy.

thickness exceeds a certain threshold value called the *critical thickness*. The critical thickness, in turn, depends on the lattice parameter mismatch. Representative values for some of the materials discussed in this volume are shown in Fig. 2. The critical thickness can range from 10 Å for the growth of Ge on Si to hundreds of microns for the growth of AlAs on GaAs.

The growth of lattice-mismatched heterostructures by liquid-phase epitaxy has been largely abandoned for two reasons. One is the difficulty of the nucleation of epitaxial layers with even a modest amount of mismatch relative to the substrate. The second is the inability to grow epitaxial layers reproducibly with a thickness of less than 1000 Å. These considerations have driven researchers who continue to use LPE to consider heterostructures in which the lattice mismatch could be kept very small, such as AlGaAs/GaAs or GaInAsP/InP (Antypas *et al.*, 1973).

II. Some Principles of Strained-Layer Epitaxy

A. Critical Layer Thickness

The basic principle of strained-layer epitaxy is that a certain amount of elastic strain can be accomodated by any material without generating dislocations or defects. Whereas this statement may appear trivial on the macroscopic level, it also applies to crystal growth on the atomic level. It takes energy to accomodate an epitaxial layer of lattice-mismatched material. The energy depends on both the thickness and the size of the lattice mismatch. Of course, it also takes energy to create a dislocation that will relieve the lattice mismatch strain. If the thickness of the epitaxial layer is kept low enough to maintain the elastic strain energy below the energy of dislocation formation, the strained-layer structure will be thermodynamically stable against dislocation formation. Of course the unstrained state of the lattice-mismatched layer is energetically most favorable, but the strained structure is stable against transformation to the unstrained state by the energy barrier associated with the generation of enough dislocations to relieve the strain (Frank and van de Merwe, 1949). The idea of a critical thickness beyond which the energy required to accomodate elastic strain exceeds the fixed energy of dislocation formation was then further developed and quantified (van der Merwe, 1963, 1972; Matthews and Blakeslee, 1974, 1975).

The treatment by Matthews and Blakeslee has found wide application to a number of strained systems. In this model, the critical thickness, h_C, depends on the lattice parameter mismatch or strain, ε, and the magnitude of the Burgers vector **b** for slip dislocation

$$h_C = (1/\varepsilon)[\mathbf{b}(1 - v\cos^2\theta)/8\pi(1 + v)\cos\lambda] \cdot [\ln(4h_C/\mathbf{b})]. \tag{1}$$

The quantity v is Poisson's ratio, which is calculated from the elastic constants; θ and λ are the angles between the slip planes and the crystal surfaces. However, most existing models of critical thickness, like the Matthews–Blakeslee model, are based on equilibrium considerations, while it is clear that kinetics play an important role as well.

B. Planar Nucleation and Growth in Strained-Layer Epitaxy

The idea of critical thickness depends in part on the assumption that it is possible to grow epitaxial layers whose thickness can be controlled and maintained to a uniform value over a useable surface area of several square millimeters. That is, understanding and using the ideas of critical thickness depend on the ability to grow planar epitaxial strained layers. In Fig. 2, there are some strained-layer systems, such as Ge–Si/Si, for which the notion of critical thickness specifies layer thicknesses on the order of only a few atomic monolayers. As shown in Fig. 3, growth of these structures can be achieved by a two-dimensional Frank–van der Merwe mechanism (Bean et al., 1984). Under less favorable conditions, provoked by a combination of kinetic forces involving strain, substrate temperature, and orientation, growth can occur by a three-dimensional mode or a hybrid Stranski–Krastanov process. The

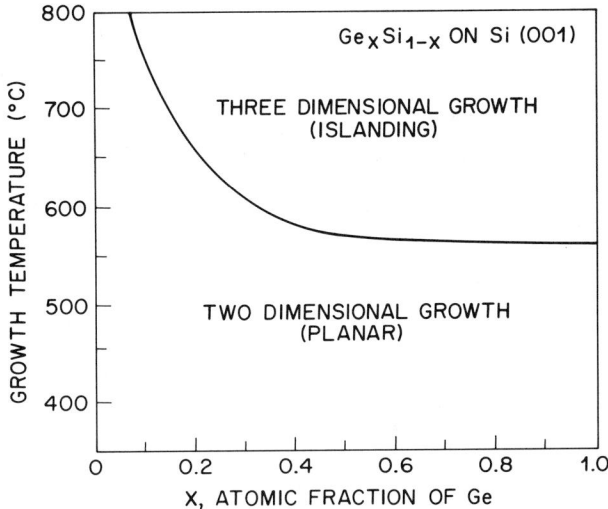

FIG. 3. The nucleation of Ge–Si strained layers can be either two-dimensional and planar, or three-dimensional, as shown in this work by Bean *et al.* (1984)[Reprinted with permission from the American Institute of Physics, Bean, J.C., Feldman, L.C., Fiory, A.T., Nakahara, S., and Robinson, I.K. (1984). *J. Vac. Sci. Technol.* **A2**, 435.]

growth of GaAs and InAs, symmetrically strained on InP, is characterized by this process over a wide range of conditions. It should be easy to recognize that growth proceeding along these lines is at odds with the notion of extended planar epitaxy of layers of only a few atoms in thickness. These considerations are presented in detail by Hull and Bean in Chapter 6 of Volume 33.

During growth by LPE, the transition between 2D and 3D nucleation occurs generally when the lattice mismatch exceeds 1%. However, using molecular-beam epitaxy (MBE), extended planar layering can still be achieved with strains as large as 4%. This kind of growth is shown in Fig. 4 for Ge–Si alloys grown on Ge (001) substrates (Pearsall *et al.*, 1989a). This example underscores some of the enhanced power of MBE to benefit from properly chosen kinetic effects (as opposed to the idea of epitaxy under strict equilibrium conditions).

C. Achieving Uniform Compositional Growth—Clustering

In addition to defining critical thickness and extended planar growth of epitaxial layers that respect this thickness, a third important consideration is maintaining a uniform composition within the strained layer. This is a concern principally when one of the strained layers is an alloy, rather than an elemental or compound material. The issue here is clustering. A significant

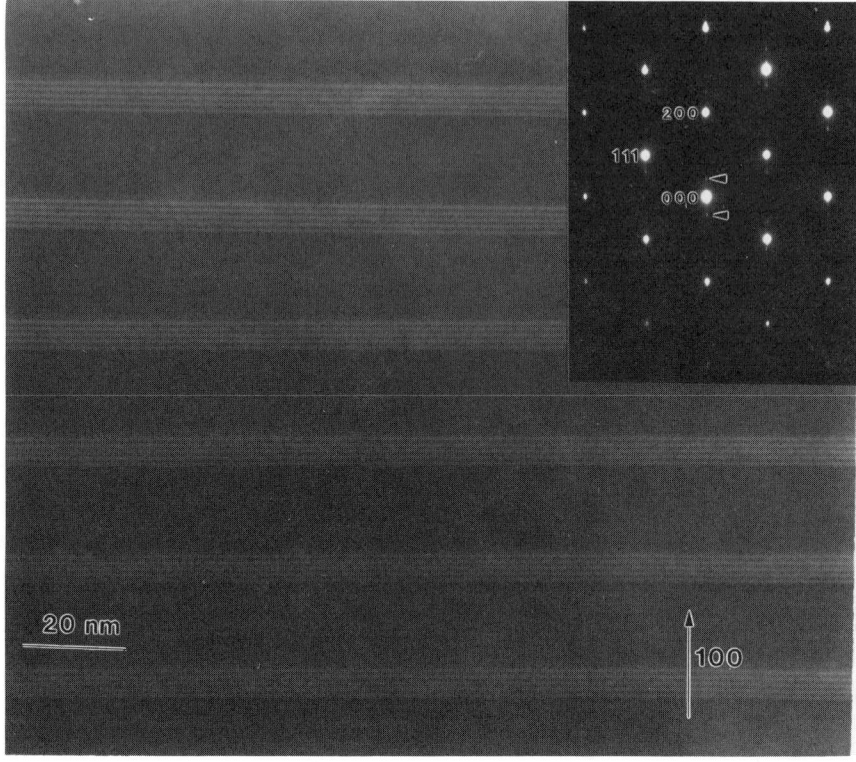

FIG. 4. High-resolution transmission electron micrograph taken by R. H. Hull of a structure consisting of four atomic monolayers of strained Si interposed with six atomic monolayers of unstrained Ge. The Ge layers are strain free because the structure is grown epitaxially on a Ge substrate. The strained layers of Si show both uniform thickness and extended planarity. [Reprinted with permission from the American Physical Society, Pearsall, T.P., Vandenberg, J.M., Hull, R., and Bonar, J.M. (1989). *Phys. Rev. Lett.* **63**, 2104.]

contribution to the understanding of clustering in strained-layer growth has been achieved through molecular-dynamic simulations (Grabow and Gilmer, 1987). This remarkable work, in which nucleation and growth at a three-dimensional interface is visualized, has shown that a strained-layer film will, in almost all cases, be unstable against clustering. Thus, while clusters may form dynamically during nucleation, they are driven to break up. This principle can be understood by considering the ratio of the free energy of a strained homogeneous film to that of a clustered film (Feldman *et al.*, 1988)

$$R = \frac{1 + n \cdot (E/y)}{1 + (n/x) \cdot (4n_S/N)} \tag{2}$$

In this equation, n is the number of atoms per unit of surface area, while N is the number of atoms per unit volume. E is the strain energy density, and y is the surface energy density. When R is set equal to 1, a critical cluster dimension, X_C, is defined:

$$X_C = (4n_s/N)/(E/y). \tag{3}$$

For the growth of Ge on Si, clustering becomes energetically favorable when the cluster diameter is about 200 Å or about 20 times the critical thickness. The work of Grabow and Gilmer establishes that growth of strained-alloy films will result in epitaxy of layers with uniform composition (absence of clustering) under almost all experimental conditions. For example, if the layer thickness were to exceed the critical thickness, strain would be eventually reduced through dislocation formation. This lowers the strain energy density and avoids the conditions for cluster formation as outlined in Eqs. (2) and (3).

The requirements of superlattice growth are beyond the capabilities of liquid-phase epitaxy when the superlattice period is less than 500 Å. The study of strained-layer heterostructures has depended heavily on the codevelopment of MBE. MBE is a procedure that makes use of nonequilibrium, in addition to equilibrium, growth mechanisms. It is the ability to control these nonequilibrium aspects of the growth process that leads to the extraordinary control over epitaxial layer nucleation and morphology. These features of strained-layer epitaxy are discussed by Kasper and Schäffler in Chapter 9, Volume 33 for the case of Group IV compounds, by Partin for narrow-gap IV–VI materials in Chapter 10, Volume 33, and by Gunshor, Kolodziejski, Nurmikko, and Otsuka for wide-bandgap II–VI compounds in Chapter 11, Volume 33.

III. Applications of Strained-Layer Epitaxy

A. STRAIN AS A TOOL

The study of lattice-mismatched epitaxy took on a new dimension when Osborn proposed that strained-layer structures might display new electronic and optical properties not seen in the unstrained-constituent materials, (Osbourn, 1982; 1986). Before this conceptual breakthrough, strained-layer growth was perceived as a compromise between a desire to produce semiconductor heterostructures and simultaneously to avoid misfit dislocations in these structures. The work of Osbourn introduced the notion that this strain could be a tool for modifying the band structure of semiconductors in a useful and predictable fashion.

The effect of strain on the band structure of semiconductors is well understood from numerous measurements made under conditions of externally imposed stress (Bir and Pikus, 1974). The studies have shown that substantial levels of strain, close to the elastic limit, are required to produce changes in the band structure that are easily measured at 295 K. In the case of strained-layer growth, strains of this magnitude are easily created. Moreover, the strain is imposed internally, as a consequence of lattice-mismatch and it may be compressive or tensile. These effects are discussed by Pollak, Chapter 2, and by People and Jackson, in Chapter 4, Volume 32.

The condition of large strain (i.e., those that will modify the band structure by more than 100 meV) implies strained-layer structures with a lattice mismatch greater than about 2%. From Fig. 2, it can be seen that the critical thickness will be less than a hundred angstroms for heterostructures made of typical semiconductors. This in turn means that careful control over layer thickness is required to produce extended strained structures that meet the critical thickness requirement. Control over the composition of the structure is essential not only for respecting the limits imposed by critical thickness, but also for ensuring a uniform level of strain throughout the structure itself. This means that in order to grow structures with enough built-in strain to be interesting for physics or device applications, it is a necessary requirement to be able to grow thin ($\varepsilon < 100$ Å) planar layers of uniform composition. Such structures are commonly known as superlattices, and so the study of strained-layer structures is in many cases the study of strained-layer superlattices.

B. Strained-Layer Devices

Semiconductor devices using strained-layer epitaxy have been proposed for most major applications: lasers, photodetectors, bipolar transistors, and field-effect transistors (FET). In the case of lasers and photodetector design, strain is used to modify the electronic energy-level structure. The strain tensor imposed by lattice-mismatched epitaxy has a uniaxial component that decouples the top of the valence band. The strain tensor will also modify the electronic bandgap energy. Marzin, Gérard, Voisin, and Brum discuss these effects in III-V semiconductors in Chapter 3, Volume 32. The change in bandgap energy can be used to tune laser emission wavelength and photodetector spectral response. This latter example is discussed in detail by People and Jackson. In addition to modification of laser emission wavelength, decoupling of the valence band by strain can lead to a lower valence-band density of states and thus a reduction in laser threshold current. This possibility was first proposed in 1986 (Yablonovitch and Kane, 1986; Adams, 1986) and is discussed by Tasker, Schaff, Foisy, and Eastman in Chapter 7, Volume 33.

Most of the experimental work in transistor design that involves strained-layer structures has sought to introduce heterostructures into the active region of the device in question. In these designs, large levels of strain are

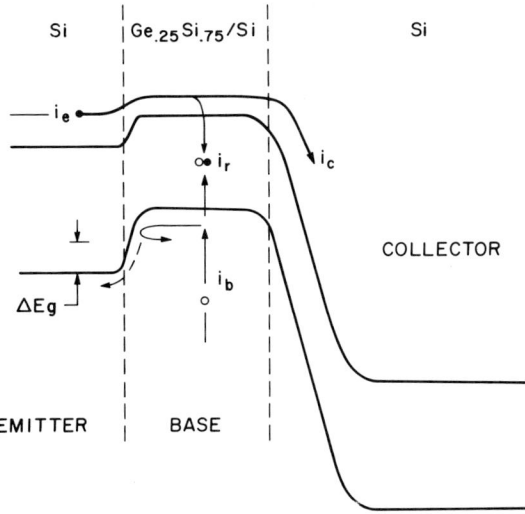

FIG. 5. Schematic energy-level diagram for a heterojunction bipolar transistor. The dotted line shows a potential leakage path for holes from the base into the emitter. This leakage current is reduced exponentially by the heterojunction offset, ΔE_g, thereby increasing emitter injection efficiency. The wider bandgap emitter permits much higher doping in the emitter for the same level of injection efficiency at the base emitter junction. This leads directly to a higher-gain bandwidth product and improved device performance.

tolerated as a condition for benefits derived from these heterostructures. The heterojunction bipolar transistor (Shockley, 1952; Kroemer, 1982) exploits a heterojunction barrier ΔE_g between the wider gap emitter and lower bandgap base. This barrier (as shown in Fig. 5) lowers the leakage rate of majority carriers from the base into the emitter. This structure permits much higher doping levels in the emitter (N_e) relative to that (P_b) in the base, raising the gain bandwidth product without degrading device performance. The gain bandwidth product can be approximated by

$$G \approx A(N_e/P_b)\cdot\exp(\Delta E_g/kT). \qquad (4)$$

Si bipolar transistors are used by IBM in its computers because of the superior combination of speed and transconductance of bipolar transistors compared to FETs. The development of strained-layer Ge_xSi_{1-x}/Si heterostructures means that Si-based bipolar transistors can now be built with performance superior to conventional Si devices. Significant progress along these lines is discussed by Kasper and Schäffler in Chapter 9, Volume 33.

The first application of strained-layer epitaxy to see commercial development is the **s**trained-layer **mo**dulation-**d**oped **f**ield-**e**ffect **t**ransistor (S-MODFET). As in the preceding example, strained-layer growth is used to create a heterostructure in the transistor that helps to increase the confinement of carriers, at the expense of residual strain. In a MODFET, charge carriers in the source–drain circuit are confined by energy barriers to regions of the device where doping is low and the mobility, consequently, is high (Pearsall, 1984). As the energy of these carriers increases, the confinement barriers become rapidly less effective, leading to degradation of the mobility. The S-MODFET (Rosenberg et al., 1985) used strained-layer epitaxy to increase the confinement energy. In the case of GaInAs/AlGaAs S-MODFETs, which are discussed in some detail by Tasker, Schaff, Foisy, and Eastman, the energy barrier is approximately doubled by use of strained-layer growth. This technique is so effective at improving performance that it has become a standard part of high-performance FET fabrication.

IV. Strained-Layer Superlattices—Opportunities for Materials Science

The astonishing progress in strained-layer epitaxy of planar heterostructures of only a few atomic layers in thickness is due in a large part to the development and improvement of materials analysis methods. These two fields of study should continue to advance together, because strained-layer superlattice structures have dimensions and features that challenge the latest improvements in analysis technology.

Among some of the important parameters that need characterization in strained-layer structures are:

Parameter	*Method*
composition dislocation density thickness	Rutherford backscattering (RBS)
strain periodicity interfacial roughening long-range ordering	high-resolution x-ray diffraction (HRXRD)
layer thickness dislocations interfacial flatness short-range ordering	transmission electron microscopy (TEM)

From this list, it can be seen that the principal methods of analysis each have unique capabilities as well as those that overlap or complement the capabilities of some other methods.

Rutherford backscattering is the tool of choice to measure the critical layer thickness limit in strained-layer structures. Such a result is shown in Fig. 6 for the case of strained Ge grown on (001)Si (Bevk *et al.*, 1986). Here the dramatic growth of dislocation concentration can be seen when the Ge-layer thickness exceeds six atomic monolayers. Use of RBS is presented and discussed in detail by Picraux, Doyle, and Tsao in Chapter 8, Volume 33.

X-ray diffraction is perhaps the most widely used tool for materials analysis. HRXRD is a powerful extension of the x-ray technique that is based on a narrow-linewidth x-ray beam (Vandenberg *et al.*, 1988). Through precise measurement of lattice parameters, HRXRD can measure the level of strain in samples of only a few atomic layers in thickness. In Fig. 7, we show another application of HRXRD measurements. The pattern of peaks measures results from a periodic structure of 45 Å of Ge–Si strained-layer superlattice and a 307 Å Si spacer between each alloy layer. The diffraction peaks give the 352 Å periodicity for this structure. The lower curve in Fig. 7 shows a kinematic

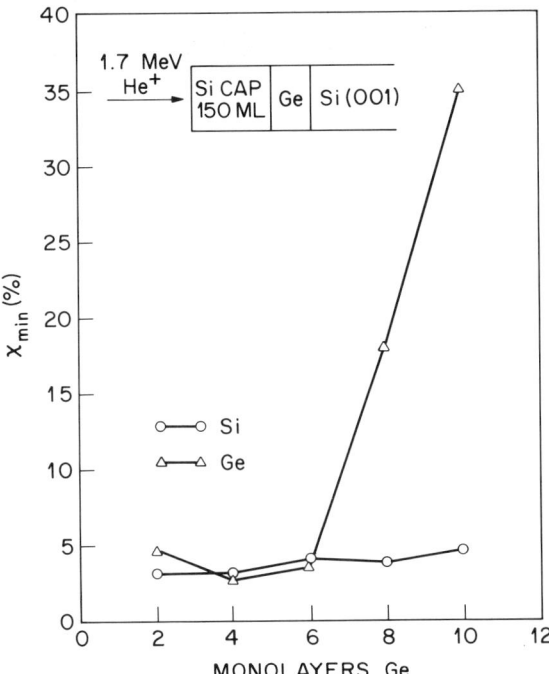

FIG. 6. Rutherford backscattering is used to determine the critical layer thickness for strained-layer materials. Here, it can be seen that the sharp increase in backscattered intensity occurs when the number of Ge monolayers grown on Si exceeds 6, corresponding to about 10 Å. [Reprinted with permission from the American Institute of Physics, Bevk, J., Mannaerts, J.P., Feldman, L.C., Davidson, B.A., and Ourmazd, A. (1986). *Appl. Phys. Lett.* **49**, 286.]

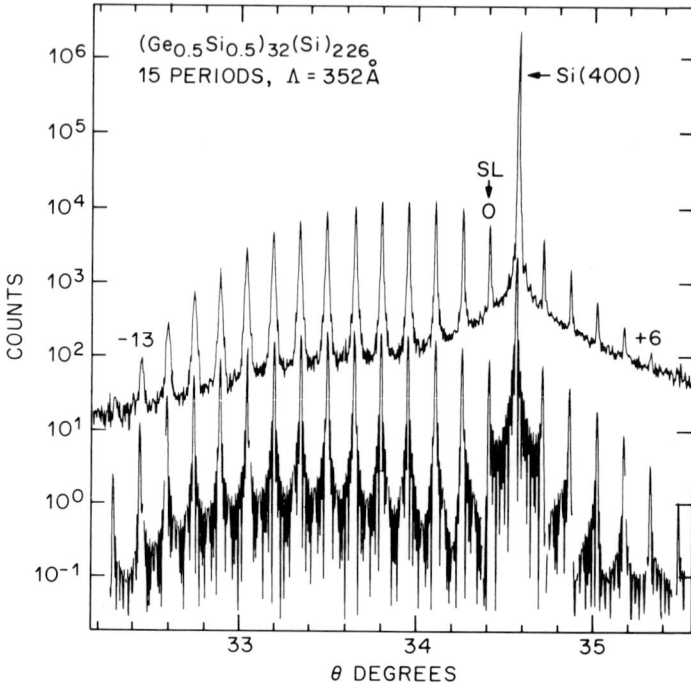

FIG. 7. High-resolution x-ray diffraction can be used as a sensitive measure of superlattice period. Here the repeat period of 250 Å appears as fine structure around a principal lattice diffraction peak from a HRXRD spectrum taken by J. M. Vandenberg. In the upper curve: experimental high-resolution X-ray diffraction results taken around the Si (400) reflection. In the lower curve: a kinematical simulation of x-ray scattering from this structure. [Reprinted with permission from the American Institute of Physics, Vandenburg, J., Bean, J.C., Hamm, R.A., and Hull, R. (1988). *Appl. Phys. Lett.* **52**, 1152.]

simulation of x-ray scattering from a Si–Ge strained-layer structure of these dimensions. Computer modelling of the diffraction pattern is used to determine the lattice mismatch, superlattice periodicity, strain in the alloy layer, and the thickness of the alloy and spacer regions.

TEM has long been recognized as the tool of choice for direct imaging of dislocations in semiconductors. The presence of dislocations is resolved in a TEM plan view by imaging the fine, nonperiodic structure associated with a diffracted beam. The size of the aperture used to select this beam is inversely proportional to the spatial resolution, and the magnification used in forming the image is inversely proportional to the field of view. These two experimental parameters—spatial resolution and field of view—will limit the dislocation density that can be resolved in TEM to about 10^7 cm^{-2} or greater. It is not uncommon in discussions of strained-layer epitaxy to refer to material having fewer than 10^7 cm^{-2} dislocations as "dislocation-free." This evaluation is usually criticized by device engineers, but supported by crystal growers. There is some truth in both views. At a level of 10^8 dislocations per cm^2, the average spacing between dislocation lines is one micron, which is too close for most devices to work properly. On the other hand, fewer than one in 5000 atom sites is occupied by a dislocation. This figure is far smaller than the concentration of dislocations that would need to be generated to relieve a 1% strain typical in strained-layer epitaxy.

Continuing improvement in transmission electron microscope technology has reduced the spatial resolution below 1.5 Å, or about one interatomic spacing for many semiconductors. Fourier recomposition of the crystal lattice by imaging several diffracted beams is now a technique commonly used to characterize superlattice structure. The sample is examined in cross section, and the structure of fine-period superlattices can be resolved with near-atomic resolution. Such an image, prepared by A. Ourmazd is shown in Fig. 8. The sample contains a strained Ge–Si superlattice composed of alternating four monolayers of Ge and Si. The resolution in the imaging mode shown here is 3 Å, so only pairs of atoms are resolved as points. Each point is the result of coherent diffraction of a column of electron pairs, 20 to 25 monolayers thick. Hence, such cross sections do not show the resolution of a single atom or atom pair, but rather the collective properties of a group of atoms. The resolution of a point or interfacial contrast—both clearly apparent in Fig. 8—give a measure of the coherent ordering of atoms and the definition of the Si–Ge interface.

The resolution capabilities of TEM that allow imaging of the lattice as shown in Fig. 8 also result in higher-quality direct-mode transmission images

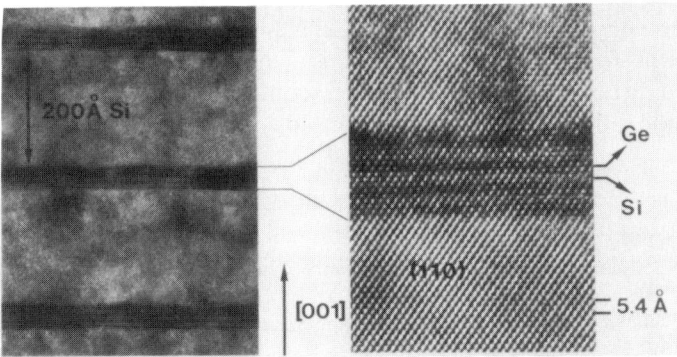

FIG. 8. Fourier reconstruction imaging of the superlattice structure. In this image, taken by A. Ourmazd, the dark bands correspond to regions of Ge, whereas the lighter bands correspond to regions of Si. The contrast is built up through about 10–20 monolayers thickness, and it indicates the presence of extended layering in the superlattice sample. [Reprinted with permission from the American Physical Society, Pearsall, T.P., Bevk, J., Bean, J.C., Bonar, J., Mannaerts, J.P., and Ourmazd, A. (1989). *Phys. Rev.* **B39**, 3741.]

formed from the structure in the undiffracted beam. Such an image, taken by R. H. Hull, of a Ge_6Si_4 strained-layer structure grown on Ge is shown in Fig. 4. In this picture, TEM shows at a glance that strained-layer structures are grown with good structural definition and planarity over an extended range of the sample. The transmission diffraction pattern from this same sample is shown as an inset to Fig. 4. Each diffraction spot has easily resolved satellites whose position is determined by the relative periods of the Ge–Si superlattice and the Ge substrate. The presence of these satellites is compelling evidence for the formation of a well-ordered superlattice with a period similar to that of a unit cell of the bulk constituent materials. Many of these characterization techniques are described in more detail in the chapters of Hull and Bean, as well as those of Picraux, Doyle, and Tsao in Vol. 33, while the effect of ordering on the electronic levels in the superlattice are presented by Jaros in Chapter 5 of this volume.

The field of strained-layer superlattices is growing rapidly because of constructive collaborations between groups of scientists who are using tools for crystal growth, characterization, and analysis that did not yet exist only a few years ago. Rather than being a review of past work, this volume of *Semiconductors and Semimetals* should serve both as a survey of the field in 1990 and as a guide to fruitful avenues of research in the decade ahead. It is clear now that it is possible both to produce and to measure a continuous planar epitaxial layer of semiconductor of only a few atoms in thickness with lateral dimensions many thousands of atoms in extent. There is a palpable synergy between the improving capabilities of materials scientists to characterize material on an atomic scale and to grow heterostructures with better definition on the atomic scale. It is producing new physics, new devices, and better technology for growth and analysis. It is the hope of the editor that this volume helps to convey a sense of this synergy to those in any of the interrelated disciplines who would like to contribute to this rapidly growing area of semiconductor science and technology.

References

Adams, A. R. (1986). *Electron Lett.* **22**, 249.
Antypas, G. A., Moon, R. L., James, L. W. Edgecombe, J., and Bell, R. L. (1973). *Symp. GaAs and Related Compounds*, London, Inst. of Phys., p. 48.
Bean, J. C., Feldman, L. C., Fiory, A. T., Nakahara, S., and Robinson, I. K. (1984). *J. Vac. Sci. Technol.* **A2**, 436.
Bevk, J., Mannaerts, J. P., Feldman, L. C. Davidson, B. A., and Ourmazd, A. (1986). *Appl. Phys. Lett.* **49**, 286.
Bir, G. L., and Pikus, G. E. (1974). "Symmetry and Strain-Induced Effects in Semiconductors," New York, J. Wiley & Sons.
Butler, J. F., Calawa, A. R., Phelen, R. J. Harman, T. C., Strauss, A. J., and Rediker, R. H. (1964). *Appl. Phys. Lett.* **5**, 75.

Casey, H. C., Jr., and Panish, M. B. (1978). "Heterostructure Lasers," New York, Academic Press.
Feldman, L. C., Bevk, J., Davidson, B. A., Gossmann, H.-J., Ourmazd, A., Pearsall, T. P., and Zinke-Allmang, M. (1988). *Mat. Res. Soc. Symp.* **102**, 405.
Frank, F. C., and van der Merwe, J. H. (1949). *Proc. Roy. Soc. London, A* **198**, 216–225.
Grabow, M. H., and Gilmer, G. H. (1987). "Initial Stages of Epitaxial Growth," Pittsburgh, Materials Research Society, p. 15.
Kroemer, H. (1982). *Proc. IEEE* **70**, 13.
Matthews, J. W., and Blakeslee, A. E. (1974). *J. Cryst. Growth*, **27**, 118.
Matthews, J. W., and Blakeslee, A. E. (1975). *J. Cryst. Growth*, **29**, 273.
Matthews, J. W., and Blakeslee, A. E. (1976). *J. Cryst. Growth*, **32**, 265.
Osbourn, G. C. (1982). *J. Appl. Phys.* **53**, 1586.
Osbourn, G. C. (1986). *IEEE J. Quant. Electron.* **QE-22**, 1677.
Pearsall, T. P., Miller, B. I., Capik, R. J., and Bachmann, K. J. (1976a). *Appl. Phys. Lett.* **28**, 449–450.
Pearsall, T. P., Nahory, R. E., and Pollack, M. A. (1976b). *Appl. Phys. Lett.* **28**, 403.
Pearsall, T. P. (1984). *Surface Science* **142**, 529.
Pearsall, T. P., Bevk, J., Feldman, L. C., Bonar, J. M. Mannaerts, J. P., and Ourmazd, A. (1987). *Phys. Rev. Lett.* **58**, 729.
Pearsall, T. P., Vandenberg, J. M. Hull, R., and Bonar, J. M. (1989a). *Phys. Rev. Lett.* **63**, 2104.
Pearsall, T. P., Bevk, J., Bean, J. C., Bonar, J., Mannaerts, J. P., and Ourmazd, A. (1989b). *Phys. Rev. B* **39**, 3741.
Rosenberg, J. J., Benlamri, M., Kirchner, P. D., Woodall, J. M., and Pettit, G. D. (1985). *IEEE Electron. Dev. Lett.* **EDL-6**, 491.
Shockley, W. (1951). U.S. Patent 2,569,347.
Vandenberg, J., Bean, J. C., Hamm, R. A., and Hull, R. (1988). *Appl. Phys. Lett.* **52**, 1152.
van der Merwe, J. H. (1963). *J. Appl. Phys.* **27**, 123.
van der Merwe, J. H. (1972). *Surface Science* **31**, 198.
van der Merwe, J. H. (1978). "The Role of Lattice Misfit in Epitaxy," *CRC Critical Reviews of Solid-State Materials Science* **7**, 209–231.
Yablonovitch, E., and Kane, E. O., (1986). *IEEE J. Lightwave Technol.* **LT-4**, 961.

CHAPTER 2

Effects of Homogeneous Strain on the Electronic and Vibrational Levels in Semiconductors

Fred H. Pollak

DEPARTMENT OF PHYSICS
BROOKLYN COLLEGE OF THE CITY UNIVERSITY OF NEW YORK
BROOKLYN, NEW YORK

I.	INTRODUCTION	17
II.	EFFECTS OF HOMOGENEOUS DEFORMATION ON ELECTRONIC ENERGY LEVELS	18
	A. *Critical Points near* $\mathbf{k} = 0$	19
	B. *Bands at* $\mathbf{k} \neq 0$	35
III.	EFFECTS OF EXTERNAL STRESS ON QUANTUM STATES	40
IV.	INFLUENCE OF HOMOGENEOUS DEFORMATION ON $\mathbf{q} \approx 0$ OPTICAL PHONONS	43
	A. *Diamond-Type Materials*	43
	B. *Zincblende-Type Materials*	46
V.	SUMMARY	48
	APPENDICES	49
	REFERENCES	52

I. Introduction

The introduction of a homogeneous strain in a solid produces changes in the lattice parameter and in some cases, in the symmetry of the material. These in turn produce significant changes in the electronic band structure and vibrational modes. All configurations of homogeneous strain can be divided into two contributions: the isotropic or hydrostatic components, which give rise to a volume change without disturbing the crystal symmetry, and the anisotropic component, which in general reduces the symmetry present in the strain-free lattice. For the electronic states, energy gaps are altered, and in some cases, degeneracies are removed. Effective masses are affected by the variations in energy gaps as well as by changes in interband matrix elements. The strain dependence of electronic levels can be characterized by deformation potentials, i.e., the energy shift per unit strain, which are typically in the range from 1 to 10 eV. For vibrational states, there is a shift of phonon frequencies and also in some cases, a destruction of symmetry. The "deformation potentials" of the lattice vibrations, i.e., the relative change in phonon frequency per unit strain, are of order 1. The electronic and vibrational deformation potentials have been measured experimentally for a large number of diamond- and zincblende-type semiconductors. In addition, there have also been several theoretical calculations.

During the past several years, there has been considerable interest in strained-layer superlattices and quantum wells from both fundamental and applied points of view. Strained-layer heterostructures allow the use of lattice-mismatched materials without the generation of misfit dislocations. This freedom from the need for precise lattice matching widens the choice of compatible materials and greatly increases the ability to control the electronic and optical properties of such structures.

To fully describe the electronic energy levels of such strained heterostructures, it is important to have information about the effects of strain on the properties of the host materials. These include strain-induced changes in energy gaps, splittings due to the lowering of symmetry, and variations in effective masses. Whereas a considerable amount of work has been done on the former two areas, little has been published in the latter field. The changes in vibrational modes are extremely useful for the characterization of the distribution of strain in the components of the heterostructure.

In this article, we review the effects of homogeneous strains on the electronic states of the highest valence and lowest conduction bands of diamond- and zincblende-type materials. Special emphasis is placed on the band extrema at the center of the Brillouin zone (BZ). Electronic deformation potentials for a number of relevant bands will be summarized. In addition, the effects of an external stress on the quantum levels of quantum wells will be discussed. The properties of strained-layer superlattices with built-in strain are considered by other authors in these volumes. The effects of strain on free carriers (cyclotron resonance and free-carrier absorption) and impurity centers will not be considered in this article. These phenomenon are discussed by Bir and Pikus (1974), as well as Balslev (1972). The influence of a homogeneous strain on the optic phonons at the BZ center of these materials will be described. Also for this situation, vibrational deformation potentials will be listed.

II. Effects of Homogeneous Deformation on Electronic Energy Levels

The influence of a homogeneous strain on the electronic energy levels of critical points in diamond- and zincblende-type semiconductors has been discussed by a number of authors (Brooks, 1955; Herring and Vogt, 1956;

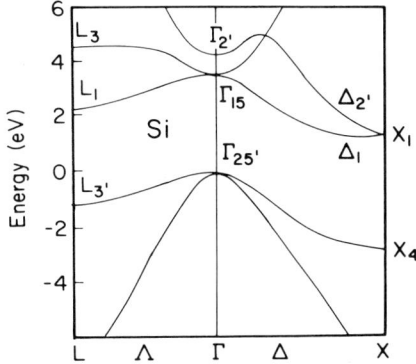

FIG. 1. Band structure of silicon along [001] and [111] [Reprinted with permission from the American Physical Society, Chelikowsky, J.R. and Cohen, M.L (1976). *Phys. Rev.* **B14**, 556.]

Kleiner and Roth, 1959; Hensel and Feher, 1963; Hasegawa, 1963; Kane, 1973; Balslev, 1972; Pollak, 1973; Bir and Pikus, 1974). In the following sections, the strain-induced changes in the conduction- and valence-band extrema at $\mathbf{k} = 0$, conduction-band minima at Δ_1 (Si, GaP, AlSb, AlAs) and L_1(Ge) as well as $\Lambda_3 - \Lambda_1$ interband transitions will be described.

For purposes of discussion, we present the band structures of Si and GaAs along the (111) and (100) directions of the BZ as calculated by the pseudopotential method (Chelikowsky and Cohen, 1976), in Figs. 1 and 2, respectively.

A. CRITICAL POINTS NEAR $\mathbf{k} = 0$

In the absence of strain or spin-orbit splitting, the valence-band edge at $\mathbf{k} = 0$ in diamond- and zincblende-type materials is a sixfold degenerate multiplet with orbital symmetry $\Gamma_{25'}$ (diamond) or Γ_{15} (zincblende). The spin–orbit interaction lifts this degeneracy into a fourfold degenerate (including spin) $P_{3/2}$ multiplet ($J = 3/2$, $M_J = \pm 3/2, \pm 1/2$ in spherical notation) and a $P_{1/2}$

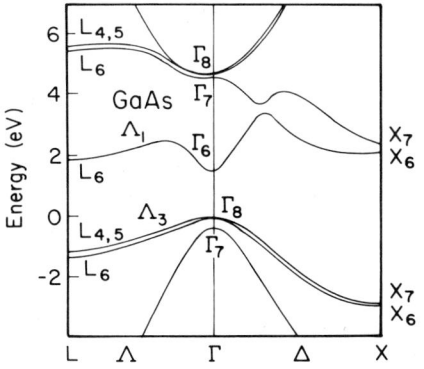

FIG. 2. Band structure of GaAs along [001] and [111]. [Reprinted with permission from the American Physical Society, Chelikowsky, J.R. and Cohen, M.L.

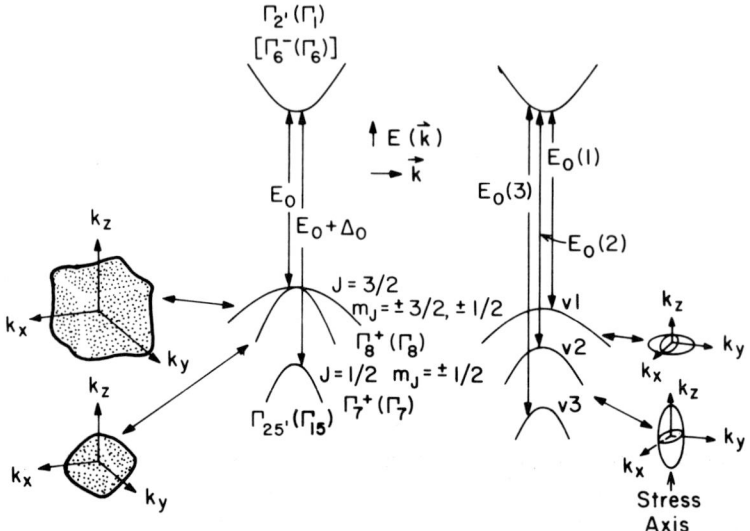

FIG. 3. Valence bands ($J = 3/2$, $M_J = \pm 3/2$, $\pm 1/2$, and $J = 1/2$, $M_J = \pm 1/2$ in spherical notation) and lowest conduction band [$\Gamma_{2'}(\Gamma_1)$] in diamond- and zincblende-type semiconductors for unstrained (left-band) and strained (right-band) crystals. Also indicated is the double group notation. [Reprinted with permission from the American Physical Society, Chandrasekhar, M., and Pollak, F.H. (1977). *Phys. Rev.* **B15**, 2127.]

multiplet ($J = 1/2$, $M_J = \pm 1/2$) as shown in detail in Fig. 3. Also given in Fig. 3 is the double group notation for the spin–orbit split bands as well as the lowest conduction bands [$\Gamma_{2'}(\Gamma_1)$] in diamond- and zincblende-type semiconductors. Because of the $J = 3/2$ degeneracy, the valence bands have warped energy surfaces (Hensel and Feher, 1963).

The fundamental direct gap, i.e., the energy difference between the $\Gamma_6^-(\Gamma_6)$ conduction and $\Gamma_8^-(\Gamma_8)$ valence bands, is denoted as E_0. The spin–orbit splitting is Δ_0, and $E_0 + \Delta_0$ is the transition energy between $\Gamma_6^-(\Gamma_6)$ and the spin–orbit split $\Gamma_7^+(\Gamma_7)$ bands.

A strain with a uniaxial component splits the $J = 3/2$ multiplet into a pair of degenerate Kramers doublets (Kleiner and Roth, 1959; Hensel and Feher, 1963; Pollak, 1973; Bir and Pikus, 1974). The three valence bands for the case of a compressive uniaxial strain is shown schematically in Fig. 3, where the bands are labelled v_1, v_2, and v_3. In addition, the hydrostatic component of the strain will shift the energy gap between the valence bands and the lowest-lying conduction band. The transition between the conduction band and the v_i valence band is denoted $E_0(i)$, where $i = 1, 2, 3$. The removal of the $J = 3/2$ degeneracy produces energy surfaces that are ellipsoids of revolution around the uniaxial strain axis (Hensen and Feher, 1963).

It has been shown that the strain Hamiltonian H_ε for a p-like multiplet can be expressed as (Pollak, 1973)

$$H_\varepsilon = H_\varepsilon^{(1)} + H_\varepsilon^{(2)}, \tag{1}$$

where $H_\varepsilon^{(1)}$ is the orbital-strain Hamiltonian and $H_\varepsilon^{(2)}$ is the strain-dependent spin-orbit Hamiltonian.

The orbital-strain Hamiltonian $H_\varepsilon^{(1)}$ of the valence bands can be written as (Pollak, 1973)

$$H_\varepsilon^{(1)} = -a_1(\varepsilon_{xx} + \varepsilon_{yy} + \varepsilon_{zz}) - 3b_1[(L_x^2 - \tfrac{1}{3}L^2)\varepsilon_{xx} + \text{cp}]$$
$$- \sqrt{3}d_1[(L_xL_y + L_yL_x)\varepsilon_{xy} + \text{cp}], \tag{2}$$

where ε_{ij} denotes the components of the strain tensor, $\mathbf{L}$ is the angular momentum operator and cp denotes cyclic permutation with respect to the indices x, y, z. The quantity a_1 represents the intraband (absolute) shift of the orbital valence bands due to the hydrostatic component of the stress (intraband or absolute hydrostatic deformation potential), while b_1 and d_1 are orbital uniaxial deformation potentials appropriate to strains of tetragonal and rhombohedral symmetries, respectively.

The stress-dependent spin-orbit Hamiltonian can be expressed as (Pollak, 1973):

$$H_\varepsilon^{(2)} = -a_2(\mathbf{L}\cdot\boldsymbol{\sigma})(\varepsilon_{xx} + \varepsilon_{yy} + \varepsilon_{zz})$$
$$- 3b_2[(L_x\sigma_x - \tfrac{1}{3}\mathbf{L}\cdot\boldsymbol{\sigma})\varepsilon_{xx} + \text{cp}]$$
$$- \sqrt{3}d_2[(L_x\sigma_y + L_y\sigma_x)\varepsilon_{xy} + \text{cp}], \tag{3}$$

where a_2, b_2, and d_2 are additional deformation potentials describing the effects of a strain on the spin–orbit interaction, and $\boldsymbol{\sigma}$ is the Pauli matrix vector.

At $\mathbf{k} = 0$ the conduction-band minima for the diamond- and zincblende-type solids (except for Si and "zero" bandgap materials such as α-Sn and HgTe) is an antibonding s-state with symmetry $\Gamma_{2'}(\Gamma_1)$. The effects of a strain is to produce a hydrostatic shift given by:

$$H_\varepsilon^{(c)} = a_c(\varepsilon_{xx} + \varepsilon_{yy} + \varepsilon_{zz}), \tag{4}$$

where a_c is the intraband (absolute) hydrostatic deformation potential of the $\Gamma_{2'}(\Gamma_1)$ conduction band.

To explore the structure of the conduction and valence bands near $\mathbf{k} = 0$, we shall use the $\mathbf{k}\cdot\mathbf{p}$ perturbation approach where the Hamiltonian is given by (Kane, 1966):

$$H_k = \left(-\frac{\hbar^2}{2m}\right)\nabla^2 + V(r) + \left(\frac{\hbar^2 k}{2m}\right) + \left(\frac{\hbar}{m}\right)\mathbf{k}\cdot\mathbf{p}. \tag{5}$$

Therefore, the total Hamiltonian can be written as:

$$H = H_k + H_{so} + H_\varepsilon^{(c)} + H_\varepsilon^{(1)} + H_\varepsilon^{(2)}, \tag{6}$$

where H_{so} is the effect of the spin–orbit interaction.

In using perturbation theory, the selection of the appropriate set of unperturbed wave functions is of great importance. The treatment is simpler when the basis functions are chosen so that the term in the Hamiltonian that gives the greatest energy contributions is already diagonal. Hence, if the energy changes produced by $H_k + H_\varepsilon$ are much smaller than the spin–orbit splitting, the most convenient set of basis functions for a perturbation expansion are those that make H_{so} diagonal. Conversely, if the spin–orbit splitting is small compared with the energy changes produced by the other terms, functions that make $H_k + H_\varepsilon$ diagonal are the most convenient set of unperturbed wave functions for the problem. Here, variations of spin–orbit splitting with **k** are neglected.

We start our study with wave functions that make H_{so} diagonal. These are functions that have the same transformation properties as the eigenfunctions of the total angular momentum operator $\mathbf{J}(=\mathbf{L}+\mathbf{S})$, i.e., $|J, M_J\rangle$. These wave functions are for the s-like $\Gamma_{2'}(\Gamma_1)$ conduction band (Kane, 1966):

$$|1/2, 1/2\rangle_c = |S\uparrow\rangle, \tag{7a}$$

$$|1/2, -1/2\rangle_c = |S\downarrow\rangle, \tag{7b}$$

while for the p-like $\Gamma_8^+(\Gamma_8)$, $\Gamma_7^+(\Gamma_7)$ valence bands, they can be written as:

$$|3/2, 3/2\rangle = (1/\sqrt{2})(X+iY)\uparrow, \tag{7c}$$

$$|3/2, 1/2\rangle = (1/\sqrt{6})(X+iY)\downarrow - (\sqrt{2/3})Z\uparrow, \tag{7d}$$

$$|3/2, -1/2\rangle = -(1/\sqrt{6})(X-iY)\uparrow - (\sqrt{2/3})Z\downarrow, \tag{7e}$$

$$|3/2, -3/2\rangle = -(1/\sqrt{2})(X-iY)\downarrow, \tag{7f}$$

$$|1/2, 1/2\rangle = (1/\sqrt{3})(X+iY)\downarrow + (1/\sqrt{3})Z\uparrow, \tag{7g}$$

$$|1/2, -1/2\rangle = (1/\sqrt{3})(X-iY)\uparrow - (1/\sqrt{3})Z\downarrow. \tag{7h}$$

In many diamond- and zincblende materials, the $\Gamma_{2'}(\Gamma_1)$ conduction band is only about 1 eV above the $\Gamma_{25'}(\Gamma_{15})$ valence-band maximum. Therefore, we shall write down explicitly the Hamiltonian matrix for the $\Gamma_{2'}(\Gamma_1)$ conduction and $\Gamma_8^+(\Gamma_8)$, $\Gamma_7^+(\Gamma_7)$ valence bands. Thus, in the following treatment, we shall group the $\Gamma_{2'}(\Gamma_1)$ conduction and $\Gamma_{25'}(\Gamma_{15})$ valence bands in Lowdin's class A and put all other states in class B. That is, we remove all interactions between class A and class B to lowest order in perturbation theory (Kane, 1966). The Hamiltonian matrix can then be written as Eq. (8)

2. EFFECTS OF HOMOGENEOUS STRAIN

$$\begin{bmatrix}
 & |S\uparrow\rangle & |S\downarrow\rangle & |3/2,3/2\rangle & |3/2,1/2\rangle & |3/2,-1/2\rangle & |3/2,-3/2\rangle & |1/2,1/2\rangle & |1/2,-1/2\rangle \\
 & E_o+H'_{11} & 0 & H_{13} & H'_{14} & (1/\sqrt{3})(H'_{13})^* & 0 & (1/\sqrt{2})(H'_{14})^* & -(\sqrt{2/3})(H'_{13})^* \\
 & 0 & E_o+H'_{11} & 0 & (1/\sqrt{3})H'_{13} & H'_{14} & (H'_{13})^* & (\sqrt{2/3})H'_{13} & (1/\sqrt{2})H'_{14} \\
 & (H'_{13})^* & 0 & H_{33} & H_{34} & H'_{35} & 0 & H'_{37} & H_{38} \\
 & (H'_{14})^* & (1/\sqrt{3})(H'_{13})^* & (H'_{34})^* & H_{44} & 0 & H'_{35} & H'_{47} & H_{48} \\
 & (1/\sqrt{3})H'_{13} & (H'_{14})^* & (H'_{35})^* & 0 & H'_{44} & H'_{34} & H'_{48} & -H'_{47} \\
 & 0 & H_{13} & 0 & (H'_{47})^* & (H'_{34})^* & H_{33} & -(H'_{38})^* & H'_{37} \\
 & (1/\sqrt{2})H'_{14} & (\sqrt{2/3})(H'_{13})^* & (H'_{37})^* & (H'_{47})^* & (H'_{48})^* & -H'_{38} & -\Delta_0+H_{77} & 0 \\
 & -(\sqrt{2/3})H'_{13} & (1/\sqrt{2})(H'_{14})^* & (H'_{38})^* & (H'_{48})^* & -(H'_{47})^* & (H'_{37})^* & 0 & -\Delta_0+H_{77} \\
\end{bmatrix} \quad (8)$$

In this treatment, we have taken the zero of energy to be Γ_8^+ (Γ_8) for $\varepsilon = 0$. The various matrix elements H_{ij} are defined in Appendix A.

1. Strain-Dependent Levels at k = 0

At $\mathbf{k} = 0$ for a strain with a uniaxial component along [001] or [111], the Hamiltonian matrix of Eq. (8) has a simple form. At the BZ center, the matrix elements connecting the $\Gamma_{2'}(\Gamma_1)$ conduction band with the $\Gamma_8^+(\Gamma_8)$ and $\Gamma_7^+(\Gamma_7)$ valence bands are zero. In these cases, Eq. (8) can be written as (Pollak, 1973):

$$\begin{bmatrix} E_0 + \delta E_{H,c}^m & 0 & 0 & 0 \\ 0 & -\delta E_{H,v}^m - \delta E_S^m & 0 & 0 \\ 0 & 0 & -\delta E_{H,v}^m + \delta E_S^m & \sqrt{2}(\delta E_S')^m \\ 0 & 0 & \sqrt{2}(\delta E_S')^m & -\Delta_0 - (\delta E_{H,v}')^m \end{bmatrix}. \quad (9)$$

In Eq. (9) m denotes either the [001] or [111] case. Since the strain does not remove the Kramers degeneracy, there is a similar expression for the $|S\downarrow\rangle$, $|J, -M_J\rangle^m$ manifold. The band v_2 corresponds to $|3/2, \pm 3/2\rangle^m$, whereas v_1

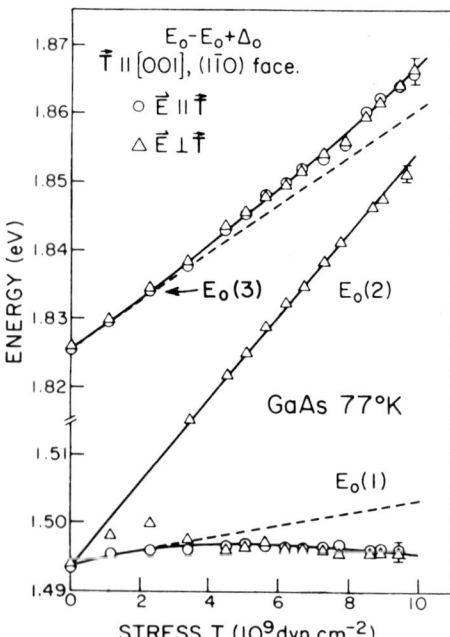

FIG. 4. Energies of the $E_0(1)$, $E_0(2)$ and $E_0(3)$ peaks of the direct-gap interband electroreflectance structure of GaAs at 77K for $\vec{T} \parallel$ [001] with light ($\vec{E}$) polarized parallel and perpendicular to the stress axis. The dashed lines represent the linearized portion of the curves. [Reprinted with permission from the American Physical Society, Chandrasekhar, M., and Pollak, F. H. (1977). Phys. Rev. **B15**, 2127.]

and v_3 are strain-induced linear combinations of $|3/2, \pm 1/2\rangle^m$ and $|1/2, \pm 1/2\rangle^m$. It can be shown that the transition $E_o(1)$ is allowed only for the polarization of the light perpendicular to the uniaxial strain axis, while $E_o(2)$ and $E_o(3)$ can be observed for both perpendicular and parallel configurations (Pollak, 1973; Chandrasekhar and Pollak, 1977). In this treatment, we have neglected the small strain-induced coupling between Γ_1 and Γ_{15} in zincblende-type semiconductors (Pollak, 1973).

The expressions for $\delta E_{H,c}$, $\delta E_{H,v}$, $\delta E'_{H,v}$, δE_S^m and $(\delta E'_S)^m$ for the case of an external stress T along ($\|$) [001] and [111] are given in Pollak (1973) and will not be repeated here.

Shown in Fig. 4 are the energies of $E_o(1)$, $E_o(2)$, and $E_o(3)$ for GaAs as a function of external stress $T \| [001]$ for light polarized parallel and perpendicular to the stress axis. The measurements were made using the electroflectance technique. The solid lines are least-squares fits to Eq. (9), making it possible to evaluate the deformation potentials $(a_c + a_1 + a_2)$, $(a_c + a_1 - 2a_2)$, b_1, and b_2. Similar studies for $T \| [111]$ have yielded values for d_1 and d_2 (Chandrasekhar and Pollak, 1977).

For strained-layer superlattices, the more relevant situation is for biaxial strain, ε, in the (001), (111), and (011) planes. In the former two cases, the form of Eq. (9) remains the same with:

$$\delta E_{H,c} = a_c(2 - \lambda^m)\varepsilon, \tag{10a}$$

$$\delta E_{H,v} = (a_1 + a_2)(2 - \lambda^m)\varepsilon = a(2 - \lambda^m)\varepsilon, \tag{10b}$$

$$\delta E'_{H,v} = (a_1 - 2a_2)(2 - \lambda^m)\varepsilon = a'(2 - \lambda^m)\varepsilon, \tag{10c}$$

$$\delta E_S^{(001)} = -(b_1 + 2b_2)[1 + \lambda^{(001)}]\varepsilon = -b[1 + \lambda^{(001)}]\varepsilon, \tag{10d}$$

$$(\delta E'_S)^{(001)} = (b_1 - b_2)[1 + \lambda^{(001)}]\varepsilon = -b'[1 + \lambda^{(001)}]\varepsilon, \tag{10e}$$

$$\delta E_S^{(111)} = -[(d_1 + 2d_2)/2\sqrt{3}][1 + \lambda^{(111)}]\varepsilon = -(d/2\sqrt{3})[1 + \lambda^{(111)}]\varepsilon, \tag{10f}$$

$$(\delta E'_S)^{(111)} = -[(d_1 - d_2)/2\sqrt{3}][1 + \lambda^{(111)}]\varepsilon = -(d'/2\sqrt{3})[1 + \lambda^{(111)}]\varepsilon, \tag{10g}$$

with

$$\lambda^{(001)} = 2C_{12}/C_{11}, \tag{10h}$$

$$\lambda^{(111)} = 2(C_{11} + 2C_{12} - 2C_{44})/(C_{11} + 2C_{12} + 4C_{44}), \tag{10i}$$

where a and a' are interband hydrostatic deformation potentials, ε is the in-plane strain, and the C_{ij} are the elastic stiffness constants. Thus, the effects of the strain is to create a hydrostatic shift and to split the degeneracy of the Γ_8^+ (Γ_8) valence band into the v_1 and v_2 bands similar to those in Fig. 3. The band v_2 corresponds to the $|3/2, \pm 3/2\rangle^m$ state, which does not couple to either the other Γ_8^+ (Γ_8) state or the Γ_7^+ (Γ_7) spin–orbit split band. However, the strain does couple the $|3/2, \pm 1/2\rangle^m$ bands to the corresponding $|1/2, \pm 1/2\rangle^m$ spin–orbit-split state.

The stress-dependent wave function for the $\mathbf{k} = 0$ valence bands can then be written as (Laude *et al.*, 1971):

$$|v_2\rangle^m = |3/2, \pm 3/2\rangle^m, \tag{11a}$$

$$|v_1\rangle^m = \alpha^m|3/2, \pm 1/2\rangle^m + \beta^m|1/2, \pm 1/2\rangle^m, \tag{11b}$$

$$|v_3\rangle^m = -\beta^m|3/2, \pm 1/2\rangle^m + \alpha^m|1/2, \pm 1/2\rangle^m, \tag{11c}$$

where the quantities α^m and β^m are given in Appendix B. The $|J, \pm M_J\rangle^{(001)}$ wave functions are given by Eqs. (7), whereas $|J, \pm M_J\rangle^{(111)}$ also are listed in Appendix B.

The strain dependence of the spin–orbit splitting in these materials is not very significant, and hence for simplicity, we shall ignore it ($a_2 = 0$) (Pollak, 1973). Under this approximation,

$$\delta E^m_{H,v} = (\delta E'_{H,v})^m = a_1(2 - \lambda^m)\varepsilon, \tag{12a}$$

$$\delta E^{(001)}_S = (\delta E'_S)^{(001)} = -b[1 + \lambda^{(001)}]\varepsilon, \tag{12b}$$

$$\delta E^{(111)}_S = (\delta E'_S)^{(111)} = -(d/2\sqrt{3})[1 + \lambda^{(111)}]\varepsilon. \tag{12c}$$

In the case where we neglect the strain dependence of the spin–orbit splitting, we can write for the strain dependence of the conduction to v_1, v_2, and v_3 bands:

$$E_0(2) = E_0 + \delta E^m_H + \delta E^m_S, \tag{13a}$$

$$E_0(1) = E_0 + \delta E^m_H - (\Delta_0 - \delta E^m_S)/2 - (1/2)[\Delta_0^2 + 2\Delta_0\delta E^m_S + 9(\delta E^m_S)^2]^{1/2}, \tag{13b}$$

$$E_0(3) = E_0 + \delta E^m_H - (\Delta_0 - \delta E^m_S)/2 + (1/2)[\Delta_0^2 + 2\Delta_0\delta E^m_S + 9(\delta E^m_S)^2]^{1/2}, \tag{13c}$$

where E_0 is the zero-strain bandgap and:

$$\delta E^m_H = (a_c + a_1)(2 - \lambda^m)\varepsilon = a(2 - \lambda^m)\varepsilon. \tag{14}$$

The relation between the Bir and Pikus deformation potentials (a, b, and d) and other notations (Kleiner and Roth, 1959; Kane, 1970) is given in Appendix C.

Listed in Table I are the experimental values of the interband hydrostatic pressure coefficient (a) and shear deformation potentials (b and d) for the $\mathbf{k} = 0$ bands of a number of diamond- and zincblende-type semiconductors. These parameters have been measured by a variety of optical and transport experiments.

Pseudopotential calculations of these parameters have recently been performed (Blacha *et al.*, 1984; Verges *et al.*, 1982). These authors also discuss other theoretical results. The determination of the intraband (absolute)

2. EFFECTS OF HOMOGENEOUS STRAIN

TABLE I

THE INTERBAND HYDROSTATIC PRESSURE DEFORMATION POTENTIAL (a) OF E_0 AND SHEAR DEFORMATION POTENTIALS (b AND d) OF THE $\Gamma_{25'}(\Gamma_{15})$ VALENCE BANDS FOR A NUMBER OF DIAMOND- AND ZINCBLENDE-TYPE SEMICONDUCTORS
(All values were taken from Landolt-Börnstein (1982), unless otherwise indicated.)

	a (eV)	b (eV)	d (eV)
Si	-5.1^a	-2.27	-5.1
		-1.92	-4.84
		-2.10	-4.85
		-2.14	-5.3
		-2.2	-5.1
Ge	-12.7	-2.4	-3.5
	-9.56^b	-2.6	-4.7
	-9.95^b	-2.86	-5.3
	-11.7^b	-2.21	-6.6
GaAs	-9.77	-1.7	-4.55
	-6.70	-2.0	-5.4
	-8.46^b	-2.0	-5.3
	-9.43^b		
	-6.36^b		
InAs	-6	-1.8	-3.6
	-5.7		
	-5.9		
	-6.9		
GaSb	-8.3	-1.8	-4.6
	-8.28	-2.0	-4.8
	-8.2^b		
	-7.9^b		
	-8.3^b		
AlSb	-5.9	-1.35	-4.3
	-2.0^b		
	-0.9^b		
InP	-6.35	-2.0	-5.0
	-6.6	-1.55	-4.2
GaP	-9.3	-1.8	-4.5
	-9.9	-1.5	-4.6
	-9.5^b		
	-9.4^b		

TABLE I (Cont.)

	a (eV)	b (eV)	d (eV)
InSb	−7.7 −7.6[b] −7.5[b] −8.0[b]	−2.05	−5.0
CdTe	−5.1	−1.2	−5.4[c] −4.8[d]
ZnTe	−5.8[e]	−1.8[f]	−4.6[f]
ZnSe	−5.4[g]	−1.2[g]	−3.8[e]
ZnS	−4.0[g]	−0.53[c] −0.7[g]	−3.7[c]

[a] Theoretical calculation from Blacha et al. (1984).
[b] Evaluated from the value of dE_0/dP using $a = -[dE_0/dP](C_{11} + 2C_{12})/3]$.
[c] A. Gavini and M. Cardona (1970). Phys. Rev. B **1**, 672.
[d] F. Meseguer, J. C. Merle, and M. Cardona (1984). Solid State Comm. **50**, 709.
[e] B. A. Weinstein, R. Zallen, M. L. Slade, and A. DeLozanne (1981). Phys. Rev. B **24**, 4652.
[f] A. A. Kaplianski and L. G. Suslina (1966). Soviet Phys. Solid State **7**, 1881.
[g] Langer et al. (1970).

hydrostatic pressure coefficients is more difficult in relation to the interband deformation potentials. This topic also is discussed by Blacha et al. (1984) and Verges et al. (1982). Nolte et al. (1987) have recently measured the absolute band-edge hydrostatic deformation potential $a_c(\Gamma_1)$ for GaAs and InP.

The situation for (011) uniaxial or biaxial strain is more complicated, since for this low symmetry direction, the $|3/2, 3/2\rangle^{(011)}$ band is coupled to both the $|3/2, -1/2\rangle^{(011)}$ and $|1/2, -1/2\rangle^{(011)}$ states by terms proportional to $(d - \sqrt{3}b)[1 - \lambda^{(011)}]\varepsilon$ (Laude et al., 1971). In the diamond- and zincblende-type semiconductors, $d \approx \sqrt{3}b$ (see Table I), and so to first order, we can neglect this coupling. Under this approximation, the form of Eqs. (9), (11), and (13) remain the same with

$$\delta E_H^{(011)} = a[2 - \lambda^{(011)}]\varepsilon, \tag{15a}$$

$$\delta E_S^{(011)} = (1/4)(b + \sqrt{3}d)[1 + \lambda^{(011)}]\varepsilon, \tag{15b}$$

$$\lambda^{(011)} = (C_{11} + 3C_{12} - 2C_{44})/(C_{11} + C_{12} + 2C_{44}). \tag{15c}$$

2. *Band Structure with $\varepsilon = 0$*

(a) *Valence bands.* In this section, we take $\varepsilon = 0$ and study the behavior of the Γ_8^+ (Γ_8) and Γ_7^+ (Γ_7) valence bands near $\mathbf{k} = 0$. In order to accomplish this, it is convenient to include the effect of the Γ_6^- (Γ_6) conduction-band implicity instead of explicity, as was done in Eq. (8). In this case, we can write:

$$\begin{array}{cccccc}
|3/2,3/2\rangle & |3/2,1/2\rangle & |3/2,-1/2\rangle & |3/2,-3/2\rangle & |1/2,1/2\rangle & |1/2,-1/2\rangle
\end{array}$$

$$\begin{bmatrix}
H_{11}^0 & H_{12}^0 & H_{13}^0 & 0 & -(1/\sqrt{2})H_{12}^0 & -\sqrt{2}H_{13}^0 \\
(H_{12}^0)^* & -H_{11}^0 & 0 & H_{13}^0 & \sqrt{2}H_{11}^0 & \sqrt{3/2}H_{12}^0 \\
(H_{13}^0)^* & 0 & -H_{11}^0 & -H_{13}^0 & \sqrt{3/2}H_{12}^0 & -\sqrt{2}H_{11}^0 \\
0 & (H_{13}^0)^* & -(H_{13}^0)^* & H_{11}^0 & \sqrt{2}(H_{13}^0)^* & -(1/\sqrt{2})(H_{12}^0)^* \\
-(1/\sqrt{2})(H_{12}^0)^* & \sqrt{2}(H_{11}^0)^* & \sqrt{3/2}(H_{12}^0)^* & \sqrt{2}H_{13}^0 & -\Delta_0+H_{55}^0 & 0 \\
-\sqrt{2}(H_{13}^0)^* & \sqrt{3/2}(H_{12}^0)^* & -\sqrt{2}(H_{11}^0)^* & -(1/\sqrt{2})H_{12}^0 & 0 & -\Delta_0+H_{55}^0
\end{bmatrix} \quad (16)$$

Again, we have chosen the zero of energy to coincide with the Γ_8^+ (Γ_8) valence band. The quantities H_{ij}^0 are defined in Appendix D.

At $\mathbf{k}=0$, there is an energy gap equal to Δ_0 between the $J=3/2$ and $J=1/2$ bands. For $\mathbf{k} \neq 0$ but small, the off-diagonal terms connecting these two multiplets will give contributions of the order $|H_{1j}^0|^2/\Delta_0$ ($j=1,2,3,$). In the limit of small $\mathbf{k}$, the quantity $|H_{1j}^0|^2/\Delta_0 \ll 1$, and hence these terms (nonparabolicities for $\mathbf{k}$) can be set equal to zero. The matrix of Eq. (16) can then be resolved into two diagonal blocks.

The lower block of Eq. (16) is already diagonal ($\mathbf{k}$ does not remove the Kramers degeneracy of the $J=1/2$ band) and yields for the energy of the $J=1/2$ band the expression [taking the zero of energy at $\Gamma_8^+(\Gamma_8)$]:

$$E_3(\mathbf{k}) = -\Delta_0 - \frac{\hbar^2}{2m} A_{so} k^2, \quad (17)$$

where A_{so} is defined in Appendix D. For the $J=3/2$ bands, Eq. (16) reduces to:

$$\begin{array}{cccc}
|3/2,3/2\rangle^{(001)} & |3/2,1/2\rangle^{(001)} & |3/2,-1/2\rangle^{(001)} & |3/2,3/2\rangle^{(001)}
\end{array}$$

$$\begin{bmatrix}
H_{11}^0 & H_{12}^0 & H_{13}^0 & 0 \\
(H_{12}^0)^* & -H_{11}^0 & 0 & H_{13}^0 \\
(H_{13}^0)^* & 0 & -H_{11}^0 & -H_{12}^0 \\
0 & (H_{13}^0)^* & -(H_{12}^0)^* & H_{11}^0
\end{bmatrix}, \quad (18)$$

which yields two double roots, since $\mathbf{k}$ preserves the Kramers degeneracy (Dresselhaus et al., 1955).

$$E_{1,2}(\mathbf{k}) = -(\hbar^2/2m)\{Ak^2 \pm [B^2k^4 + c^2(k_x^2k_y^2 + k_x^2k_z^2 + k_y^2k_z^2)]^{1/2}\}, \quad (19)$$

where A and B are defined in Appendices D and A, and $3C^2 = N^2 - 3B^2$. The plus sign corresponds to the heavy-hole band, while the minus sign corresponds to the light-hole band. The notation $E_{1,2}$ denotes the degenerate v_1 and v_2 bands. The warped energy surfaces generated by Eq. (19) are shown on the left-hand side of Fig. 3. The relation between the mass parameters A, B, and C (Dresselhaus *et al.*, 1955) and the Luttinger (1956) mass parameters γ_1, γ_2, and γ_3 is given in Appendix D.

In the intermediate range, where the energy shifts produced by **k** are comparable to Δ_0, the terms connecting the $J = 3/2$ and $J = 1/2$ manifolds cannot be neglected. In this case, a 6×6 matrix must be diagonalized, which must reduce to two 3×3 matrices because **k** does not destroy the Kramers degeneracy. It is thus possible to obtain nonparabolic terms, i.e., the k^4 dependence.

(b) Conduction bands. From Eq. (8), the dispersion relation for the $\Gamma_{2'}(\Gamma_1)$ conduction band can be written as (Hermann and Weisbuch, 1977):

$$E_c(\mathbf{k}) = \left[(1 + D'_c) + \left(\frac{P^2}{3}\right)\left(\frac{2}{E_0} + \frac{1}{E_0 + \Delta_0}\right)\right]\frac{\hbar^2 k^2}{2m}, \qquad (20)$$

where D'_c and P are defined in Appendix A.

3. *Band Structure for* $\varepsilon \neq 0$

(a) Valence bands. The band structure in the presence of an arbitrary strain is quite complicated, because momentum matrix elements as well as energy gaps are strain dependent (Aspnes and Cardona, 1978). Thus the mass parameters A, B, C, and A_{so} as well as the matrix element P will vary with strain. This topic has been treated in detail by Aspnes and Cardona for the case of GaAs (Aspnes and Cardona, 1978), using the pseudopotential method with a limited basis. In the section below, we shall discuss this subject with the understanding that these quantities may have a complex strain dependence. We shall, however, take into account explicity the strain-induced mixing between v_1 and v_3.

The simplest approach to evaluating the band properties of the v_1, v_2, and v_3 states is to use the wave functions of Eqs. (7) and (11) for [001], [111], and [011] uniaxial strain and the Hamiltonian of Eq. (5).

(001) Biaxial Strain. For this case, we find

$$k_\| = k_z \qquad \text{out-of-plane}, \qquad (21a)$$
$$k_\perp^2 = k_x^2 + k_y^2 \qquad \text{in-plane}. \qquad (21b)$$

Therefore, it can readily be shown that for the $v_2^{(001)}$ [$|3/2, \pm 3/2\rangle^{(001)}$], band the effective masses are:

$$m/m_\|^*[v_2^{(001)}] = [A_2^{(001)} - B_2^{(001)}], \qquad \text{out-of-plane mass} \qquad (22a)$$
$$m/m_\perp^*[v_2^{(001)}] = [A_2^{(001)} + (\tfrac{1}{2})B_2^{(001)}], \qquad \text{in-plane mass}. \qquad (22b)$$

Since, for the diamond- and zincblende-type semiconductor, A and B have the same sign (Landolt-Bornstein, 1982) $m_\parallel^*[v_2^{(001)}]$ is heavy-hole like, while $m_\perp^*[v_2^{(001)}]$ is light-hole like. The notation $A_2^{(001)}$ and $B_2^{(001)}$ for the mass parameters reflects the fact that A, B and N are different for the v_1, v_2 and v_3 bands due to the difference in the energies $E_0(1)$, $E_0(2)$ and $E_0(3)$ as given by Eqs. (D5)–(D8) and (13). Also, strains in different directions will produce different strain-dependent matrix-element effects.

For the v_1 and v_3 bands, the effective masses are given by:

$$m/m_\parallel^*[v_1^{(001)}] = [\alpha^{(001)}]^2[A_1^{(001)} + B_1^{(001)}] - (4/\sqrt{2})\alpha^{(001)}\beta^{(001)}B_1^{(001)}$$
$$+ [\beta^{(001)}]^2 A_{so}^{(001)}, \tag{23a}$$

$$m/m_\perp^*[v_1^{(001)}] = [\alpha^{(001)}]^2[A_1^{(001)} - (1/2)B_1^{(001)}] + \sqrt{2}\alpha^{(001)}\beta^{(001)}B_1^{(001)}$$
$$+ [\beta^{(001)}]^2 A_{so}^{(001)}, \tag{23b}$$

$$m/m_\parallel^*[v_3^{(001)}] = [\alpha^{(001)}]^2 A_{so}^{(001)} + (2\sqrt{2})\alpha^{(001)}\beta^{(001)}B_1^{(001)}$$
$$+ [\beta^{(001)}]^2[A_1^{(001)} + B_1^{(001)}], \tag{23c}$$

$$m/m_\perp^*[v_3^{(001)}] = [\alpha^{(001)}]^2 A_{so}^{(001)} - (\sqrt{2})\alpha^{(001)}\beta^{(001)}B_1^{(001)}$$
$$+ [\beta^{(001)}]^2[A_1^{(001)} - (1/2)B_1^{(001)}]. \tag{23d}$$

(111) Biaxial Strain. For this situation, we take:

$$k_\parallel = k_{\bar{z}} \quad \text{out-of-plane}, \tag{24a}$$

$$k_\perp^2 = k_{\bar{x}}^2 + k_{\bar{y}}^2 \quad \text{in-plane}, \tag{24b}$$

where $\bar{x} = (1/\sqrt{2})(x - y)$; $\bar{y} = (1/\sqrt{6})(x + y - 2z)$, and $\bar{z} = (1/\sqrt{3})(x + y + x)$.

For the $|3/2, \pm 3/2\rangle^{(111)}$ band, the masses are given by

$$m/m_\parallel^*[v_2^{(111)}] = [A_2^{(111)} - (\tfrac{1}{3})B_2^{(111)}] \quad \text{out-of-plane mass}, \tag{25a}$$

$$m/m_\perp^*[v_2^{(111)}] = [A_2^{(111)} + (\tfrac{1}{6})B_2^{(111)}] \quad \text{in-plane mass}. \tag{25b}$$

Again, since $A_2^{(111)}$ and $N_2^{(111)}$ have same sign, $m_\parallel^*$ is the heavy-hole like, while $m_\perp^*$ is light-hole like for the $|3/2, \pm 3/2\rangle^{(111)}$ state.

For the $v_1^{(111)}$ and $v_3^{(111)}$ bands, it can be shown that

$$m/m_\parallel^*[v_1^{(111)}] = [\alpha^{(111)}]^2[A_1^{(111)} + (1/3)v^{(111)}] - (2\sqrt{2}/3)\alpha^{(111)}\beta^{(111)}B_1^{(111)}$$
$$+ [\beta^{(111)}]^2 A_{so}^{(111)}, \tag{26a}$$

$$m/m_\parallel^*[v_1^{(111)}] = [\alpha^{(111)}]^2[A_1^{(111)} - (1/6)v^{(111)}] + (\sqrt{2}/3)\alpha^{(111)}\beta^{(111)}B_1^{(111)}$$
$$+ [\beta^{(111)}]^2 A_{so}^{(111)}, \tag{26b}$$

$$m/m_\parallel^*[v_3^{(111)}] = [\alpha^{(111)}]^2 A_{so}^{(111)} + (2\sqrt{2}/3)\alpha^{(111)}\beta^{(111)}B_1^{(111)}$$
$$+ [\beta^{(111)}]^2[A_1^{(111)} + (1/3)v^{(111)}], \tag{26c}$$

$$m/m_\perp^*[v_3^{(111)}] = [\alpha^{(111)}]^2 A_{so}^{(111)} - (\sqrt{2}/3)\alpha^{(111)}\beta^{(111)}B_1^{(111)}$$
$$+ [\beta^{(111)}]^2[A_1^{(111)} - (1/6)v^{(111)}]. \tag{26d}$$

(b) Conduction bands. For the conduction band mass, it can be shown that

$$m/m_\parallel^*(c) = (1 + D_c') + (P^2/3)\left\{\frac{[2(\alpha^m)^2 - (2\sqrt{2})\alpha^m\beta^m + (\beta^m)^2}{E_0(1)}\right.$$
$$\left. + \frac{[(\alpha^m)^2 + (2\sqrt{2})\alpha^m\beta^m + 2(\beta^m)^2]}{E_0(3)}\right\}, \tag{27a}$$

$$m/m_\perp^*(c) = (1 + D_c') + (P^2/6)\left\{\frac{3}{E_0(2)} + \frac{[(\alpha^m)^2 + (2\sqrt{2})\alpha^m\beta^m + 2(\beta^m)^2]}{E_0(1)}\right.$$
$$\left. + \frac{[2(\alpha^m)^2 - (2\sqrt{2})\alpha^m\beta^m + (\beta^m)^2]}{E_0(3)}\right\}, \tag{27b}$$

where $E_0(i)$ ($i = 1, 2, 3$) are given by Eq. (13), and the matrix element p and mass parameter D_c' are functions of the strain.

(c) Zero bandgap Materials. Zero bandgap materials such as α-Sn or HgTe, which are semimetals for $\varepsilon = 0$, can become semiconductors under a uniaxial strain. The lowest conduction-band minimum at $\mathbf{k} = 0$ becomes strongly anisotropic as discussed above. The valence band for a compressive external uniaxial stress has the highest maxima away from $\mathbf{k} = 0$. Cardona (1967) has obtained expressions for the toroidal energy surfaces for $\mathbf{T} \parallel [001]$ or $[111]$.

4. Exciton Effects at $\mathbf{k} = 0$

In the previous sections, the effects of strain on the band structure near $\mathbf{k} = 0$ have been discussed. However, these bands still possess spin degeneracy that cannot be removed by the strain within the framework described above. However, in the case of excitons, there are two particles involved. The lowering of the symmetry of the system by both the strain (uniaxial or biaxial) and the spin–exchange interaction results in a splitting of the spin degeneracy (Langer et al., 1970; Pollak, 1973). In this manner, it has been possible to obtain information about the spin-exchange parameter of the direct edge of a number of semiconductors.

We shall now consider as a specific example the effects of a [001] uniaxial strain on the exciton spectrum of the lowest direct transition of a diamond- or zincblende-type material. The results for [111] strain are qualitatively

2. EFFECTS OF HOMOGENEOUS STRAIN

similar. The case of wurtzite-type semiconductors has been treated in detail by Langer *et al.* (1970).

The exciton part of the Hamiltonian can be written as (Langer *et al.*, 1970; Pollak, 1973):

$$H = \text{BE} + (\tfrac{1}{2})j\boldsymbol{\sigma}_h \cdot \boldsymbol{\sigma}_e, \tag{28}$$

where BE is the exciton binding energy, and the second term describes the crystalline exchange interaction. Both BE and the exchange constant (j) are assumed to be stress independent. The operators $\boldsymbol{\sigma}_h$ and $\boldsymbol{\sigma}_e$ operate on valence-hole and conduction-electron wave functions, respectively.

The total Hamiltonian is then described by Eqs. (6) and (28). The basis functions can be written as (Langer *wet al.*, 1970; Pollak, 1973):

$$\begin{aligned}
\psi_1 &= |3/2, 3/2\rangle^{(001)}|S\downarrow\rangle, & \psi_2 &= |3/2, -3/2\rangle^{(001)}|S\uparrow\rangle, \\
\psi_3 &= |3/2, 1/2\rangle^{(001)}|S\uparrow\rangle, & \psi_4 &= |3/2, -1/2\rangle^{(001)}|S\downarrow\rangle, \\
\psi_5 &= |3/2, 3/2\rangle^{(001)}|S\uparrow\rangle, & \psi_6 &= |3/2, 1/2\rangle^{(001)}|S\downarrow\rangle, \\
\psi_7 &= |3/2, -3/2\rangle^{(001)}|S\downarrow\rangle, & \psi_8 &= |3/2, -1/2\rangle^{(001)}|S\uparrow\rangle,
\end{aligned} \tag{29}$$

where the valence-hole wave functions $|J, \pm M_J\rangle^{(001)}$ are given by Eqs. (7) and the conduction-electron function transform as $|S\uparrow\rangle$ and $|S\downarrow\rangle$.

In the following discussion, energies that equally affect all the states of Eq. (29), i.e., spin–orbit interaction and hydrostatic pressure term, will be dropped, and the interaction with the spin–orbit split levels will be neglected. The Hamiltonian is then quasidiagonal in the basis of Eq. (29) (Langer *et al.*, 1970; Pollak, 1973):

$$H = \begin{bmatrix} H^{(1,2)} & 0 & 0 & 0 \\ 0 & H^{(3,4)} & 0 & 0 \\ 0 & 0 & H^{(5,6)} & 0 \\ 0 & 0 & 0 & H^{(7,8)} \end{bmatrix}, \tag{30}$$

where

$$\begin{aligned}
H^{(1,2)} &= \begin{bmatrix} \delta E_S^{(001)} + (\tfrac{1}{2})j & 0 \\ 0 & \delta E_S^{(001)} + (\tfrac{1}{2})j \end{bmatrix}, \\
H^{(3,4)} &= \begin{bmatrix} -\delta E_S^{(001)} - (\tfrac{1}{6})j & (\tfrac{2}{3})j \\ (\tfrac{2}{3})j & -\delta E_S^{(001)} - (\tfrac{1}{6})j \end{bmatrix}, \\
H^{(5,6)} &= H^{(7,8)} = \begin{bmatrix} \delta E_S^{(001)} - (\tfrac{1}{2})j & (1/\sqrt{3})j \\ (1/\sqrt{3})j & -\delta E_S^{(001)} + (\tfrac{1}{6})j \end{bmatrix},
\end{aligned} \tag{31}$$

and $\delta E_S^{(001)}$ is given by Eq. (12b) for (001) biaxial strain, and

$$\delta E_S^{(001)} = bT/(C_{11} - C_{12}) \qquad (32)$$

for an externally applied stress $\mathbf{T} \parallel [001]$.

From Eqs. (31), the exciton energy levels are (including the hydrostatic term)

$$E_1 = E_2 = \delta E_H^{(001)} + \delta E_S^{(001)} + (\tfrac{1}{2})j, \qquad (33a)$$

$$E_3 = \delta E_H^{(001)} - \delta E_S^{(001)} + (\tfrac{1}{2})j, \qquad (33b)$$

$$E_4 = \delta E_H^{(001)} - \delta E_S^{(001)} - (\tfrac{5}{6})j, \qquad (33c)$$

$$E_{5,6} = E_{7,8} = \delta E_H^{(001)} - (\tfrac{1}{6})j$$

$$\pm (\tfrac{1}{2})\{[2\delta E_S^{(001)}]^2 - (\tfrac{8}{3})j\delta E_S^{(001)} + (\tfrac{16}{9})j^2\}^{1/2}. \qquad (33d)$$

It can be shown that of the above levels only 3, 5, and 7 (which belong to the representation F_2) are optically active at zero stress, with level 3 being allowed for $\parallel$, while 5, 7 are allowed for $\perp$. Application of a uniaxial strain splits the 3 level from 5, 7 and causes the 6, 8 level to become optically active for $\perp$.

Shown in Fig. 5 is the stress-induced splittings of the exciton lines of ZnTe, as observed in the wavelength-modulated reflectivity spectra for $\mathbf{T} \parallel [111]$ (Wardzynski and Suffczynski, 1972). These results are in agreement with the above considerations. Note that the lowest energy line is observed only at finite stresses for $\perp$. The exchange-splitting parameter has been determined from the zero-stress extrapolation.

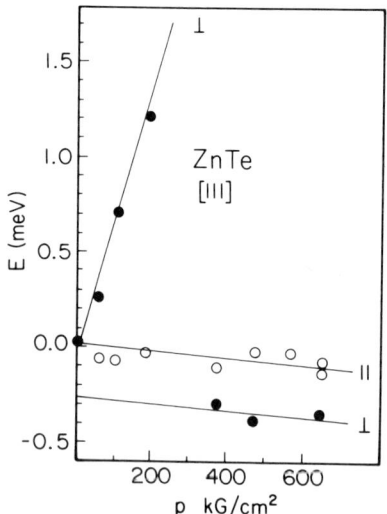

FIG. 5. Splitting of the exciton line for ZnTe for [111] uniaxial stress as observed in wavelength-modulated reflectivity for light polarized parallel and perpenducular to the stress axis. The lowest energy line is seen only at finit stress for the perpendicular configuration. (Reprinted with permission from Wardzynski and Suffczynski, *Solid State Comm.* 10, 417, 1972, Pergamon Press PLC.)

The spin-exchange parameters of the direct edge in a number of cubic and hexagonal materials has been determined by a variety of experimental techniques (Langer et al., 1970; Pollak, 1973). Rohner (1971) has calculated theoretical values of this parameter for CdS, dSe, ZnO, ZnS, and ZnSe, all of which are in good agreement with experiment. Wardzynszki and Suffczynski (1972) have observed that the exchange splitting depends exponentially on the interatomic distance of the cation and anion of the compound.

B. Bands at $\mathbf{k} \neq 0$

For band extrema or interband critical points at $\mathbf{k} \neq 0$, the shear component of the applied uniaxial stress can cause three effects (Pollak, 1973): (1) band states of different $\mathbf{k}$ vector, which are degenerate because of the symmetry of the crystal, may have their degeneracy reduced depending on the projections of their $\mathbf{k}$ vectors onto the stress direction (interband splitting), (2) a splitting of degenerate orbital bands whose $\mathbf{k}$ vectors are not parallel to the stress, and (3) a stress-induced coupling between neighboring bands [e.g., the Δ_1 conduction-band minima and nearby Δ_2', band in silicon (see Fig. 1)]. The second and third effects mentioned above are denoted as intraband splittings.

Deformation-potential theory for many-valley cubic semiconductors was first considered by Brooks (1955); Herring and Vogt (1956) as well as Kane (1970). In the notation of Brooks (1955), the stress-induced hydrostatic shift and interband splitting of a band-extrema interband critical point due to effect (1) above plus the hydrostatic pressure component are given by:

$$\Delta E = \hat{n} \cdot \{\mathscr{E}_1(\varepsilon_{xx} + \varepsilon_{yy} + \varepsilon_{zz}) + \mathscr{E}_2[\varepsilon - (\tfrac{1}{3})(\varepsilon_{xx} + \varepsilon_{yy} + \varepsilon_{zz})\mathbb{I}]\} \cdot \hat{n}, \tag{34}$$

where $\hat{n}$ is a unit vector in the direction of the band extrema or critical point in k-space, $\mathbb{I}$ is the unit diadic, and $\mathscr{E}_1$ and $\mathscr{E}_2$ are the hydrostatic and shear deformation potentials, respectively. The relation between $\mathscr{E}_1$ and $\mathscr{E}_2$ and the notation for other deformation potentials is given in Appendix C. The intraband effects will depend on the specific bands under consideration.

We shall now consider two specific examples of major interest, i.e., the conduction-band minimum along the equivalent $\langle 100 \rangle$ directions of the BZ, which occurs in such indirect gap materials as Si, $Si_{1-x}Ge_x$ ($x < 0.85$), GaP, AlAs, etc., and the L_1 conduction-band minima in Ge (equivalent $\langle 111 \rangle$ directions).

Listed in Table II are the band edge (absolute) deformation potentials $\mathscr{E}_1$ and $\mathscr{E}_2$ for Si(Δ_1), Ge(L_1), GaP(Δ_1) and AlSb(Δ_1). For the sake of completeness, also listed are values for the indirect gap interband hydrostatic pressure deformation potential ($\mathscr{E}_1 + a_1$).

TABLE II

Band Edge Deformation Potentials $\mathscr{E}_1$, $\mathscr{E}_2$, and $|\mathscr{E}_2^*|$ for the Conduction Band Minima of Several Indirect Gap Diamond- and Zincblende-type Semiconductors. For the Sake of Completeness, we also List the Indirect Gap Interband Hydrostatic Pressure Deformation Potential ($\mathscr{E}_1 + a_1$).
(All values are taken from Landolt-Börnstein (1982) unless otherwise indicated.)

	$\mathscr{E}_1$ (eV)	$\mathscr{E}_1 + a_1$ (eV)	$\mathscr{E}_2$ (eV)	$|\mathscr{E}_2^*|$ (eV)
Si(Δ)$_1$	-3^a	3.8^c	-8.77	8.0
	-7.4^b	3.1^c	-9.0	
		1.6^c	-8.1	
		1.2^c	-8.6	
		1.5^c	-9.2	
Ge(L_1)	-5.6^a	-5.7^e	-18.7	
	-4.3^d		-19.3	
	-6.1^d		-16.3	
	-12.6^b		-15.9	
GaP(Δ_1)	13.0	$1.6^{c,f}$	-6.5	
			-6.3	
AlSb(Δ_1)	4.6	2.2^c	-5.4^g	

[a] K. Murase, K. Enjouji, and E. Otsuka (1970). *Japan J. Appl. Phys.* **29**, 1255.
[b] R. D. Rode (1975). In "Semiconductors and Semimetals," (R. K. Willardson and A. C. Beer, eds.). Vol. 10, Academic Press, New York.
[c] $\Gamma_{25'}(\Gamma_{15}) - \Delta_1$ indirect gap.
[d] Evaluated from $\mathscr{E}_1 = \Xi_d + (1/3)\Xi_u$.
[e] $\Gamma_{25'} - L_1$ indirect gap.
[f] Evaluated from $\mathscr{E}_1 + a_1 = [(dE/dP)(C_{11} + 2C_{12})/3]$.
[g] This value is erroneously listed as $+5.4$ eV in Landolt-Börnstein (1982) due to an error in L. Laude, M. Cordona and F. H. Pollack (1970), *Phys. Rev.* **B1**, 1436.

1. Conduction Band Minima at Δ_1

In addition to the linear shifts and splittings given by Eq. (34), there is a nonlinear shift due to the stress-induced coupling between the neighboring Δ_1 and $\Delta_{2'}(\Delta_1)$ conduction bands (see Figs. 1 and 2). For the [100] direction in the BZ, this interaction is given by (Pollak, 1973)

$$\langle \Delta_1 | H_\varepsilon | \Delta_{2'}(\Delta_1) \rangle = 2\mathscr{E}_2^* \varepsilon_{xy}, \tag{35}$$

where $\mathscr{E}_2^*$ is a shear deformation potential. Components for the other equivalent $\langle 100 \rangle$ directions of the BZ are obtained by cyclical permutations of x, y, z.

Thus, combining Eqs. (34) and (35), we can write

$$\begin{array}{cc} |\Delta_1\rangle & |\Delta_{2'}(\Delta_1)\rangle \\ \begin{bmatrix} E_c(\Delta_1) & 2\mathscr{E}_2^* \varepsilon_{xy} \\ 2\mathscr{E}_2^* \varepsilon_{xy} & E_c[\Delta_{2'}(\Delta_1)] \end{bmatrix} \end{array} \tag{36a}$$

with

$$E_c(\Delta_1) = E_c^0(\Delta_1) + \Delta E_c(\Delta_1), \qquad (36b)$$

$$E_c[\Delta_{2'}(\Delta_1)] = E_c^0[\Delta_{2'}(\Delta_1)] + \Delta E_c[\Delta_{2'}(\Delta_1)], \qquad (36c)$$

where $E_c^0(\Delta_1)$, and $E_c^0[\Delta_{2'}(\Delta_1)]$ are the zero-strain energies of the Δ_1 and $\Delta_{2'}(\Delta_1)$ bands, and $\Delta E_c(\Delta_1)$, $\Delta E_c[\Delta_{2'}(\Delta_2)]$ are given by Eq. (34) with the appropriate deformation potentials. In silicon, $E_c^0(\Delta_1) - E_c^0(\Delta_2') \approx 0.8\,\text{eV}$ (see Fig. 1). Listed in Table II is $\mathscr{E}_2^*$ for silicon.

(a) (001) Biaxial strain. For a uniaxial strain along [001], the quantities $\varepsilon_{xy} = \varepsilon_{xz} = \varepsilon_{yz} \equiv 0$, and hence there is no intraband coupling of Δ_1 to $\Delta_{2'}(\Delta_1)$. For this situation, it can be shown that the shift of the [001] singlet band is given by

$$\Delta E^S(\Delta_1) = \mathscr{E}_1[2 - \lambda^{(001)}]\varepsilon - \mathscr{E}_2[1 + \lambda^{(001)}](2\varepsilon/3), \qquad (37a)$$

while for the [100] and (010] doublet bands, the strain-dependent energy shifts are:

$$\Delta E^D(\Delta_1) = \mathscr{E}_1[2 - \lambda^{(001)}]\varepsilon + \mathscr{E}_2[1 + \lambda^{(001)}](\varepsilon/3). \qquad (37b)$$

Since for $\langle 001 \rangle$ indirect gap materials like Si the quantity $\mathscr{E}_2 < 0$, a tensile biaxial strain in the (001) plane moves the singlet [100] conduction band below the doublet bands.

(b) (111) Biaxial strain. For this strain direction, there is no removal of the interband degeneracy of the equivalent $\langle 100 \rangle$ conduction-band minima. However, there is an intraband coupling between the Δ_1 and $\Delta_{2'}(\Delta_1)$ conduction bands. Although $\mathscr{E}_1$ and $\mathscr{E}_2$ for Δ_1 have been evaluated (see Table II), no values for the $\Delta_{2'}(\Delta_1)$ band have been measured. Thus, to first order, we neglect the strain dependence of $E_c[\Delta_{2'}(\Delta_1)]$. Equation (36a) can then be diagonalized to yield the shift of all the Δ_1 conduction-band minima.

$$E_c(\Delta_1) = \{E_c^0(\Delta_1) + E_c^0[\Delta_{2'}(\Delta_1)] + \Delta E_c(\Delta_1)\}/2$$
$$+ (\tfrac{1}{2})\{[E_c^0(\Delta_1) - E_c[\Delta_{2'}(\Delta_1)]$$
$$+ \Delta E_c(\Delta_1)]^2 + 16(\mathscr{E}_2^* \varepsilon_{xy})^2\}^{1/2} \qquad (38a)$$

with

$$\Delta E_c(\Delta_1) = \mathscr{E}_1[2 - \lambda^{(111)}]\varepsilon, \qquad (38b)$$

$$\varepsilon_{xy} = \varepsilon_{xz} = \varepsilon_{yz} = -[1 + \lambda^{(111)}](\varepsilon/3). \qquad (38c)$$

2. Conduction-Band Minima at L_1

Another situation of considerable interest is the degenerate conduction-band minima along $\langle 111 \rangle$, such as occurs for Ge(L_1). Since there are not other bands near L_1, we neglect intraband-mixing terms analogous to Eq. (35) for the $\Delta_2 (\Delta_1)$ bands. Thus for the L_1 conduction-band minima, only Eq. (34) applies. Again, we consider only the case of biaxial strain, since the situation for external stress T has already been considered (Balslev, 1972).

(a) (001) Biaxial strain. This type of strain does not remove the degeneracy of the equivalent L_1 minima, and hence there is only a hydrostatic pressure shift given by

$$\Delta E(L_1) = \mathscr{E}_1 [2 - \lambda^{(001)}] \epsilon. \tag{39}$$

(b) (111) Biaxial strain. This strain direction makes the [111] conduction band minima (singlet) inequivalent to the remaining L_1 states (triplet). The energy shifts are thus from Eq. (34):

$$\Delta E^S(L_1) = \mathscr{E}_1 [2 - \lambda^{(111)}]\varepsilon - 2\mathscr{E}_2 [1 + \lambda^{(111)}]\varepsilon, \tag{40a}$$

$$\Delta E^T(L_1) = \mathscr{E}_1 [2 - \lambda^{(111)}]\varepsilon + 2\mathscr{E}_2 [1 + \lambda^{(111)}](\varepsilon/3). \tag{40b}$$

3. *Interband Transitions along* $\langle 111 \rangle$

Another important set of bands are the Λ_1 conduction and Λ_3 valence bands, which occur along the equivalent $\langle 111 \rangle$ directions of the BZ zone. The orbital degeneracy of Λ_3 is removed by the spin–orbit splitting. The spin–orbit split transitions between Λ_3 and Λ_1 are denoted as E_1 and $E_1 + \Delta_1$, where Δ_1 is the spin–orbit splitting of Λ_3 (see Fig. 2).

A [001] uniaxial strain does not remove the **k**-space degeneracy of these bands (no interband splitting) but does cause an intraband effect, i.e., a strain dependence of the separation between the Λ_3 orbital bands. A [111] uniaxial strain produces both interband and intraband splitting; the [111] band is split off from the remaining three bands ($[1\bar{1}\bar{1}]$, $[\bar{1}1\bar{1}]$, $[\bar{1}\bar{1}1]$), and there is an intraband effect on this latter group whose **k** vector does not lie along the uniaxial strain direction.

(a) (001) Biaxial strain. For a (001) biaxial strain, the energies of the E_1 and $E_1 + \Delta_1$ transitions are given by (neglecting exciton effects, which are considered in the next section):

$$\Delta E_1 = \left(\frac{\Delta_1}{2}\right) + \mathscr{E}_2 [2 - \lambda^{(001)}]\varepsilon$$

$$- (\tfrac{1}{2})\{\Delta_1^2 + (\tfrac{8}{3})(D_3^3)^2 [1 + \lambda^{(001)}]^2 \varepsilon^2\}^{1/2}, \tag{41a}$$

$$\Delta (E_1 + \Delta_1) = \left(\frac{\Delta_1}{2}\right) + \mathscr{E}_2 [2 - \lambda^{(001)}]\varepsilon$$

$$+ (\tfrac{1}{2})\{\Delta_1^2 + (\tfrac{8}{3})(D_3^3)^2 [1 + \lambda^{(001)}]^2 \varepsilon^2\}^{1/2}. \tag{41b}$$

The quantity D_3^3 is an intraband deformation potential for the Λ_3 valence band for a [001] uniaxial strain.

(b) (111) Biaxial strain. A uniaxial strain along the [111] direction preferentially selects out the [111] direction (singlet) while making equal angles with the other six ⟨111⟩ directions (triplet). This gives rise to an interband splitting between the singlet and triplet states. In addition, there is an intraband effect for the triplet states.

The strain-dependent energy eigenvalues for the E_1, $E_1 + \Delta_1$ singlet states are thus from Eq. (34):

$$\Delta E_1^S = \mathscr{E}_1[2 - \lambda^{(111)}]\varepsilon + \mathscr{E}_2[1 + \lambda^{(111)}]\left(\frac{\varepsilon}{3}\right), \quad (42a)$$

$$\Delta(E_1 + \Delta_1)^S = \mathscr{E}_1[2 - \lambda^{(111)}]\varepsilon + \mathscr{E}_2[1 + \lambda^{(111)}]\left(\frac{\varepsilon}{3}\right). \quad (42b)$$

By using the wave functions with the proper transformation properties, it can be shown that for the triplet state, the energy shifts caused by the strain are given by (Chandrasekhar and Pollak, 1977):

$$\Delta E_1^T = \left(\frac{\Delta_1}{2}\right) + \mathscr{E}_1[2 - \lambda^{(111)}]\varepsilon - \mathscr{E}_2[1 + \lambda^{(111)}]\left(\frac{\varepsilon}{9}\right)$$
$$- (\tfrac{1}{2})\{\Delta_1^2 + (\tfrac{8}{27})(D_3^5)^2[1 + \lambda^{(111)}]^2\varepsilon^2\}^{1/2}, \quad (43a)$$

$$\Delta(E_1 + \Delta_1)^T = \left(\frac{\Delta_1}{2}\right) + \mathscr{E}_1[2 - \lambda^{(111)}]\varepsilon - \mathscr{E}_2[1 + \lambda^{(111)}]\left(\frac{\varepsilon}{9}\right)$$
$$+ (\tfrac{1}{2})\{\Delta_1^2 + (\tfrac{8}{27})(D_3^5)^2[1 + \lambda^{(111)}]^2\varepsilon^2\}^{1/2}, \quad (43b)$$

where D_3^5 is an intraband deformation potential appropriate to a [111] uniaxial strain.

In the above cases, $\mathscr{E}_1$ and $\mathscr{E}_2$ are interband ($\Lambda_3 - \Lambda_1$) deformation potentials. Listed in Table III are $\mathscr{E}_1$, $\mathscr{E}_2$, D_3^3, and D_3^5 for silicon, germanium, and GaAs.

4. Exciton Effects at $\mathbf{k} \neq †$

In the above section on the E_1, $E_1 + \Delta_1$ transitions, we have neglected exciton-exchange effects. For example, a [001] uniaxial strain does not remove the equivalence of the ⟨111⟩ bands, and hence only two polarization-independent transitions given by Eq. (40) are predicted. That is, the Kramers degeneracy of the Λ_1 and Λ_3 bands is not removed in the one-electron picture. However, inclusion of the spin-exchange effect (Chandrasekhar and Pollak, 1977) does destroy this degeneracy as in the case of the bands at $\mathbf{k} \neq 0$.

TABLE III

INTERBAND ($\mathscr{E}_1$ AND $\mathscr{E}_2$) AND INTRABAND VALENCE (D_3^3 AND D_3^5) DEFORMATION POTENTIALS OF THE $\Lambda_3 - \Lambda_1$ TRANSITION IN SEVERAL DIAMOND- AND ZINCBLENDE-TYPE SEMICONDUCTORS

	$\mathscr{E}_1$ (eV)	$\mathscr{E}_2$ (eV)	D_3^3 (eV)	D_3^5 (eV)
Si	-4.6^a	8.7^a	5^a	4^a
Ge	-5.0^b	9.8^b	5.85^b	6.1^b
GaAs	-4.3^b	8.0^b	3.45^b	0^b
InSb	-3.25^c	6.4^c		
GaSb	-4.0^c	6.4^c		

[a]F. H. Pollak and G. W. Rubloff (1972). *Phys. Rev. Lett.* **29**, 789.
[b]Chandrasekhar and Pollak (1977).
[c]T. Tuomi, M. Cardona, and F. H. Pollak (1970). *Phys. Stat. Sol.* **40**, 227.

For the case of uniaxial strain along [001], it can be shown that the strain-dependent E_1 and $E_1 + \Delta_1$ energies are given by (Chandrasekhar and Pollak, 1977):

$$\Delta E_1 = \left(\frac{\Delta_1}{2}\right) + \delta E_H^{(001)} - (\tfrac{1}{2})[\Delta_1^2 + 4(\delta_j \pm \delta_S)^2]^{1/2}, \tag{44a}$$

$$\Delta(E_1 + \Delta_1) = \left(\frac{\Delta_1}{2}\right) + \delta E_H^{(001)} + (\tfrac{1}{2})[\Delta_1^2 + 4(\delta_j \pm \delta_S)^2]^{1/2}, \tag{44b}$$

where

$$\left. \begin{aligned} \delta E_H^{(001)} &= \mathscr{E}_1 T/(C_{11} + 2C_{12}) \\ \delta_S &= (\sqrt{2/3})D_3^3 T/(C_{11} - C_{12}) \end{aligned} \right\} \quad T \parallel [001], \tag{44c, 44d}$$

$$\left. \begin{aligned} \delta E_H^{(001)} &= \mathscr{E}_1[2 - \lambda^{(001)}]\varepsilon \\ \delta_S &= (\sqrt{2/3})D_3^3[1 + \lambda^{(001)}]\varepsilon \end{aligned} \right\} \quad \text{biaxial (001) strain,} \tag{44e, 44f}$$

$$\delta_j = p(0)j, \tag{44g}$$

where $p(0)$ is the probability that the electron and the hole are in the same lattice site, and j is the exchange interaction between Wannier functions.

III. Effects of External Stress on Quantum States

The effects of external stress along [001] and [111] on the electronic structure and optical properties of various superlattices have been investigated both experimentally (Jagannath *et al.*, 1986; Sooyakumar *et al.*, 1987; Lee *et al.*, 1988; Gil *et al.*, 1988; Collins *et al.*, 1987) and theoretically (Sanders and Chang, 1985; Mailhoit and Smith, 1987; Lee and Vassell, 1988a, 1988b).

Lee et al. (1988) have studied the influence of a stress **T** ∥ [100] on the photoluminescence excitation spectra at 5 K of a $GaAs/Ga_{0.7}Al_{0.3}As$ multiple quantum well (MQW) grown along the [001] axis (Lee et al., 1988). Thus the strain is perpendicular to the quantum axis. Shown in Fig. 6 are the uniaxial stress dependence of higher energy exciton transitions with respect to the fundamental heavy-hole to conduction-subband transition. The

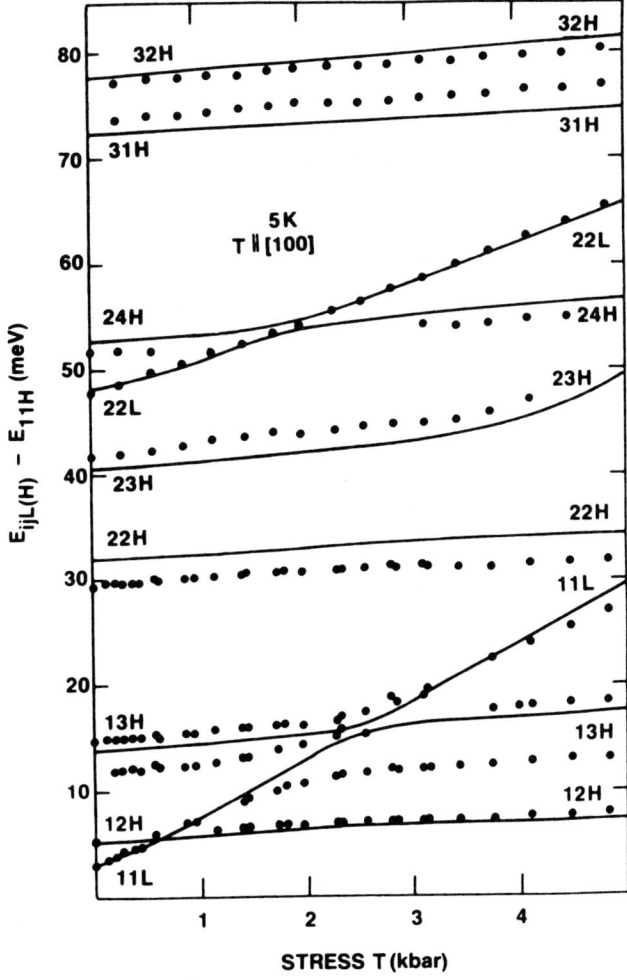

FIG. 6. Uniaxial stress dependence of higher exciton transitions of a $GaAs/Ga_{0.7}Al_{0.3}As$ multiple quantum well. The energies are taken with respect to the fundamental heavy-hole to conduction subband transition (11H). The stress is perpendicular to the growth (quantum) axis. The solid circles are experimental data and the solid lines represents the theoretical evolution [Reprinted with permission from the American Physical Society, Lee, J., Jagannath, C., Vessell, M.O., and Koteles, E. (1988). *Phys. Rev.* **B37**, 4164.]

notation $mnH(L)$ denotes transitions from the mth conduction subband to the nth valence subband of heavy (H)- or light (L)-hole character. The solid circles are the experimental data, and the solid lines represent the predictions of the theoretical model. The valence-band mixing is a consequence of the fact that **T** is perpendicular to the quantum axis. The work of Lee *et al.* (1988) used only a 4×4 valence-band Hamiltonian corresponding to the $J = 3/2$ states. Their analysis does not include the strain-induced wave-function mixing of the $|3/2, \pm 3/2\rangle$ and $|1/2, \pm 1/2\rangle$ valence-band states.

Mailhoit and Smith (1987) have presented a theoretical analysis of the effects of a compressive uniaxial stress on the electronic structure of

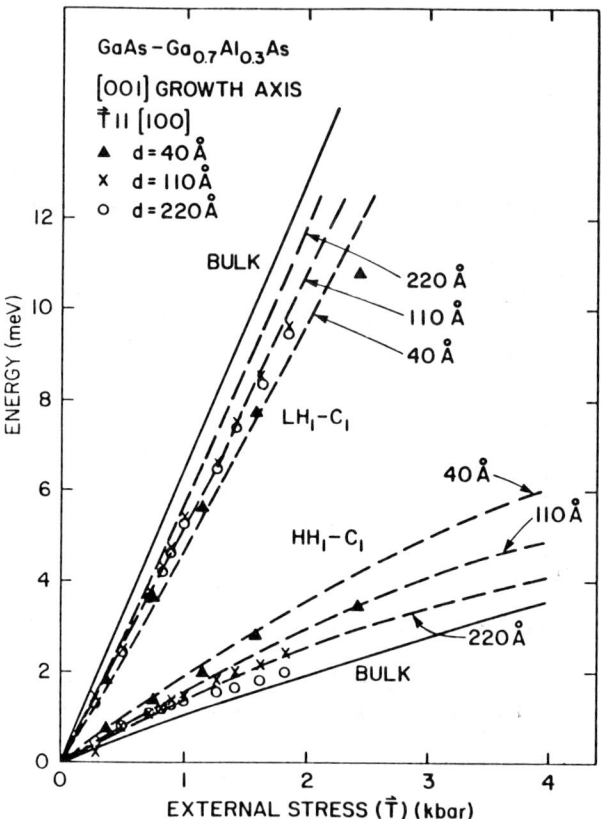

FIG. 7. Stress dependence of the fundamental heavy-hole to conduction ($HH_1 - C_1$) subband transitions for several GaAs/Ga$_{0.7}$Al$_{0.3}$As multiple quantum wells with different well widths. The stress is perpendicular to the growth axis. The dashed lines are theoretical calculations, which includes the coupling between the v_1 and v_3 bands. The solid lines are the strain dependence of $E_0(1)$ and $E_0(2)$ of bulk GaAs. (D. L. Smith and C. Mailhoit, Reviews of Modern Physics **62**, 173, 1990.) [Reprinted with permission from the American Physical Society]

GaAs/Ga$_{1-x}$Al$_x$As quantum wells grown along the [001] axis. They have considered the cases where the external stress **T** is parallel (**T** ∥ [001]) and perpendicular (**T** ∥ [100]) to the [001] growth (quantum) axis. Their approach utilizes the full 6 × 6 valence-band Hamiltonian so that nonlinear effects and wave-function mixing are taken into account. Shown in Fig. 7 are the experimental results of Jagannath *et al.* (1986) for the fundamental heavy-hole to conduction (HH$_1$–C$_1$) and fundamental light-hole to conduction (LH$_1$–C$_1$) subband transitions of GaAs/Ga$_{0.7}$Al$_{0.3}$As MQWs of different well thickness (*d*). This configuration also has **T** perpendicular to the growth axis. The dashed lines are theoretical results for the MQWs, whereas the solid lines are the theoretical behavior of $E_0(1)$ and $E_0(2)$ of bulk GaAs (Mailhiot, 1989).

IV. Influence of Homogeneous Deformation on q ≈ 0 Optical Phonons

For diamond- and zincblende-type materials, the **q** ≈ 0 optical phonons have symmetry $\Gamma_{25'}(\Gamma_{15})$. Therefore, to determine the effects of a strain on these phonons, we can take the same approach as used in Section II.A to compute the effect of strain on the $\Gamma_{25'}(\Gamma_{15})$ electronic valence bands at **k** = 0. For the Raman-active phonons (**q** ≈ 0 optical phonons), we can therefore use an effective Hamiltonian of the form (Cerdeira, 1971):

$$H_\varepsilon = [(p + 2q)/6\omega_0](\varepsilon_{xx} + \varepsilon_{yy} + \varepsilon_{zz})$$
$$- [(p - q)/2\omega_0][(L_x^2 - \tfrac{1}{3}L^2)\varepsilon_{xx} + \text{cp}]$$
$$- (r/\omega_0)[L_x L_y + L_y L_x]\varepsilon_{xy} + \text{cp}], \tag{45}$$

where p, q, and r are "deformation potentials" that describe the changes in the restoring force constants (Cerdeira, 1971; Cerdeira *et al.*, 1972; Anastassakis, 1980; Weinstein and Cardona, 1972; Balslev, 1974; Anastassakis *et al.*, 1988), and ω_0 is the frequency of the unperturbed **q** ≈ 0 optical phonon. The quantity $(p + 2q)/6\omega_0^2 = -\gamma$, where γ is the mode Gruneisen parameter. In Anastassakis (1980), the parameters p, q, and r are denoted as K_1, K_2, and K_3, respectively, whereas in Anastassakis *et al.* (1988) $K_{11} - K_{12} = (p - q)/2\omega_0^2$, and $K_{44} = r/\omega_0^2$.

A. Diamond-Type Materials

For these semiconductors, the $\Gamma_{25'}$ **q** ≈ 0 optical phonons are triply degenerate. In this case, the total Hamiltonian is given by:

$$H = \omega_0 + H_\varepsilon. \tag{46}$$

Using the $\Gamma_{25'}$ the wave functions X, Y, and Z, this Hamiltonian can be written in matrix form as:

$$\begin{array}{ccc} |X\rangle & |Y\rangle & |Z\rangle \\ \begin{bmatrix} [p+2q]/6\omega_0](\varepsilon_{xx}+\varepsilon_{yy}+\varepsilon_{zz}) \\ -[(p-q)/6\omega_0](\varepsilon_{yy}+\varepsilon_{zz}-2\varepsilon_{xx}) \\ +\omega_0 \end{bmatrix} & (r/\omega_0)\varepsilon_{xy} & (r/\omega_0)\varepsilon_{xz} \\ (r/\omega_0)\varepsilon_{xy} & \begin{matrix}[(p+2q)/6\omega_0](\varepsilon_{xx}+\varepsilon_{yy}+\varepsilon_{zz}) \\ -[(p-q)/6\omega_0](\varepsilon_{xx}+\varepsilon_{zz}-2\varepsilon_{yy}) \\ +\omega_0 \end{matrix} & (r/\omega_0)\varepsilon_{yz} \\ (r/\omega_0)\varepsilon_{xz} & (r/\omega_0)\varepsilon_{yz} & \begin{matrix}[(p+2q)/6\omega_0](\varepsilon_{xx}+\varepsilon_{yy}+\varepsilon_{zz}) \\ -[(p-q)/6\omega_0](\varepsilon_{xx}+\varepsilon_{yy}-2\varepsilon_{zz}) \\ +\omega_0 \end{matrix} \end{array} \quad (47)$$

For a uniaxial strain along the [001] or [111] directions, the matrix of Eq. (47) can be easily diagonalized. In this case, the threefold degeneracy of the $\mathbf{q} \approx 0$ optical phonons is split into a singlet (Ω_s) with eigenvector parallel to the strain axis, and a doublet (Ω_d) with eigenvectors located in the plane perpendicular to the strain axis. For these two directions of strain, the frequency of the $\mathbf{q} \approx 0$ optical phonons is given by:

$$\Omega_s = \omega_0 + \Delta\Omega_H + (2/3)\Delta\Omega_S, \tag{48a}$$

$$\Omega_d = \Omega_0 + \Delta\Omega_H - (1/3)\Delta\Omega_S. \tag{48b}$$

The expressions for $\Delta\Omega_H$ and $\Delta\Omega_S$ in the cases of an external stress $\mathbf{T} \parallel [001]$ and $\mathbf{T} \parallel [111]$ have been given in Cerdeira et al. (1972) and Anastassakis (1980) and will not be repeated here.

For the situation in which there is an in-plane strain, ε, in the (001) or (111) planes, these quantities can be written as:

$$\Delta\Omega_H^{(001)} = [(p + 2q)/6\omega_0][2 - \lambda^{(001)}]\varepsilon \tag{49a}$$
$$\Delta\Omega_S^{(001)} = [(p - q)/2\omega_0][1 + \lambda^{(001)}]\varepsilon, \tag{49b}$$
(001) strain

$$\Delta\Omega_H^{(111)} = [(p + 2q)/6\omega_0][2 - \lambda^{(111)}]\varepsilon, \tag{50a}$$
$$\Delta\Omega_S^{(111)} = [r/\omega_0][1 + \lambda^{(111)}]\varepsilon \tag{50b}$$
(111) strain

where $\lambda^{(001)}$ and $\lambda^{(111)}$ are given by Eqs. (10h) and (10i), respectively.

For a uniaxial strain along [011], the situation is more complex, since this strain direction splits the triply-degenerate phonons into three components given by:

$$\Omega_1 = \omega_0 + \Delta\Omega_H^{(011)} - \{[(p-q)/12\omega_0] - (r/4\omega_0)\}[1 + \lambda^{(011)}]\varepsilon, \tag{51a}$$

$$\Omega_2 = \omega_0 + \Delta\Omega_H^{(011)} - [(p-q)/6\omega_0][1 + \lambda^{(011)}]\varepsilon, \tag{51b}$$

$$\Omega_3 = \omega_0 + \Delta\Omega_H^{(011)} - \{[(p-q)/12\omega_0] + (r/4\omega_0)\}[1 + \lambda^{(011)}]\varepsilon, \tag{51c}$$

where

$$\Delta\Omega_H^{(011)} = [(p + 2q)/6\omega_0][2 - \lambda^{(011)}]\varepsilon, \tag{51d}$$

and $\lambda^{(011)}$ is given by Eq. (15c).

The vibrations of Ω_1, Ω_2, and Ω_3 occur along the axes x_1, x_2, and x_3 given by:

$$x_1 = (1/\sqrt{2})(y - z), \tag{52a}$$

$$x_2 = x, \tag{52b}$$

$$x_3 = (1/\sqrt{2})(y + z). \tag{52c}$$

B. ZINCBLENDE-TYPE MATERIALS

Although the strain Hamiltonian of the valence bands at $\mathbf{k} = 0$ is the same for the diamond- and zincblende-type materials, this is not the case for the $\mathbf{q} \approx 0$ phonons. For these phonons in the polar zincblende-type materials, there is an additional perturbation, i.e., the long-range Coulomb force, which causes a splitting between the longitudinal (LO) and transverse (TO) optic phonons at the center of the BZ. This splitting depends on the effective charge, e^*. Thus, we have to add into our strain Hamiltonian a term H_{ε,e^*} that takes into account the possible strain dependence of e^*. Thus, the Hamiltonian to be diagonalized is

$$H = H_\varepsilon + H_{e^*} + H_{\varepsilon,e^*}, \tag{53}$$

where H_{e^*} is the term responsible for the LO–TO splitting.

We consider only those situations that have been investigated experimentally (Cerdeira et al., 1972; Anastassakis, 1980; Weinstein and Cardona, 1972; Balslev, 1974; Anastassakis et al., 1988). If the uniaxial strain is along either [001] or [111], then H_ε has cylindrical symmetry around the strain direction, and any coordinate system that has this direction as one of its axes diagonalizes H_ε. The second term in Eq. (52) always has cylindrical symmetry around $\mathbf{q}$ (for the small values of $\mathbf{q}$ under consideration). Then, in the situation in which $\mathbf{q}$ lies in a plane perpendicular to the stress direction, the term $H_\varepsilon + H_{e^*}$ can be simultaneously diagonalized. The third term H_{ε,e^*} in Eq. (53) has to preserve the symmetry planes defined by the strain axis and $\mathbf{q}$. Therefore, in a coordinate system that has the stress direction along one of its axes and $\mathbf{q}$ along another H_{ε,e^*} must also be diagonal. That means that in such an experimental situation (i.e., uniaxial strain along [001] or [111] and $\mathbf{q}$ perpendicular to the strain axis), the results derived in the previous paragraph are still valid, and we only have to add the TO–LO splitting to the expressions for the frequency shifts of Eqs. (48), (49), and (50). Notice that one member of the stress doublet is now the LO phonon, while the singlet (TO_s) and the other stress doublet component (TO_d) are TO phonons. The equations for the frequency changes are:

LO phonon: $\quad \Omega_{\text{LO}} = \omega_{\text{LO}} + \Delta\Omega_H - (1/3)\Delta\Omega_S, \tag{54a}$

TO phonons: $\quad \Omega_d = \omega_{\text{TO}} + \Delta\Omega_H - (1/3)\Delta\Omega_S, \tag{54b}$

$$\Omega_s = \omega_{\text{TO}} + \Delta\Omega_H + (2/3)\Delta\Omega_S,$$

where ω_{TO} and ω_{LO} are the unperturbed TO and LO frequencies, and $\Delta\Omega_H$ and $\Delta\Omega_S$ are given by Eqs. (49) and (50) (with ω_0 replaced by either ω_{TO} or ω_{LO}). The contribution of the term H_{ε,e^*} (stress dependence of the effective charge) may be accounted for by using two sets of values for p, q, and r in Eqs.

TABLE IV

VALUES OF THE MODE GRUNEISEN PARAMETER $\gamma = -[(p+2q)/6\omega_0^2]$, $(p-q)$ AND r FOR THE $\mathbf{q} \approx 0$ OPTIC PHONONS OF SEVERAL DIAMOND- AND ZINCBLENDE-TYPE SEMICONDUCTORS.
(All values are taken from Landolt-Börnstein (1982). unless otherwise indicated.) For completeness, we have also listed the value of ω_0^2

	γ	$p-q$ (10^{27} sec^{-2})	r (10^{27} sec^{-2})	ω_0^2 (10^{27} sec^{-2})
Si	0.9 1.02	6.0^a	-6.3^a	9.7
Ge	1.12 0.89^a	1.4 1.47^a	-2.7 $-3.5^{a,b}$	3.2
GaAs	1.23^c 1.39^d	1.7	-1.5	3.0^c 2.6^d
InAs	0.85^a	1.72	-1.3	2.1^c 1.7^d
GaP	0.95^c 1.16^c 1.1^c 1.09^d 1.19^d 1.07^d	3.9	-2.6	5.7^c 4.8^d
InP	1.19^{c,e} 0.82^{c,e} 1.24^{c,e} 1.48^{d,e} 0.96^{d,e} 1.44^{d,e}	10^{c,e} 8.1^{c,e} 4.6^{d,e} 2.8^{d,e}	$-0.8^{c,e}$ $-1.6^{d,e}$ $-1.3^{d,e}$	4.2^c 3.3^d
AlSb	1.15^{c,f} 1.21^{d,f}	0.97^{c,f} 0.55^{d,f}	0.34^{c,f} $-0.71^{d,f}$	4.1^{c,f} 3.6^{d,f}
GaSb	1.21^c 1.23^d	0.82	-2.0	2.0^c 1.9^d
InSb	1.5^g	1.2^g	-0.8^g	1.4^c 1.2^d
ZnSe	1.8 1.7	1.8	-0.6	1.6^d

^{a}Cerdeira et al. (1972).
^{b}There is a factor of 10 error for r/ω_0^2 for Ge in Cerdeira et al. (1972).
^{c}LO phonon.
^{d}TO phonon.
^{e}Anastassakis et al. (1988).
^{f}E. Anastassakis and M. Cardona (1987). Solid State Comm. 63, 893.
^{g}E. Anastassakis, F. H. Pollak, and G. W. Rubloff (1972). Proc. ††th Int. Conf. Physics of Semiconductors, Warsaw (PWN-Polish Scientific Publishers, Warsaw) p. 1188.

(49) and (50), one for LO phonons and another for TO phonons. The strain dependence of e^* has been discussed by Anastassakis and Cardona (1985).

When the strain direction lies along the [011] axis, the requirement that $\mathbf{q}$ be perpendicular to the strain axis is not sufficient to ensure diagonalization of the Hamiltonian of Eq. (53). If we take $\mathbf{q}$ so as to coincide with one of the principal axes of H_ε [given by Eq. (52)] in the plane perpendicular to the strain direction, then the previous arguments may be repeated. For instance, if the strain axis is along [011] and $\mathbf{q} \parallel [011]$, we have Ω_{LO} given by Eq. (51a), with ω_0 replaced by ω_{LO}. For the TO phonon, Ω_d is given by Eq. (51b), Ω_s is equal to (51d), and ω_0 is replaced by ω_{TO}.

However, if $\mathbf{q}$ is chosen in an arbitrary direction in a plane perpendicular to the stress axis, the shifts of the LO and TO_d modes will be nonlinear due to stress-induced mixing of these two modes. These cases have not been investigated experimentally.

Listed in Table IV are experimental values of the mode Gruneisen parameter γ, $(p-q)$, and r for several diamond- and zincblende-type semiconductors. For the sake of completeness, we also list ω_0^2. Theoretical calculations for p, q, and r are discussed in Cerdeira (1971), Cerdeira et al. (1972), and Bell (1972).

V. Summary

This article has reviewed the effects of homogeneous strain, either externally applied or internally generated, on the electronic band structure and vibrational levels (phonons) of diamond- and zincblende-type semiconductors. The changes in these states can be described in terms of "deformation potentials." To fully describe the electronic energy levels of strained heterostructures, it is necessary to understand the influence of strain on the properties of the host materials. These include changes in energy gaps, splittings due to lowering of symmetry, and variations in effective masses. A considerable amount of information has been gained in the former two areas from external stress measurements of the optical and transport properties of these materials. In the latter category, we have considered the effects of stress-induced valence-band mixing, although other less-studied interactions are also significant. Electronic deformation potentials for a number of relevant bands have been summarized. In addition, some recent works on the impact of external stress on the quantum levels of GaAs/GaAlAs multiple quantum wells has been discussed. The influence of strain on the $\mathbf{q} \approx 0$ optic phonons has been described and vibrational "deformation potentials" have been listed. The strain dependence of these phonons is extremely useful for the characterization of the built-in strain in strained-layer superlattices.

2. EFFECTS OF HOMOGENEOUS STRAIN

Appendix A

$$H'_{11} = (1 + D'_c)(\hbar^2 k^2/2m) + a_c(\varepsilon_{xx} + \varepsilon_{yy} + \varepsilon_{zz}). \tag{A1}$$

$$H'_{13} = (\hbar/2\sqrt{m})[iP(k_x + ik_y)]. \tag{A2}$$

$$H'_{14} = -(\hbar/\sqrt{3m})iPk_z. \tag{A3}$$

$$H'_{33} = [A'k^2 - (B'/2)(2k_z^2 - k_x^2 - k_y^2)](\hbar^2/2m) \tag{A4}$$
$$- (a_1 + a_2)(\varepsilon_{xx} + \varepsilon_{yy} + \varepsilon_{zz})$$
$$- [(b_1 + 2b_2)/2](2\varepsilon_{zz} - \varepsilon_{xx} - \varepsilon_{yy}).$$

$$H'_{34} = -[(1/\sqrt{3})N'(k_x k_z - ik_y k_z)](\hbar^2/2m) + (d_1 + 2d_2)(\varepsilon_{xz} - \varepsilon_{yz}). \tag{A5}$$

$$H'_{35} = -[(\sqrt{3}/2)B'(k_x^2 - k_y^2) + (1/\sqrt{3})N'k_x k_y](\hbar^2/2m) \tag{A6}$$
$$- [\sqrt{3}(b_1 + 2b_2)/2](\varepsilon_{xx} - \varepsilon_{yy})$$
$$- i(d_1 + 2d_2)\varepsilon_{xy}.$$

$$H'_{37} = [(1/\sqrt{6})N'(k_x k_z - ik_y k_z)](\hbar^2/2m) - (d_1 - d_2)(\varepsilon_{xz} - i\varepsilon_{yz}). \tag{A7}$$

$$H'_{38} = [(\sqrt{3}/2)B'(k_x^2 - k_y^2) - i(\sqrt{2/3})N'k_x k_y](\hbar^2/2m) \tag{A8}$$
$$+ (\sqrt{3/2})(b_1 - b_2)(\varepsilon_{xx} - \varepsilon_{yy})$$
$$+ i\sqrt{2}(d_1 - d_2)\varepsilon_{xy}.$$

$$H'_{44} = [A'k^2 + (B'/2)(2k_z^2 - k_x^2 - k_y^2)](\hbar^2/2m) \tag{A9}$$
$$- (a_1 + a_2)(\varepsilon_{xx} + \varepsilon_{yy} + \varepsilon_{zz})$$
$$+ [(b_1 - b_2)/2](2\varepsilon_{zz} - \varepsilon_{xx} - \varepsilon_{yy}).$$

$$H'_{47} = [(B'/\sqrt{2})(2k_z^2 - k_x^2 - k_y^2)](\hbar^2/2m) \tag{A10}$$
$$- [(b_1 - b_2)/\sqrt{2}](2\varepsilon_{zz} - \varepsilon_{xx} - \varepsilon_{yy}).$$

$$H'_{48} = -[(1/\sqrt{2})N'(k_x k_z - ik_y k_z)](\hbar^2/2m) \tag{A11}$$
$$+ (\sqrt{3/2})(d_1 - d_2)(\varepsilon_{xz} - i\varepsilon_{yz}).$$

$$H'_{77} = A'_{so}\hbar^2 k^2/2m - (a_1 - 2a_2)(\varepsilon_{xx} + \varepsilon_{yy} + \varepsilon_{zz}), \tag{A12}$$

where:

$$P = -i\sqrt{2/m}\langle s|p_x|X\rangle. \tag{A13}$$

$$D'_c = (2/m) \sum_j^{\Gamma_{15},\Gamma_{25'}} \frac{|\langle S|p_x|u_j\rangle|^2}{(E_c - E_j)}. \tag{A14}$$

$$F' = (2/m) \sum_j^{\Gamma_1, \Gamma_{2'}} \frac{|\langle X|p_x|u_j\rangle|^2}{(E_v - E_j)}. \tag{A15}$$

$$G = (1/m) \sum_j^{\Gamma_{12}, \Gamma_{12'}} \frac{|\langle X|p_x|u_j\rangle|^2}{(E_v - E_j)}. \tag{A16}$$

$$H_1 = (2/m) \sum_j^{\Gamma_{15}, \Gamma_{15}} \frac{|\langle X|p_x|u_j\rangle|^2}{(E_v - E_j)}. \tag{A17}$$

$$H_2 = (2/m) \sum_j^{\Gamma_{25}, \Gamma_{25}} \frac{|\langle X|p_y|u_j\rangle|^2}{(E_v - E_j)}. \tag{A18}$$

$$L' \equiv F' + 2G. \tag{A19}$$

$$M \equiv H_1 + H_2. \tag{A20}$$

$$N = F' - G + H_1 - H_2. \tag{A21}$$

The above units of P, D'_c, F', G, H_1 and H_2 are somewhat different from those of Kane (1966) in order to make the units of energy consistent in the H'_{ij}.

The sums in Eqs. (A14)–(A18) are over all states u_j transforming like Γ_j as indicated above the summation sign. The Γ on the left is for zincblende and on the right for diamond.

$$A' = (1/3)(L' + 2M) + 1. \tag{A22}$$

$$B' = (1/3)(L' - M). \tag{A23}$$

$$A'_{so} = (1/3)(L'_{so} + 2M_{so}) + 1, \tag{A24}$$

where L'_{so} and M_{so} are L' and M given above with $E_v \to E_v - \Delta_0$.

Appendix B

The coefficients α^m and β^m for [001] and [111] are listed in Laude *et al.* (1971). If the strain dependence of the spin–orbit splitting is neglected ($\delta E'_H = \delta E_H$ and $\delta E'_S = \delta E_S$), these factors can be written as:

$$\alpha^m = [2\sqrt{2}|\delta E_S^m|]/q_m, \tag{B1}$$

$$\beta^m = (n_m - p_m)(\delta E_S^m)/q_m|\delta E_S^m|, \tag{B2}$$

where

$$p_m = \Delta_0 + \delta E_S^m, \tag{B3}$$

$$n_m = [p_m^2 + 8(\delta E_S^m)^2]^{1/2}, \tag{B4}$$

$$q_m = [2n_m(n_m - p_m)]^{1/2}. \tag{B5}$$

The situation for [011] is more complex, since the $|3/2, 3/2\rangle^{(011)}$ band is coupled to both the $|3/2, -1/2\rangle^{(011)}$ and $|1/2, -1/2\rangle^{(011)}$ states. The complete expression for the v_1, v_2, and v_3 wave functions are given in Laude et al. (1971). However, if this small coupling is neglected, the [011] case is qualitatively similar to [001] and [111] as given above.

The $|J, M_J\rangle^{(111)}$ wave functions are given by:

$$|3/2, 3/2\rangle^{(111)} = (1/2)|[(X - Y) + (i/\sqrt{3})(X + Y - 2Z)]\uparrow\rangle \tag{B6}$$

$$|3/2, 1/2\rangle^{(111)} = (1/6)|[\sqrt{3}(X - Y) + i(X + Y - 2Z)]\downarrow$$
$$- [2\sqrt{2}(X + Y + Z)\uparrow]\rangle. \tag{B7}$$

$$|3/2, -1/2\rangle^{(111)} = -(1/6)|[\sqrt{3}(X - Y) - i(X + Y - 2Z)]\uparrow$$
$$+ [2\sqrt{2}(X + Y + Z)\downarrow]\rangle. \tag{B8}$$

$$|3/2, -3/2\rangle^{(111)} = -(1/2)|[(X - Y) - (i/\sqrt{3})(X + Y - 2Z)]\downarrow\rangle. \tag{B9}$$

$$|1/2, 1/2\rangle^{(111)} = (1/3)|[(\sqrt{2/3})(X - Y) + (i/\sqrt{2})(X + Y - 2Z)]\downarrow$$
$$+ (X + Y + Z)\uparrow\rangle. \tag{B10}$$

$$|1/2, -1/2\rangle^{(111)} = (1/3)|[(\sqrt{2/3})(X - Y) - (i/\sqrt{2})(X + Y - 2Z)]\uparrow$$
$$- (X + Y + Z)\downarrow\rangle. \tag{B11}$$

Appendix C

At $\mathbf{k} = 0$, the relations between the Bir and Pikus (1974), Kleiner and Roth (1959), and Kane (1970) deformation potentials are

Bir–Pikus	Kleiner–Roth	Kane
a	D_d^v	$D_1^1/\sqrt{3}$
b	$-(2/3)D_u$	$D_3/\sqrt{3}$
d	$-(2/\sqrt{3})D_{u'}$	$D_5/\sqrt{2}$

At $\mathbf{k} \neq 0$, the relations between the Brooks (1955), Herring and Vogt (1956), and Kane (1970) deformation potentials are

Brooks	Herring–Vogt	Kane	
$\mathscr{E}_1$	$\Xi_d + (1/3)\Xi_u$	$D_1^1/\sqrt{3}$	
$\mathscr{E}_2$	Ξ_u	$(\sqrt{2/3})D_1^3$	$\mathbf{k}$ along $\langle 001 \rangle$
$\mathscr{E}_2$	$\Xi_u/2$	$(\sqrt{3/2})D_1^5$	$\mathbf{k}$ along $\langle 111 \rangle$

Appendix D

$$H^0_{11} = [Ak^2 - (B/2)(2k_z^2 - k_x^2 - k_y^2)](\hbar^2/2m). \tag{D1}$$

$$H^0_{12} = -[(1/\sqrt{3})N(k_x k_z + i k_y k_z)](\hbar^2/2m). \tag{D2}$$

$$H^0_{13} = -[(\sqrt{3}/2)B(k_x^2 - k_y^2) + (i/\sqrt{3})N k_x k_y](\hbar^2/2m). \tag{D3}$$

$$H^0_{55} = A_{so}\hbar^2 k^2/2m. \tag{D4}$$

where

$$A = A' + P^2/3E_0. \tag{D5}$$

$$B = B' + P^2/3E_0. \tag{D6}$$

$$N = N' + P^2/3E_0. \tag{D7}$$

$$A_{so} = A'_{so} + P^2/3(E_0 + \Delta_0). \tag{D8}$$

The relations between the valence band mass parameter A, B, and C and the Luttinger mass parameters (γ_1, γ_2, and γ_3) are

$$\gamma_1 = -A. \tag{D9}$$

$$\gamma_2 = -(1/2)B. \tag{D10}$$

$$\gamma_3 = -(1/6)N = -(1/6)[3C^2 + 9B^2]^{1/2}. \tag{D11}$$

References

Anastassakis, E. (1980). in "Dynamical Properties of Solids" (G. K. Horton and A. A. Maradudin, eds.). North Holland, New York.
Anastassakis, E., and Cardona, M. (1985). *Phys. Stat. Sol.* **129** (b), 101.
Anastassakis, E., Raptis, Y. S., Hunermann, M., Richter, W., and Cardona, M. (1988). *Phys. Rev.* **B38**, 7702.
Aspnes, D. E., and Cardona, M. (1978). *Phys. Rev.* **B17**, 726.
Balslev, I. (1972). "Semiconductors and Semimetals" (R. K. Willardson and A. C. Beer, eds.), Vol. 9. Academic Press, New York.
Balslev, I. (1974). *Phys. Stat. Sol.* **61** (b), 201.
Bell, M. I. (1972). *Phys. Stat. Sol.* **53(b)**, 675.
Bir, G. L., and Pikus, G. E. (1974). In "Symmetry and Strain-Induced Effects in Semiconductors," John Wiley, New York.
Blacha, A., Presting, H., and Cardona, M. (1984). *Phys. Stat. Sol.* **126** (b), 11.
Brooks, H. (1955). In "Advances in Electronics and Electron Physics" (L. Marton, ed.), Vol. 8. Academic Press, New York.
Cardona, M. (1967). *Solid State Comm.* **5**, 233.
Cerdeira, F. (1971). Ph.D. Thesis, Brown University (unpublished).
Cerdeira, F., Buchenauer, C. J., Pollak, F. H., and Cardona, M. (1972). *Phys. Rev.* **B5**, 580.
Chandrasekhar, M., and Pollak, F. H. (1977). *Phys. Rev.* **B15**, 2127.
Chelikowsky, J. R., and Cohen, M. L. (1976). *Phys. Rev.* **B14**, 556.

2. EFFECTS OF HOMOGENEOUS STRAIN

Collins, R. T., Vina, L., Wang, W. I., Maihiot, C., and Smith, D. L. (1987). In "Proceedings of the Society of Photo-Optical Instrumentation Engineers," Vol. 792. SPIE, Bellingham.
Dresselhaus, G., Kip, A. F., and Kittel, C. (1955). *Phys. Rev.* **98**, 36.
Gil, B., Lefebvre, P., Mathieu, H., Platero, G., Alterelli, M., Fukunaga, T., and Nakashima, H. (1988). *Phys. Rev.* **B38**, 1215.
Hasegawa, H. (1963). *Phys. Rev.* **129**, 1029.
Hensel, J. C., and Feher, G. (1963). *Phys. Rev.* **129**, 1041.
Hermann, C., and Weisbuch, C. (1977). *Phys. Rev.* **B15**, 823.
Herring, C., and Vogt, E. (1956). *Phys. Rev.* **101**, 933.
Jagannath, C., Koteles, E. S., Lee, J., Chen, Y. J., Elman, B. S., and Chi, J. Y. (1986). *Phys. Rev.* **B34**, 7027.
Kane, E. O. (1966). "Semiconductors and Semimetals" (R. K. Willardson and A. C. Beer, eds.) Vol. 1. Academic Press, New York.
Kane, E. O. (1970). *Phys. Rev.* **178**, 1368.
Kleiner, W. H., and Roth, L. M. (1959). *Phys. Rev. Letts.* **2**, 334.
Landolt-Börnstein (1982). In "Numerical Data and Functional Relationships in Science and Technology" (O. Madelung, M. Schulz and H. Weiss, eds.), Vols. 17a and 17b. Springer, New York.
Langer, D. W., Euwama, R. N., Era, K., and Koda, T. (1970). *Phys. Rev.* **B2**, 4005.
Laude, L., Pollak, F. H., and Cardona, M. (1971). *Phys. Rev.* **B3**, 2623.
Lee, J., Jagannath, C., Vassell, M. O., and Koteles, E. S. (1988). *Phys. Rev.* **B37**, 4164.
Lee, J., and Vassell, M. O. (1988a). *Phys. Rev.* **B37**, 8855.
Lee, J., and Vassell, M. O. (1988b). *Phys. Rev.* **B37**, 8861.
Luttinger, J. M. (1956). *Phys. Rev.* **102**, 1030.
Mailhiot, C., and Smith, D. L. (1987). *Phys. Rev.* **B36**, 2942.
Mailhiot, C., and Smith, D. L. (1988). *Solid State Comm.* **66**, 859.
Mailhiot, C. (1989). Private communication.
Nolte, D. D., Walukiewicz, W., and Haller, E. E. (1987). *Phys. Rev. Letts.* **59**, 501.
Pollak, F. H. (1973). *Surface Science* **37**, 863.
Rohner, P. G. (1971). *Phys. Rev.* **B3**, 433.
Sanders, G. D., and Chang, Y. C. (1985). *Phys. Rev.* **B32**, 4282.
Sooryakumar, A., Pinczuk, A. C., Gossard, D. S., Chemla, and Sham, L. J. (1987). *Phys. Rev. Lett.* **58**, 1150.
Verges, J. A., Glotzel, D., Cardona, M., and Anderson, O. K. (1982). *Phys. Stat. Sol.* **113** (b), 519.
Wardzynski, W., and Suffczynski, M. (1972). *Solid State Comm.* **10**, 417.
Weinstein, B. A., and Cardona, M. (1972). *Phys. Rev.* **B5**, 3120.

CHAPTER 3

Optical Studies of Strained III–V Heterolayers

J. Y. Marzin and J. M. Gérard*

CENTRE NATIONAL D'ETUDES DES TÉLÉCOMMUNICATIONS
BAGNEUX, FRANCE

P. Voisin

DÉPARTEMENT DE PHYSIQUE ECOLE NORMALE SUPÉRIEURE
PARIS, FRANCE

J. A. Brum†

IBM THOMAS J. WATSON RESEARCH CENTER
YORKTOWN HEIGHTS, NEW YORK

I. INTRODUCTION	56
II. STRUCTURAL ASPECTS	57
A. Thermodynamic Approach to the Critical Layer Thickness	58
B. Elastic Properties of Strained-Layer Heterostructures	62
III. EFFECTS OF STRAIN ON THE BAND STRUCTURES	64
A. Bulk Semiconductors	64
B. Superlattices	66
C. Three-Band Envelope Function Model for Strained-Layer Superlattices	68
D. In-Plane Valence Sub-bands of Externally Stressed GaAs–$Ga_{0.7}Al_{0.3}As$ Quantum Wells	72
IV. MODERATELY STRAINED SYSTEMS	82
A. $In_xGa_{1-x}As$–$GaAs$	83
B. $In_xGa_{1-x}As$–$In_yAl_{1-y}As$	91
C. $In_xAl_{1-x}As$–$GaAs$ on $GaAs$	94
D. $GaSb$–$AlSb$	95
E. $In_xGa_{1-x}As$–$In_yGa_{1-y}As$ on InP	101

*Member of the Direction des Rechaches, Etudes et Techniques, French Ministry of Defense, Paris.
†Permanent address: Depto. de Física do Estado Sólido e Ciencia dos Materiais, Universidade Estadual de Campiras (SR), Brazil.

V.	LARGE-STRAIN SYSTEMS.	101
	A. *InAs on GaAs*.	102
	B. *InAs–GaAs on InP*.	107
VI.	CONCLUSION	114
	ACKNOWLEDGMENT.	115
	REFERENCES.	115

I. Introduction

Since the pioneering work of Esaki and Tsu (1970), considerable efforts have been directed toward the growth of semiconductor heterolayers such as quantum wells (QW) and superlattices (SL) (for a recent review, see *Surface Science* **196**, *J. Phys.* (Paris) **C5**, 1987). For a long time, the attention focussed on lattice-matched materials until Osbourn (Osbourn, 1982) pointed out that the growth of thin layers (typical thickness 100 Å), which is necessary to achieve sizeable quantum confinement effects, in fact relaxes the requirement of lattice matching between host materials inasmuch as any lattice mismatch could be elastically accommodated by strains of the host layers. Since then, a considerable body of work has been devoted to the strained heterolayers due to the more flexible tailorability of their electronic properties, which arises from the competition of quantum-size effects and strain-induced effects (Bir and Pikus, 1974).

The relative lattice mismatch $\delta a_0/a_0$ between the host materials of heterolayers varies from almost zero (GaAs–Ga$_{0.7}$Al$_{0.3}$As) to values as large as several percentages (7% between InAs and GaAs, for example). Usually, the strains are treated by using the elasticity theory, which is essentially a linear response formalism. It is thus quite reasonable for many systems of actual interest, e.g., InAs–GaSb ($\delta a_0/a_0 = 0.62\%$), GaSb–AlSb ($\delta a_0/a_0 = 0.65\%$) and obviously GaAs–Ga$_{1-x}$Al$_x$As ($\delta a_0/a_0 = 0.12\%$). It may have difficulties in handling the large mismatches such as those found between InAs–GaAs ($\delta a_0/a_0 = 7\%$). The latter system, however, is of great potential interest, because if the growth of short-period superlattices (also termed pseudoalloys) is achievable, it may constitute a suitable substitute to the ternary random alloy In$_{0.53}$Ga$_{0.47}$As, which is lattice matched to InP. The pseudoalloy would display similar electronic properties (bandgap in the 1.5 μm range) as found in the alloy without being plagued by the alloy scattering. Until very recently, attempts to grow high-quality short-period superlattices have failed, and essentially, the alloy scattering was replaced by a harmful interface roughness scattering.

To interpret the electrical and optical measurements performed on strained-layer materials, energy-level calculations have to take into account the strain state of the materials. Quite often, the envelope function formalism (see, e.g., Altarelli, 1985; Bastard, 1988), suitably modified to include the strain effects (Bir and Pikus, 1974; Pollak and Cardona, 1968), has been used (Marzin, 1985; Voisin, 1985) and proven successful. In the following, we shall exclusively use this formalism referring the reader to recent reviews (Bastard, 1988; Marzin, 1985; Voisin, 1985) for discussions. However, the envelope function formalism has obvious limits, such as its inability to handle interaction between the states derived from different extrema in the hosts' Brillouin zone. Such interactions necessarily take place, for instance, when one describes the conduction states of Si–Ge strained-layer superlattices (see Jaros, this volume), in which case the envelope function formalism must be replaced by more microscopic descriptions such as the empirical tight-binding method (see, e.g., Schulman and Chang, 1985) or the empirical pseudopotential formalism (Jaros et al., 1985; Jaros, this volume). The latter models may also prove necessary in handling short-period (a few monolayers) superlattices, especially of highly strained materials (e.g., InAs–GaAs).

The paper will be organized as follows. Section II will recall some features of the elasticity theory and present a discussion of the critical layer thickness beyond which the lattice mismatch can no longer be accommodated by biaxial tensile strains but gives rise to the formation of dislocation networks. Section III will present, within the envelope function formalism, a discussion of (i) the subband extrema (i.e., the energy levels corresponding to a zero in-plane wave vector) of strained heterolayers when one uses the three-band Kane model (Kane, 1957) (suitably modified to include strain effects) to describe the hosts' energy states, and (ii) the valence-subband dispersions of externally and uniaxially stressed GaAs–Ga$_{0.7}$Al$_{0.3}$As quantum wells. Finally, Sections IV and V will be devoted to a presentation of some results of the basic optical properties (absorption, photoluminescence) of a variety of III–V strained heterolayers emphasizing the diversity that results from the interplay of strain and confinement effects.

II. Structural Aspects

Consider two perfect crystals having the same structure but different lattice parameters a^A and a^B, and try to connect them along a plane interface normal to the z-axis. Two limiting situations are easily imagined: (i) exert biaxial stresses of opposite signs on materials A and B until they have the same in-plane lattice parameter $a^A_{x,y} = a^B_{x,y} = a_{x,y}$, and then let the atoms establish their chemical bonds across the interface; this is the strained-layer regime, characterized by a pseudomorphic or coherent interface, and an excess of energy equal to the elastic energy stored in the A and B layers; (ii) keep the unperturbed lattices, and let the atoms establish as many chemical bonds as possible across the interface: This is the relaxed-layer regime, characterized by an areal density of dangling bonds equal to the difference of the areal density of atoms on each side of the interface, $n_D \approx (1/a^A)^2 - (1/a^B)^2$. This number would be as large as 2.6×10^{11} cm^{-2} for the GaAs–Al$_{0.3}$Ga$_{0.7}$As interface!

To be specific, consider a layer of material B of thickness d^B, grown on a semi-infinite substrate of material A. In the strained-layer regime, the excess of energy is proportional to the epilayer thickness d^B, whereas in the relaxed-layer regime, this excess of energy is essentially a property of the interface, and it does not depend on the epilayer thickness. Thus, the two regimes will cross for some critical layer thickness d_c. This, however, is an oversimplified view, as the point defects in the relaxed-layer regime usually rearrange in a network of dislocations that have long-range strain fields. The important problem of calculating d_c is still a matter of investigation. Indeed, there are several models based on approximate calculations of the thermodynamic or mechanical equilibrium (Van der Merwe, 1972; Matthews and Blakeslee, 1974), but they seem to predict values of d_c definitely smaller than those found experimentally (People and Bean, 1985; Kasper, 1986). Furthermore, it was recently found, in the case of the Ge$_x$Si$_{1-x}$–Si system, that d_c depends on the growth temperature (Kasper, 1986), which tends to indicate that actual critical layer thicknesses are dominated by kinetic or thermally activated processes. We discuss in the following one possible thermodynamic approach.

A. THERMODYNAMIC APPROACH TO THE CRITICAL LAYER THICKNESS

The geometry of misfit dislocations in III–V heterostructures grown along the (001) axis has been studied in detail in the pioneering work of Matthews and Blakeslee (Matthews and Blakeslee, 1974, 1975, 1976). These dislocations where found to be straight lines lying along the (110) directions of the interface plane, with a Burger vector b parallel to the (101) or (011) directions, $|b| = a_0/\sqrt{2} = 4$ Å. Such a dislocation contributes to the accommodation of the lattice mismatch in the (001) plane by $a_0/2\sqrt{2}L$, if L is the length of the sample perpendicular to the dislocation line.

In text books, one can find that in a semi-infinite homogeneous material, a dislocation parallel to the free surface is attracted towards this surface with a force-per-unit length (Landau and Lifshitz, 1967; Nabarro, 1967):

$$F = \frac{\mu b^2}{4\pi K d}, \qquad (1)$$

where $1/K = \cos^2 \Psi + \sin^2 \Psi/(1-\nu)$, Ψ being the angle between the dislocation line and its Burger vector, $\mu = 1/S_{44}$ is the shear modulus,

$v = -S_{12}/S_{11}$ is Poisson's ratio (the S_{ij} are the elastic compliance constants), and d is the distance to the free surface. We immediately derive the energy-per-unit length E_d associated with the creation of such a dislocation:

$$E_d = \frac{\mu b^2 (\ln(d/b) + \Theta)}{4\pi K},\qquad(2)$$

where the constant $\Theta \approx 1$ stands for the core energy. According to Colonetti's theorem (Landau and Lifshitz, 1967), there is no interference term, in the expression of the total elastic energy, between the dislocation strain field and the homogeneous strain (which can be considered as resulting from an external stress). Thus, Eqs. (1) and (2) remain correct for a complex system of alternate layers where the inhomogeneous built-in strain is piecewise constant along the growth axis z. We now consider a rectangular sample consisting of an $L_x \times L_y$ substrate with a large thickness d_s, on top of which is an epilayer of a lattice-mismatched material of thickness d, as sketched in Fig. 1. The accommodation of the lattice mismatch $\delta a_0/a_0$ in the direction parallel to $L_{x[or\,y]}$ is supposed to be shared between a homogeneous strain $e_{xx[yy]}$ and a network of $n_{y[x]}$ dislocations parallel to $y[x]$:

$$|\delta a_0/a_0 - e_{xx}| = n_y \frac{a_0}{2\sqrt{2}L_x}; \qquad |\delta a_0/a_0 - e_{yy}| = n_x \frac{a_0}{2\sqrt{2}L_y}. \qquad(3)$$

If we neglect the interaction between dislocations, the excess of energy E associated with this configuration of the epilayer is the sum of the homogeneous elastic energy E_{el} and of the dislocation energy $(n_x L_x + n_y L_y) E_d$,

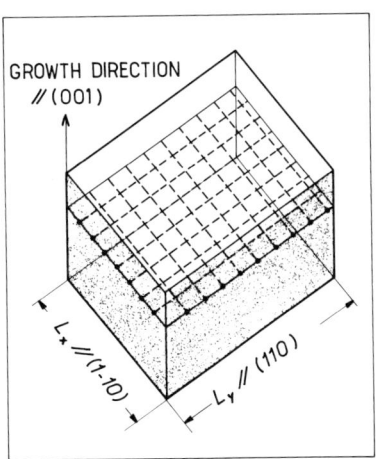

FIG. 1. Schematic representation of a square network of dislocations at the interface between a semi-infinite substrate and a partly relaxed epilayers.

where:

$$E_{el} = \left\{L_x L_y d \frac{\mu}{1-v}\right\} \{e_{xx}^2 + e_{yy}^2 + 2v e_{xx} e_{yy}\}. \quad (4)$$

Writing that $\partial E_{el}/\partial e_{xx} = \partial E_{el}/\partial e_{yy} = 0$, and and setting in these equations $e_{xx} = e_{yy} = \delta a_0/a_0$ then gives the critical layer thickness, at which the first misfit dislocation will be generated:

$$d_c = \left\{b(\delta a_0/a_0)^{-1} \frac{1 - v/4}{4\pi(1 + v)}\right\} \{\ln(d_c/b) + \Theta\}. \quad (5)$$

This expression is exactly one fourth of that deduced from the mechanical equilibrium model in the case of a superlattice where the lattice mismatch is equally shared by the two materials, as also discussed by Matthews and Blakeslee (1976). The thermodynamic approach described here is thus equivalent to their mechanical model. The critical layer thicknesses d_c versus the lattice mismatch $\delta a_0/a_0$ are plotted in Fig. 2. A representative figure is $d_c = 90$ Å for $\delta a_0/a_0 = 1\%$. It is clear that this calculation is not exact, because Eq. (1) applies to a semi-infinite homogeneous material, the approximation $\Theta = 1$ is a rough estimation of the core energy, and so on. However, these approximations are not likely to account for the discrepancy between experimental data (People and Bean, 1985; Kasper, 1986) and the prediction

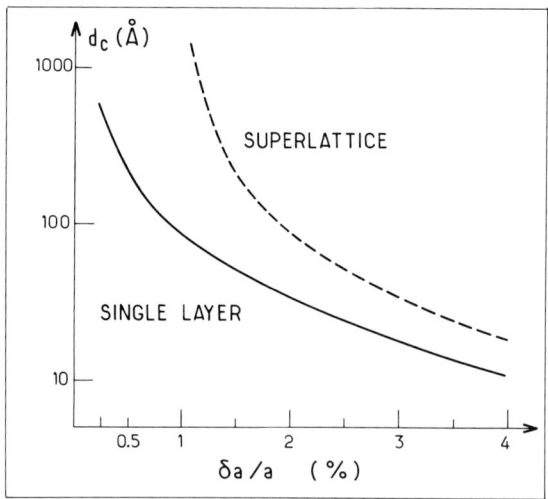

FIG. 2. Thermodynamic critical layer thickness d_c for a single layer grown on a thick substrate (solid line), and for a free-standing superlattice with equal layer thicknesses (dashed line) as a function of the lattice mismatch between the two materials.

of Eq. (5). In the model that we discuss here, the work required to bring the dislocation line from the surface to a distance αd ($\alpha \leqslant 1$) is:

$$W(\alpha d) = \left(b^2 \frac{\mu}{4\pi K}\right)\left(\ln\left(\frac{\alpha d}{b}\right) + \Theta\right) - b\mu e_{xx(yy)} \alpha d \frac{(1+\nu)}{(1-\nu)}. \qquad (6)$$

This quantity exactly vanishes for $\alpha = 0$ (dislocation outside the sample), and for $\alpha = 1$, at the thermodynamic equilibrium. In between, there is a potential barrier that has to be overcome. This accounts, at least qualitatively, for the thermally activated character of the experimental data. Note that W depends explicitly on the elastic constant μ, whereas the thermodynamic critical thickness d_c does not. W decreases with increasing lattice misfits, and therefore systems with large misfits are likely to evolve closer to the thermodynamic equilibrium than those with small mismatch.

For an epilayer grown along another direction, the situation is modified in two ways: (i) The geometry of the dislocation lines cannot obey the same conditions anymore, and (ii) the amount of lattice mismatch accommodated in the interface plane by one dislocation is changed. In particular, for a growth axis of low symmetry, the plastic relaxation probably shows some anisotropy. It turns out that the (001) growth axis is the most favorable with respect to these two points.

Finally, it is also of interest to examine the way the plastic relaxation occurs for a layer thickness d larger than d_c. We consider again the case of the (001) growth axis; for $d > d_c$, the lattice mismatch is shared between the homogeneous strain $e_{xx} = e_{yy} = e$ and the network of dislocations $n_x/L_y = n_y/nL_x = 2\sqrt{2}(\delta a_0/a_0 - e)/a_0$. We get:

$$\frac{e}{(\delta a_0/a_0)} = \left(\frac{d_c}{d}\right) \frac{\ln(d/b) + \Theta}{\ln(d_c/b) + \Theta}. \qquad (7)$$

If we neglect the logarithmic dependence in Eq. (7), the residual strain e relaxes to zero as d_c/d, that is rather slowly, and very large layer thicknesses are required if one wants to ensure a quasiperfect plastic relaxation. This is particularly important when a superlattice is grown on a buffer layer designed to match the equilibrium in-plane lattice parameter of the SL, and has to accommodate the misfit with respect to the substrate. An example of such a situation is illustrated in Fig. 3, which shows the x-ray double-diffraction spectrum obtained in a $Al_{0.9}In_{0.1}As(177\text{ Å})$–$GaAs(195\text{ Å})$ SL grown on top of a 5500-Å-thick $Al_{0.96}In_{0.04}As$ buffer layer (Sauvage et al., 1986). The quantitative fit of this diffraction spectrum, also shown in Fig. 3, indicates that the amount of plastic relaxation of the buffer layer is only 42%, which explains why the actual strain of the SL–GaAs layers (0.165%) is significantly different from the designed value (0.41%).

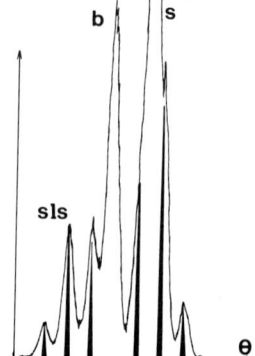

FIG. 3. Experimental and theoretical x-ray rocking curves in an $Al_{1-x}In_xAs$–GaAs superlattice (sls), grown on top of an $Al_{1-y}In_yAs$ buffer layer (b), which, in this case, presents a partial plastic relaxation with respect to the GaAs substrate (s).

To conclude this section, let us remark that for layer thicknesses small enough compared to the "thermodynamic" critical thickness discussed here, the heterostructure is certainly stable and can a priori be used to fabricate devices subjected to heating (field effect transistors, lasers, etc.). For layer thicknesses larger than d_c, it is perhaps still possible to get strained-layer structures, but clearly, these will be metastable and probably will have poorer device performances.

B. Elastic Properties of Strained-Layer Heterostructures

We now consider an A, B superlattice in the strained-layer regime. Two situations may be encountered: (i) The SL is in self-mechanical equilibrium, i.e., its in-plane lattice parameter $a_{x,y}$ is equal to the lattice parameter of the substrate (or buffer layer), a_s; (ii) the SL is strained as a whole to conform with the lattice of the substrate, which often occurs when the misfit is small and/or when one of the SL components is the substrate material itself (e.g., GaAs–AlGaAs or InGaAs–GaAs). In this case, the overall SL thickness is limited to some critical value, and the crystal is bent by the finite torque of the biaxial stress; the radii of curvature are of the order of meters, and this effect (though easily observed using x-Ray topography) does not change significantly the strain distribution in the epilayer. In the following, we suppose that the SL is in self-mechanical equilibrium. The layers in the SL exert biaxial tensile and compressive stresses on each other. A 1% strain involves stresses of the order of 10 kbar. With the usual definition of the strain tensor, $e_{ij} = \frac{1}{2}(\partial u_i/\partial x_j + \partial u_j/\partial x_i)$, where u is the displacement vector, the elastic properties of cubic materials are described by the simple matricial equation

$$\begin{bmatrix} \sigma_{xx} \\ \sigma_{yy} \\ \sigma_{zz} \\ \sigma_{xy} \\ \sigma_{yz} \\ \sigma_{zx} \end{bmatrix} = \begin{bmatrix} C_{11} & C_{11} & C_{12} & 0 & 0 & 0 \\ C_{12} & C_{11} & C_{12} & 0 & 0 & 0 \\ C_{12} & C_{12} & C_{11} & 0 & 0 & 0 \\ 0 & 0 & 0 & C_{44} & 0 & 0 \\ 0 & 0 & 0 & 0 & C_{44} & 0 \\ 0 & 0 & 0 & 0 & 0 & C_{44} \end{bmatrix} \begin{bmatrix} e_{xx} \\ e_{yy} \\ e_{zz} \\ 2e_{xy} \\ 2e_{yz} \\ 2e_{zx} \end{bmatrix}. \qquad (8)$$

The σ_{ij}s are the components of the stress tensor, and the C_{ij} are the elastic constants. The matrix of the compliances S_{ij} is the inverse of matrix C. The C_{ij} of the various III–V's are remarkably similar (Landolt-Bornstein, 1982), but not identical, and below, we keep the dependence on the C_{ij} explicit.

The analysis of the strain distribution is particularly simple for systems grown along the (001) axis, as the strain tensor reduces to $e_{xx} = e_{yy}$ and e_{zz}. From the absence of external stress on the (001) free surface, we immediately get for A [or B] material:

$$e_{zz}^{A[B]} = -\frac{2C_{12}^{A[B]}}{C_{11}^{A[B]}} e_{xx}^{A[B]} \approx -e_{xx}^{A[B]}. \qquad (9)$$

By minimizing the total elastic energy $E_{el} = \frac{1}{2}(d^A e_{ij}^A \sigma_{ij}^A + d^B e_{ij}^B \sigma_{ij}^B)$ with respect to the strain distribution, taking into account $e_{xx}^A - e_{yy}^B = \delta a_0/a_0$, one readily gets the equilibrium in-plane lattice parameter $a_{x,y}$ of the superlattice:

$$a_{x,y} = \frac{a^A d^A \xi^A + a^B d^B \xi^B}{d^A \xi^A + d^B \xi^B}, \qquad (10)$$

where $\xi^{A[B]} = (C_{11} + C_{12} - 2C_{12}^2/C_{11})^{A[B]}$.

The best experimental method to determine the strain state of a given heterostructure is certainly the quantitative analysis of x-ray double-diffraction rocking curves (Quillec et al., 1984). However, it is sometimes necessary to cross-check such structural analysis with other experimental data: For example, the strain changes the effective spring constants between the atoms, and therefore it shifts the frequency of the LO phonons (Jusserand et al., 1985; Abstreiter et al., 1986), which can be determined by Raman scattering. A representative figure for this effect is a shift of $3.5\,\text{cm}^{-1}$ for a strain $e_{xx} = 1\%$ (Jusserand et al., 1985), which means that the accuracy of this interesting method is relatively poor compared to x-ray data. Finally, the strain affects considerably the band structure of the host materials, and consequently the energy levels of the SL or QW structure.

III. Effects of Strain on the Band Structures

In the first two subsections of this part, we discuss the effects of the strains on the bulk materials and the strained-layer superlattices (SLS) band structure qualitatively. A more detailed effective mass model is discussed in Section III.C, whereas Section III.D deals with the in-plane dispersion curves of the valence subbands, calculated by means of effective mass approximations. In this latter part, a large number of stress situations are examined and illustrated by theoretical results on externally stressed GaAs–Ga$_x$Al$_{1-x}$As quantum wells.

A. BULK SEMICONDUCTORS

A detailed analysis of the effects of a strain on the band structure of bulk semiconductors can be found in Bir and Likus (1974), Pollak and Cardona (1968), and we will discuss here only the specific case of biaxial strains encountered in SLSs grown in the [001] orientation. As described in Section II, when the curvature of the ideally strained SLS is neglected, the strains in its constitutive sublayers are quadratic with z-axis, where z is the growth axis. The only nonvanishing components of the strain tensor $(e_{nm})^i$ in each material i are:

$$e^i_{xx} = e^i_{yy} = \frac{a_{x,y} - a^i}{a^i}, \qquad e^i_{zz} = -2\frac{C^i_{12}}{C^i_{11}} e^i_{xx}, \qquad (11)$$

where the C^i_{nm} and a^i are the elastic constants and bulk lattice parameter of material i; and $a_{x,y}$ is the in-plane lattice parameter of the whole superlattice.

The strain lowers the symmetry of the semiconductor from T_d to D_{2d} for III–V compounds. This symmetry change is the same that we have when going from a bulk material to a quantum well or superlattice, and has the same effect: The Γ_8 valence-band states are split into Γ_6 and Γ_7 states of D_{2d}, while the split-off band symmetry is Γ_7. The Γ_6 states correspond to $|3/2, \pm 3/2\rangle_z$, while the $|3/2, \pm 1/2\rangle_z$ and $|1/2, \pm 1/2\rangle_z$ valence-band states are coupled by the strain. In this basis, the strain Hamiltonian at $k = 0$ for the valence band reduces to two 3×3 blocks for $+m_j$ or $-m_j$:

$$\begin{bmatrix} |3/2, +3/2\rangle_z & |3/2, +1/2\rangle_z & |1/2, +1/2\rangle_z \\ -a\text{Tr}(e) + b(e_{zz} - e_{xx}) & 0 & 0 \\ 0 & -a\text{Tr}(e) - b(e_{zz} - e_{xx}) & \sqrt{2}b'(e_{zz} - e_{xx}) \\ 0 & \sqrt{2}b'(e_{zz} - e_{xx}) & \begin{pmatrix} -\Delta \\ -a'\text{Tr}(e) - b'(e_{zz} - e_{xx}) \end{pmatrix} \end{bmatrix},$$

(12)

where a, a', b, and b' are deformation potentials (in general, one assumes $b' = b$ and $a' = a$), $\text{Tr}(e) = e_{xx} + e_{yy} + e_{zz}$ and Δ is the spin–orbit splitting. The variation of the Γ_6 zone-center conduction states energy is:

$$E_c(e) - E_c(e = 0) = C \, \text{Tr}(e). \tag{13}$$

From Eqs. (12) and (13), it follows that the hydrostatic part of the strain modifies the bandgap by $(C + a)\text{Tr}(e)$, whereas the remaining (with zero trace) part splits the valence band by $2b(e_{zz} - e_{xx})$ (when Δ is large) and couples light-hole and split-off bands. The variations of the extrema of the valence bands with respect to the conduction-band edges are plotted in Fig. 4, as a function of the strain $e_{xx} = e_{yy}$ for GaAs and InAs.

The valence-band dispersion curves are also deeply affected by the strains. The most important effect is the so-called mass reversal. Whereas the effective masses along the z-axis are not strongly changed in the strained material, the in-plane masses are. The band that corresponds to $|3/2, \pm 3/2\rangle_z$ ("heavy holes") at $k = 0$, may display a light in-plane mass, and that at $k = 0$, corresponding to $|3/2, \pm 1/2\rangle_z$ a heavy in-plane mass. This effect can be simply understood when considering the Luttinger (Luttinger, 1956) $k \cdot p$ terms in the valence-band Hamiltonian: When the valence-band splitting becomes large enough, the in-plane dispersions are governed by the diagonal k-dependent terms alone. Taking for the numerical values the example of GaAs, the masses are then: For the "heavy hole" (at $k = 0$) band $m_z = 1/(\gamma_1 - 2\gamma_2) = 0.38 m_0$, $m_X = m_Y = 1/(\gamma_1 + \gamma_2) = 0.11 m_0$, and for the "light hole" (at $k = 0$) band $m_z = 1/(\gamma_1 + 2\gamma_2) = 0.09 m_0$, $m_X = m_Y = 1/(\gamma_1 - \gamma_2) = 0.21 m_0$, where γ_1, γ_2 are the two first Luttinger coefficients of GaAs.

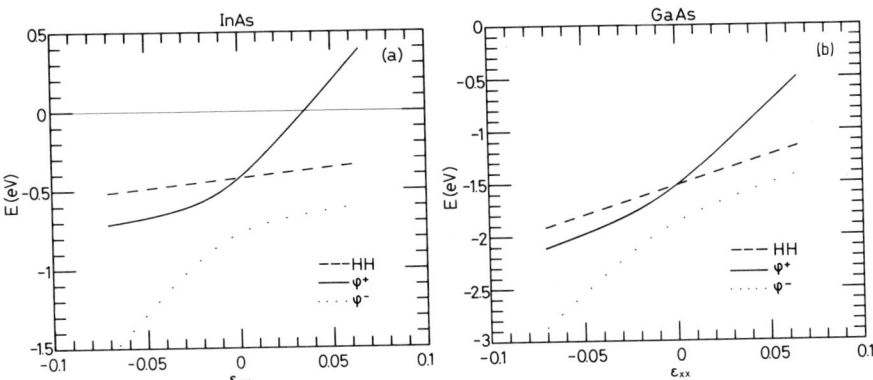

FIG. 4. Valence bands-to-conduction band zone-center energy separation for InAs and GaAs under biaxial strain. ϕ_+ and ϕ_- originate from the mixing between light-hole and spin–orbit split-off bands.

These effects on the zone-center band structure and on the dispersion curves of the stained-bulk materials will affect the superlattices band structures, as discussed in the next sections.

B. SUPERLATTICES

The most important feature resulting from the form of the strain Hamiltonian Eq. (12) is that the $|3/2, \pm 3/2\rangle_z$ heavy-hole states are left decoupled at $k_x = k_y = 0$ from the light-particle states, so that the effective mass treatment of the $k_x = k_y = 0$ states of the SLS is very similar to that of unstrained superlattices. The validity of such effective mass approach is questionable because of the difference in lattice parameter, but can be sorted out, and we will discuss this point in Section III.C. As in unstrained superlattices (Bastard, 1981), light-particle states are described in a three-band envelope function model, which includes the strain effects on the bulk materials, and the $|3/2, \pm 3/2\rangle_z$ "heavy"-hole states are treated separately (both for $k_x = k_y = 0$).

Schematically, the superlattice states for electrons, "light" holes and "heavy" holes, are obtained as the eigenvectors of three independent problems: Each of these particles is moving along the z-axis in the periodic potential arising from the corresponding strained-bulk material band-extrema modulation in the superlattice. Deviations from this simple picture are due to nonparabolicity and to the strain-induced mixing of the $|3/2, \pm 1/2\rangle_Z$ and $|1/2, \pm 1/2\rangle_Z$ valence-band states. The effective masses along the z-axis are merely affected by the strain (if the strain effects are small as compared to the bandgap and spin–orbit splitting Δ), so that the main effect of this built-in strain is to change the superpotential profiles. Depending on the strain sign in each material and to the band discontinuity, a large number of band configurations may be obtained in SLS. Some of them are shown in Fig. 5. In particular, the ordering of the valence-band states of the SLS is greatly affected by the strains. If we take the example of the situation of Fig. 5a or e, where the smaller bandgap material is a potential well for the three quasiparticles, we can have the following cases, depending on the sign of the strain in this small gap material:

(i) Unstrained: "heavy" and "light" holes are confined in the same potential. Due to difference in their effective masses, the first valence-band level is always HH_1. The energy difference between HH_1 and LH_1 depends on the potential depth and on the geometrical parameters.

(ii) Under biaxial compression (Fig. 5a, the strain-induced splitting of the bulk valence band pushes the "heavy"-hole band to higher energy than the "light"-hole one, and the energy separation between HH_1 and LH_1 is increased by this splitting. HH_1 is still the first SLS valence level at zone

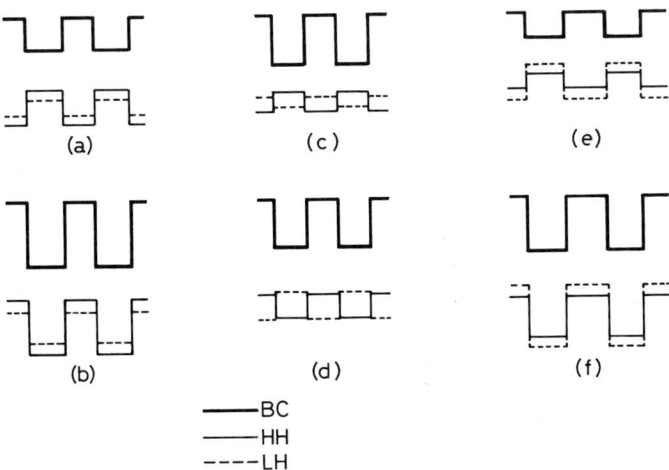

FIG. 5. Possible band-extrema configurations for strained-layer superlattices.

center, and the corresponding subband has a (with either 5(a) or (c)) light in-plane mass. $In_xGa_{1-x}As$–GaAs displays this strain configuration; the mass reversal effect has been observed experimentally (Schirber et al., 1985) and used in p-type modulation-doped $In_xGa_{1-x}As$–GaAs multiquantum-well (MQW) samples (Zipperian et al., 1988).

(iii) Under biaxial tension (Fig. 5e, the difference in confinement effects for HH_1 and LH_1 and the strain-induced splitting are of opposite signs. For large strains and thick wells, the first level can be LH_1 with a high in-plane mass (as in the bulk well material under tensile strain), as first shown in the GaSb–AlSb system (Voisin et al., 1984). Reducing the well thickness (or decreasing the strain in the well) decreases the energy difference between HH_1 and LH_1, which may even become negative.

The ordering of the valence subbands is very important because of the mixing of HH and LH states for nonzero k_x or k_y (see, e.g., Altarelli, 1985; Alterelli et al., 1985). Strain, either built-in or externally applied, thus appears as an additional and quasi-independent way to tune the in-plane dispersion relations of the valence band (see Section III.D).

When the smaller bandgap material is in compression, the situation of Fig. 5c may be encountered, where the superpotentials tend to confine the electrons and "heavy" holes in one material, whereas the light-hole states are confined in the other. Although there is a controversy on this subject, we have suggested (Marzin et al., 1985) that the $In_xGa_{1-x}As$–GaAs system, strained on GaAs substrates, displays this original feature.

In $In_xGa_{1-x}As-In_yAl_{1-y}As$ system, the sign of the strain in the smaller bandgap material $In_xGa_{1-x}As$ can be chosen, and we will discuss the (ii) and (iii) situations.

In these III–V SLS for built-in strains of 1% (which are easily obtained), the splitting of valence bands and the change in the bandgaps are of the order of 80 meV. This figure falls in the same range as the confinement energies: Strain effects cannot in general be neglected and deeply affect the optical properties of these structures, as we will see in Sections V and VI.

A last remark must be made on the SLS that concerns the band offsets. Tight-binding self-consistent calculations (Priester et al., 1988) clearly show that there is some dependence of the band discontinuities on the strain. Some care must be thus taken when comparing the same system in different strain situations (due to the substrate nature, for example) and, for the systems including alloys, when trying to extrapolate from one composition to another.

C. Three-Band Envelope Function Model for Strained-Layer Superlattices

In the application conditions of the effective mass approximation, we find that the perturbation potential has to be a slowly varying function, as compared with the unit cell size. Furthermore, as it is a perturbation theory, this potential has to remain small as compared to the unperturbed potential. In a strained material, neither of these two conditions is fulfilled. Starting from

$$H_0 = \frac{p_L^2}{2m_0} + V_0(r_L) + V_{so}(r_L) \tag{14}$$

and straining the material, the perturbed Hamiltonian is

$$H = \frac{p_E^2}{2m_0} + V(r_E) + V'_{so}(r_E). \tag{15}$$

$V_0(r) - V(r)$ is not slowly varying and can be very large due to the change in the atomic positions. The trick used in the bulk homogeneously strained materials (Bir and Likus, 1974) is to make a change of coordinates for the perturbed problem from Euler coordinates r_E to Lagrange r_L ones. The periodicity of H_0 is then restored for H. This change of coordinates is given by:

$$r_E = r_L + u(r_L); \tag{16}$$

and for a given function $f_E(r_E) = f_L(r_L)$, we have:

$$p_L f_L = (I + e) p_E f_E, \tag{17}$$

where e is the strain tensor. The Brillouin zones are changed by the change in coordinates and k_L corresponds to $(I + e)k_E$. To keep the scalar product unchanged in the transformation, one has to use:

$$\langle f_L, g_L \rangle = \int f_L^*(r_L) G_L(r_L) \det(I + e) d^3 r_L. \qquad (18)$$

The rewriting of H in these coordinates yields e-dependent terms in the kinetic energy on the one hand and allows us to write, to first order in e,

$$V_0(r_L) + V_{so}(r_L) - V(r_L) - V'_{so}(r_L) = D_{ij} e_{ij}, \qquad (19)$$

where the D_{ij} terms are the deformation potentials and where the k-dependent terms coming from the spin–orbit terms have been dropped.

A three-band Kane model description of the homogeneously strained components of the superlattice (Marzin, 1987) yields the following dispersion relation for the light particles at $k_x = k_y = 0$:

$$E\{[E + E_g + (C + a)\text{Tr}(e) - b\delta e][E + E_g + \Delta + (C + a')\text{Tr}(e)] - 2(b'\delta e)^2\}$$

$$= (1 - 4ezz) P_0^2 \hbar^2 k_z^2 \left[E + E_g + 2\frac{\Delta}{3} + \left(C + 2\frac{a'}{3} + \frac{a}{3} \right) \text{Tr}(e) \right.$$

$$\left. + \left(4\frac{b'}{3} - \frac{b}{3} \right) \delta e \right], \qquad (20)$$

where E_g and Δ are the bandgap and spin–orbit splitting of the unstrained material, respectively. P_0 is the Kane matrix element for the unstrained material, k_z is the wave vector corresponding to the Lagrange coordinates. $\text{Tr}(e) = e_{xx} + e_{yy} + e_{zz}$ and $\delta e = e_{zz} - e_{xx}$. a, a', b, and b' are the deformation potentials of Eq. (12).

In the strained-layer superlattices, an additional problem comes from the spatial variation of the strain tensor. These superlattices can be considered as built from two hydrostatically strained materials having the lattice constant corresponding to their in-plane lattice constant in the superlattice. The two starting (hydrostatically strained) materials are assumed to have the same Kane matrix elements $P'_0 = (1 - e_{xx})P_0$. This situation is taken as being the reference described in Lagrange coordinates. The two materials are then further strained with a strain tensor with only one nonvanishing component, $e'_{zz} = e_{zz} - e_{xx} = \delta e$, to build the superlattice. e'_{zz} depends only on z and is constant in each material. The dispersion relations for the constituents become, with the origin at the bottom of the strained-material conduction band,

$$E\{[E + E_g + (C + a)\text{Tr}(e) - b\delta e][E + E_g + \Delta + (C + a')\text{Tr}(e)] - 2(b'\delta e)^2\}$$

$$= (1 - 4\delta e) P_0'^2 \hbar^2 k_z^2 \left[E + E_g + 2\frac{\Delta}{3} + \left(C + 2\frac{a'}{3} + \frac{a}{3} \right) \text{Tr}(e) + \left(4\frac{b'}{3} - \frac{b}{3} \right) \delta e \right].$$

$$(21)$$

The steplike function δe (and, as in the case of unstrained superlattices, the band discontinuity) has to be further considered as a slowly varying function with respect to the bulk material lattice constants to be treated in effective-mass-type models.

Similarly to the unstrained case (Bastard, 1981), trial solutions have the form:

$$\Psi(r_L) = \sum_n \frac{F_n(r_L)U_{n0}(r_L)}{1 + \delta e/2}, \qquad (22)$$

where the sum is over the six light-particle bands of the Kane model, and where the U_{n0} are the zone-center wave functions corresponding to these six bands, in the hydrostatically deformed materials. They are assumed to be the same in the two materials.

Assuming that F, e, and its derivative with respect to the growth axis are slowly varying, one obtains, if $n = 1$ corresponds to the conduction band, the requirement that

$$(1 - \delta e)F_1(rL)$$

$$(1 + \delta e)\left[\frac{1}{m(z, E)}\right]\partial F_1/\partial z(r_L) \qquad (23)$$

are continuous across the interfaces. In Eq. (23), $m(z, E)$ is obtained by writing the relation in the form:

$$E = \frac{\hbar^2 k^2}{2m(E)} \qquad (24)$$

in each material. In these crude approximations, the superlattice dispersion relations for the light particles at $k_x = k_y = 0$ are given by the following expression:

$$\cos(x_a d_a)\cos(x_b d_b) - 0.5\left(w + \frac{1}{w}\right)\sin(x_a d_a)\sin(x_b d_b) = \cos(q(d_a + d_b)), \qquad (25)$$

where the thicknesses d_a and d_b are those of the deformed A and B layers, and w is given by:

$$w = \frac{(1 + \delta e_a)k'_a m'_b(E)}{(1 + \delta e_b)k'_b m'_a(E)}. \qquad (26)$$

One recognizes here the same formula as in the case of the unstrained superlattices except for the expression of w and for the dispersion relation giving $m'(E)$ in each material, which is

$$E\{[E + E_g + (C + a)\text{Tr}(e) - b\delta e][E + E_g + \Delta + (C + a')\text{Tr}(e)] - 2(b'\delta e)^2\}$$

$$= (1 - 2\delta e)P_0'^2 \hbar^2 k_z'^2 \left[E + E_g + 2\frac{\Delta}{3} + \left(C + 2\frac{a'}{3} + \frac{a}{3}\right)\text{Tr}(e) \right.$$

$$\left. + \left(4\frac{b'}{3} - \frac{b}{3}\right)\delta e \right]. \quad (27)$$

The main strain effects are contained in the modification of the dispersion relation (27). Some care should be taken in this respect when dealing with the valence-band states. The three-band Kane model currently used for calculation of the band structure of superlattices has the unpleasant feature of giving an asymptote at $E = -E_g - 2\Delta/3$ for the light-hole band in the unstrained material. As illustrated in Fig. 6, the energy distance between this asymptote and the top of the light-hole band is strain dependent: With respect to the unstrained case, it increases for a material under biaxial tension and decreases for a biaxial compression. Much care has to be taken to define the energy range where the $k_x = k_y = 0$ states of the superlattice can be correctly obtained by using the above approach. However, the model describes correctly the strain-induced changes in the conduction-band effective masses and their nonparabolicity, and thus the conduction states of the superlattice. In most moderately strained superlattices, it allows one to obtain the near-bandgap energies of the optical transitions.

To go further with the analysis of the valence-band states, it is possible, as in unstrained superlattices, to include k^2 terms in the valence-band Hamiltonian (see Fig. 6) for the constitutive materials in order to take into account the interaction with the remote bands. In this framework, the treatment of the

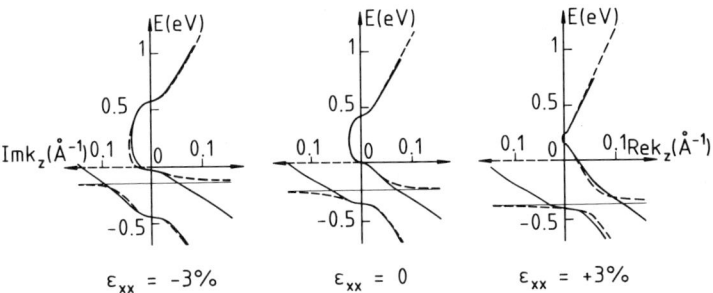

FIG. 6. InAs light-particle dispersion curves for various biaxial strain states, as calculated in a three-band Kane model (dashed line) and corrected to include the coupling with remote bands (solid line). One can note the marked effect of large strains on InAs bandgap and effective masses, and the good description of conduction states in the simple Kane model.

superlattice is quite similar to the calculation of the in-plane dispersion curves of the superlattice, which is the subject of the next part.

D. IN-PLANE VALENCE SUB-BANDS OF EXTERNALLY STRESSED GaAs–$Ga_{0.7}Al_{0.3}As$ QUANTUM WELLS

The in-plane valence subbands of unstrained quantum wells are complicated due to the degenerate Γ_8 nature of the valence-band extrema in the bulk materials (with the inversion asymmetry being neglected). As a result, the $\mathbf{k} \cdot \mathbf{p}$ valence Hamiltonian is not a vector but a 4×4 matrix if one assumes that the spin–orbit splitting Δ between Γ_8 and Γ_7 edges is large compared with the hole kinetic energy. In heterostructures, this means that the hole confinement energies should be small compared to Δ, and if stress is taken into account, that the strain-induced shifts remain also small as compared to Δ. These conditions are better realized in antimonides ($\Delta \approx 0.7\,\text{eV}$). They are reasonably fulfilled in arsenides, provided the wells are not too small and the stress (external and/or built-in) is moderate. They are on the other hand strongly invalid in phosphides ($\Delta \approx 0.1\,\text{eV}$) and in InAs–GaAs superlattices. For externally stressed GaAs–$Ga_{0.3}Al_{0.7}As$ quantum wells, the restriction of the valence Hamiltonian to the Γ_8 subspace is, we believe, an excellent approximation.

The study of the valence subbands and optical properties of externally stressed GaAs–$Ga_{0.3}Al_{0.7}As$ has already received some attention (Koteles et al., 1986; Jagannath et al., 1986; Collins et al., 1987; Sanders and Chang, 1985; Platero and Altarelli, 1987; Lee and Vassel, 1988). The 4×4-valence Hamiltonian of a uniaxially stressed (stress amplitude X applied along the $[abc]$ crystallographic direction) single-quantum well is written as:

$$H_v(X \parallel [abc]) = H_0 + M_{[abc]} \qquad (28)$$

where, in the basis $|3/2, 3/2\rangle_z$, $|3/2, -1/2\rangle_z$, $|3/2, 1/2\rangle_z$, $|3/2, -3/2\rangle_z$:

$$H_0 = \begin{bmatrix} V(z) + a_+ & c & b & 0 \\ c^* & V(z) + a_- & 0 & -b \\ b^* & 0 & V(z) + a_- & c \\ 0 & -b^* & c^* & V(z) + a_+ \end{bmatrix}, \qquad (29)$$

$$a_{\pm} = -\frac{1}{2m_0} P_z(\gamma_1 \mp 2\gamma_2) P_z - \frac{\hbar^2}{2m_0}(\gamma_1 \pm \gamma_2)(k_x^2 + k_y^2), \qquad (30)$$

$$b = \sqrt{3}\,\frac{\hbar}{2m_0} k_-(\gamma_3 P_z + P_z \gamma_3), \qquad (31)$$

$$c = \frac{\sqrt{3}}{2}\frac{\hbar^2}{m_0}[\gamma_2(k_x^2 - k_y^2) - 2i\gamma_3 k_x k_y]. \qquad (32)$$

$V(z)$ describes the difference in the valence-band maxima when going from one material to the other (in the absence of stress), and the γ are the Luttinger (Luttinger, 1956) coefficients of the valence band. $M_{[abc]}$ depends on the direction of the applied stress. It is given below for [001] (quantization axis for J), [100], and [110].

$$M_{[001]} = \begin{bmatrix} -\delta E_H - \delta E_{[001]}/2 & 0 & 0 & 0 \\ 0 & -\delta E_H + \delta E_{[001]}/2 & 0 & 0 \\ 0 & 0 & -\delta E_H + \delta E_{[001]}/2 & 0 \\ 0 & 0 & 0 & -\delta E_H - \delta E_{[001]}/2 \end{bmatrix},$$
(33)

$$M_{[100]} = \begin{bmatrix} -\delta E_H + \delta E_{[100]}/2 & \sqrt{3}\delta E_{[100]}/2 & 0 & 0 \\ \sqrt{3}\delta E_{[100]}/2 & -\delta E_H - \delta E_{[100]}/2 & 0 & 0 \\ 0 & 0 & -\delta E_H - \delta E_{[100]}/2 & \sqrt{3}\delta E_{[100]}/2 \\ 0 & 0 & \sqrt{3}\delta E_{[100]}/2 & -\delta E_H + \delta E_{[100]}/2 \end{bmatrix},$$
(34)

$$M_{[110]} = \begin{bmatrix} -\delta E_H + \delta E_{[110]}/2 & i\delta E_d & 0 & 0 \\ -i\delta E_d & -\delta E_H - \delta E_{[110]}/2 & 0 & 0 \\ 0 & 0 & -\delta E_H - \delta E_{[110]}/2 & i\delta E_d \\ 0 & 0 & -i\delta E_d & -\delta E_H + \delta E_{[110]}/2 \end{bmatrix},$$
(35)

where

$$\delta E_H = a(S_{11} + 2S_{12})X, \tag{36}$$

$$\delta E_{[001]} = 2\delta E_{[100]} = 2\delta E_{[110]} = 2b(S_{11} - S_{12})X, \tag{37}$$

$$\delta E_d = \frac{d}{2} S_{44} X. \tag{38}$$

a, b, and d are the hole deformation potentials.

The in-plane dispersion relations are obtained by projecting Eq. (28) on the basis spanned by the bound solution of its diagonal part (Bastard and

Brum, 1986). The stress dependence of V (if any) has been ignored, and the hole deformation potentials taken to be identical in GaAs–Ga$_{0.3}$Al$_{0.7}$As. In Eq. (28), the total angular momentum $\mathbf{J}(J = 3/2)$ is quantized along the growth axis. At $\mathbf{k}_\perp = (k_x, k_y) = \mathbf{0}$, and zero in-plane stress, the heavy ($J_z = \pm 3/2$) and light ($J_z = \pm 1/2$) states are decoupled. At finite in-plane $\mathbf{k}_\perp$ and/or nonvanishing in-plane stress, they become admixed. A convenient method to express this admixture (Bastard and Brum, 1986) is to plot the $\mathbf{k}_\perp$ dependence of $\sqrt{\langle J_z^2 \rangle}$ where $\langle J_z^2 \rangle$ is the J_z^2 average over the various states of Eq. (28). Unadmixed heavy- (light-) hole states corresponds to $\sqrt{\langle J_z^2 \rangle} = 3/2(1/2)$. Figures 7–15 illustrate the effect of stress on the in-plane dispersion relations of two quantum wells with $L = 220$ Å and $L = 130$ Å, respectively, for stress applied parallel to [001] (growth axis), [100], and [110]. We have taken $\gamma_1 = 6.85$ (3.45), $\gamma_2 = 2.1$ (0.68) and $\gamma_3 = 2.90$ (1.29) for GaAs (AlAs), the γ alloy parameters being linearly interpolated between GaAs and AlAs. In addition, the following values (Landolt-Bornstein, 1982) were used in the calculation: $S_{11} = 1.175 \times 10^{-9} Pa^{-1}$, $S_{12} = -0.366 \times 10^{-9} Pa^{-1}$, $S_{44} = 1.68 \times 10^{-9} Pa^{-1}$, $a = -7.93$ eV, $b = -2$ eV, and $d = -5.3$ eV.

At zero stress and $\mathbf{k}_\perp = \mathbf{0}$, the ground state is the heavy-hole state H$_1$. The mass-reversal effect arises from the fact that the diagonal terms of the Luttinger matrix correspond to light (heavy) in-plane dispersions for states that are heavy (light) along the growth axis. Without the $\mathbf{k}_\perp$-dependent off-diagonal terms, the mass-reversal effect would make the heavy-hole branches to cross the light-hole ones. These crossings are actually replaced by anticrossings (see, e.g., Figs. 7a and 13a), which, eventually, lead to camelback-shaped dispersion relations and which correspond to a strong admixture of the heavy- and light-hole character in the eigenstates at finite $\mathbf{k}_\perp$ (see, e.g., Fig 7b, 13b). When the stress is applied parallel to the growth axis (Figs. 7–10), it merely changes the energy distances between the $\mathbf{k}_\perp = \mathbf{0}$ eigenstates, pushing the light-hole levels upward and the heavy-hole levels downward. These changes are obtained in strained-layer quantum wells for a biaxial tensile stress. This strain-induced splitting goes against size quantization and, if the strain is large enough, the topmost valence band becomes L$_1$. The L$_1$ admixture with heavy holes decreases with increasing stress as witnessed by its $\sqrt{\langle J_z^2 \rangle}$ dependence upon $\mathbf{k}_\perp$ (compare Figs. 7b and 8b). When the stress is applied in the layer plane (Figs. 12–15), heavy- and light-hole states are admixed even at $\mathbf{k}_\perp = \mathbf{0}$. The heavy-hole states are pushed upward, and the overall level pattern is fairly complicated. In particular, one notices that in-plane stresses markedly increase the in-plane anisotropy of the dispersion relations (compare, for example, Figs. 9a and 12a). With increasing quantum well thickness, the size quantizations become more and more

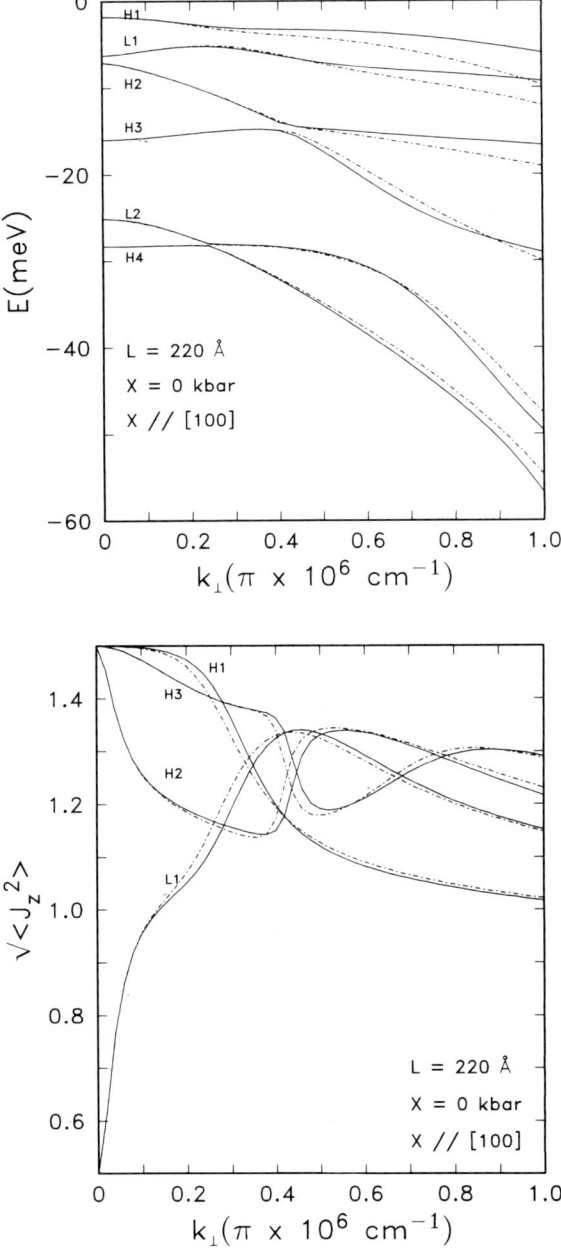

FIG. 7. (a) Calculated in-plane dispersion of the valence subbands in a 220-Å-thick unstrained GaAs–Ga$_{0.7}$Al$_{0.3}$As single-quantum well. Solid lines: $k_\perp \parallel [100]$; dashed lines: $k_\perp \parallel [110]$. (b) Calculated admixture of the valence subbands (i.e., $k_\perp$ dependence of $\sqrt{\langle J_z^2 \rangle}$) in this structure for $k_\perp \parallel [100]$ (solid line) and $k_\perp \parallel [110]$ (dashed line).

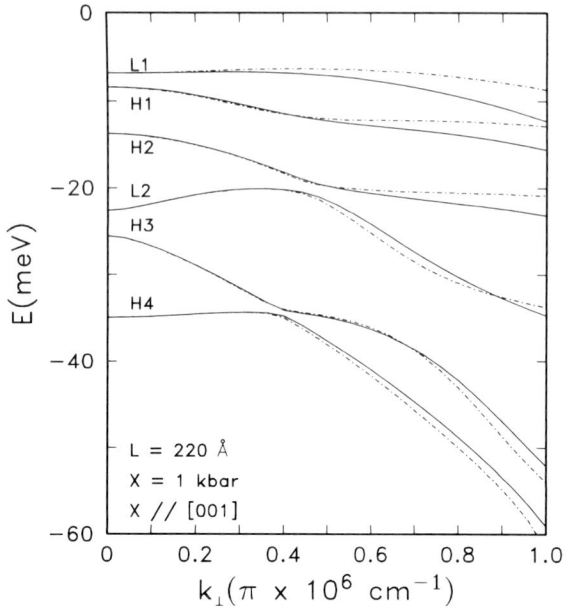

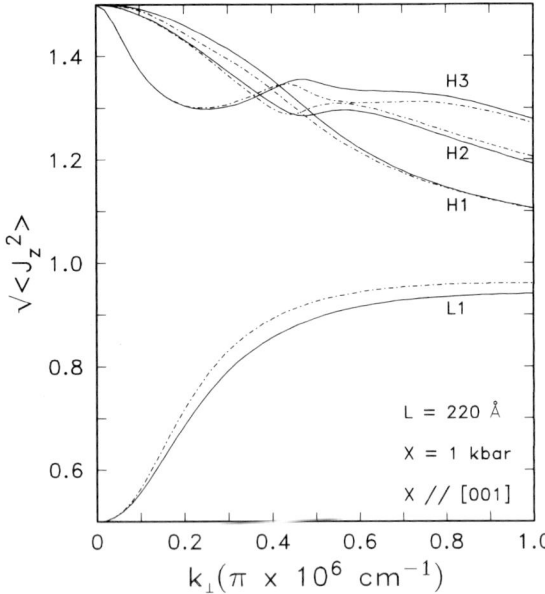

FIG. 8. (a) and (b). Same as Fig. 7(a) and (b), but for a 1 kbar stress applied parallel to [001].

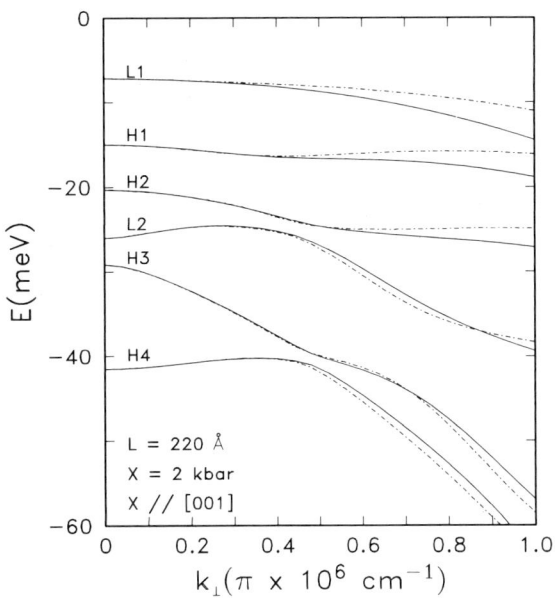

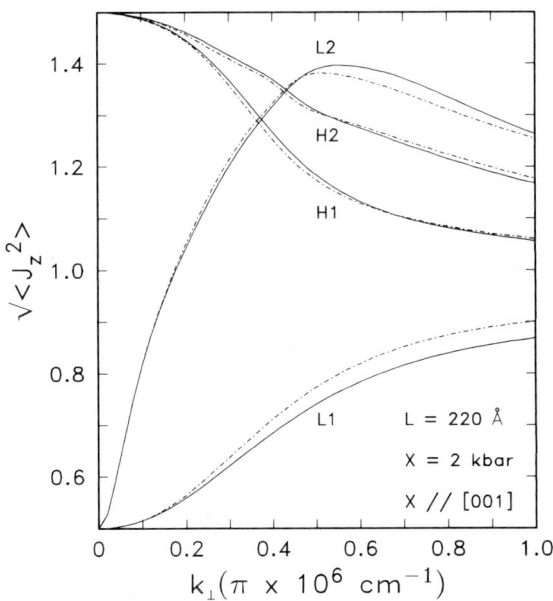

FIG. 9. (a) and (b). Same as Fig. 7(a) and (b), but for a 2 kbar stress applied parallel to [001].

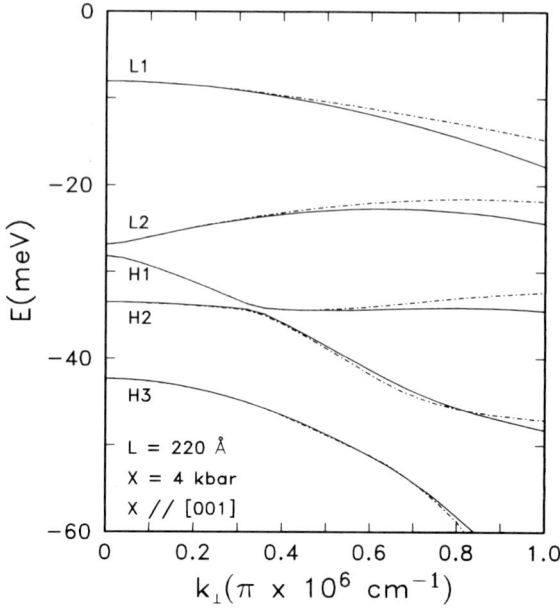

FIG. 10. Same as Fig. 7(a), but for a 4 kbar stress applied parallel to [001].

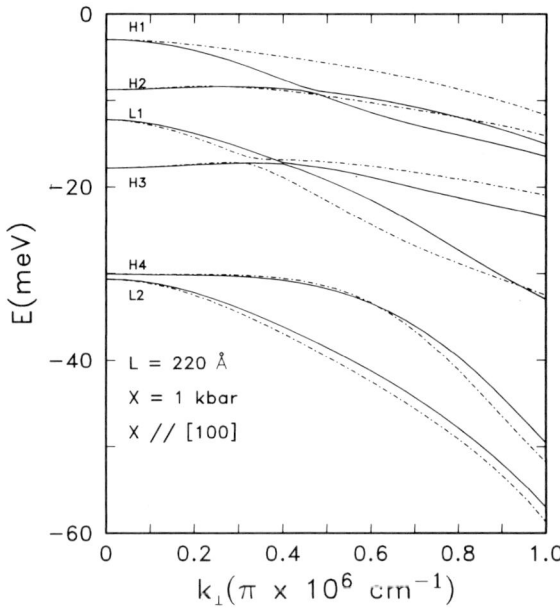

FIG. 11. Calculated in-plane dispersion of the valence subbands in a 220-Å-thick unstrained GaAs–Ga$_{0.7}$Al$_{0.3}$As single-quantum well. A 1 kbar stress is applied parallel to [100]. Solid lines: $k \parallel [100]$; dashed lines: $k \parallel [010]$.

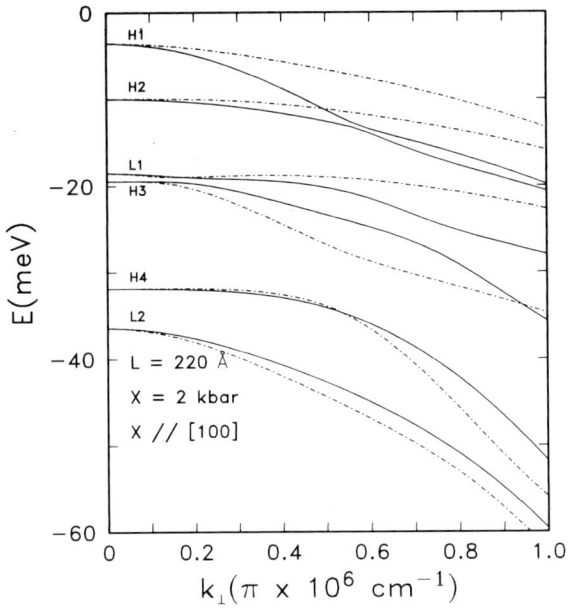

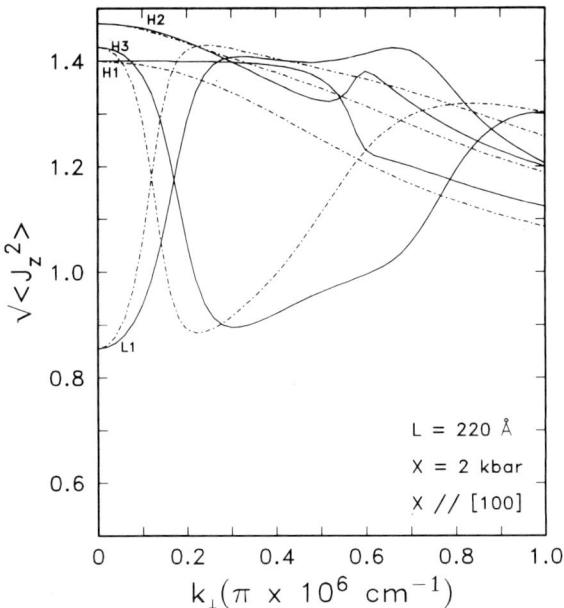

FIG. 12. (a) Same as Fig. 11, but for a 2 kbar stress. (b) $\sqrt{\langle J_z^2 \rangle}$ is plotted against k for the same parameters as used in Fig. 12(a).

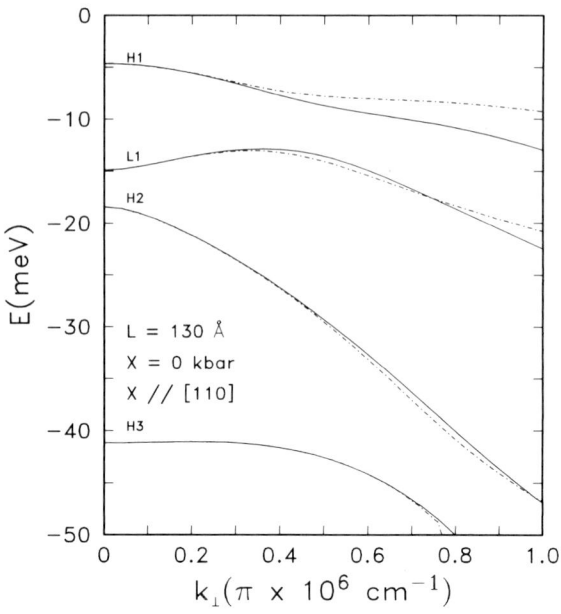

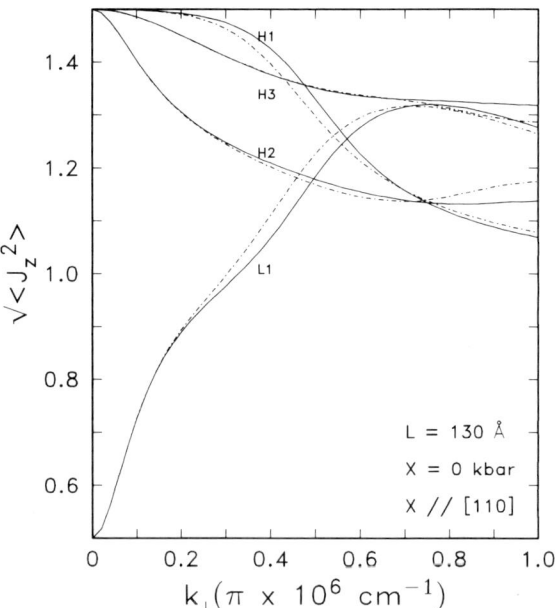

FIG. 13. (a) Same as Fig. 7(a), but for a well width $L = 130$ Å. Solid lines: $k \parallel [100]$; dashed lines: $k \parallel [110]$. (b) Same as Fig. 7(b), but for $L = 130$ Å.

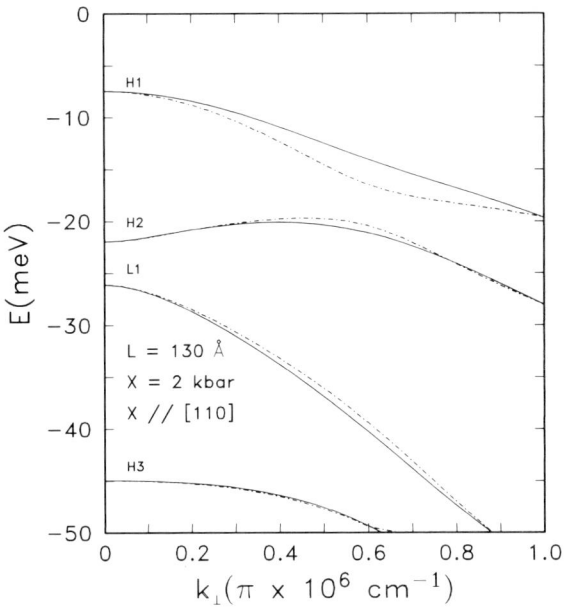

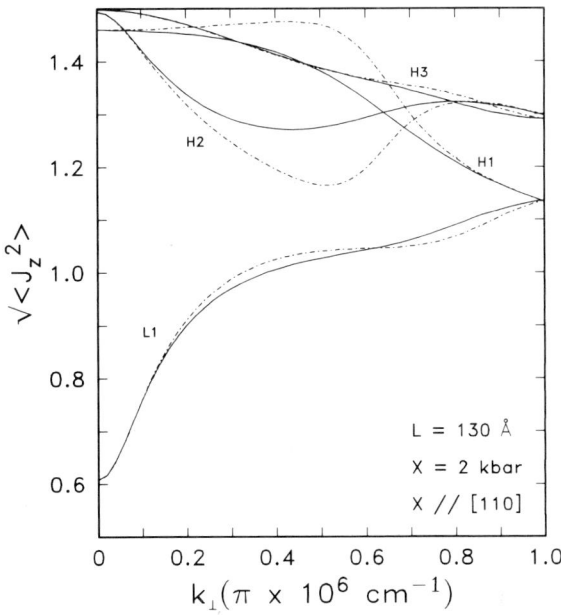

FIG. 14. (a) and (b). Same as Fig. 13(a) and (b), but for a 2 kbar stress applied parallel to [110].

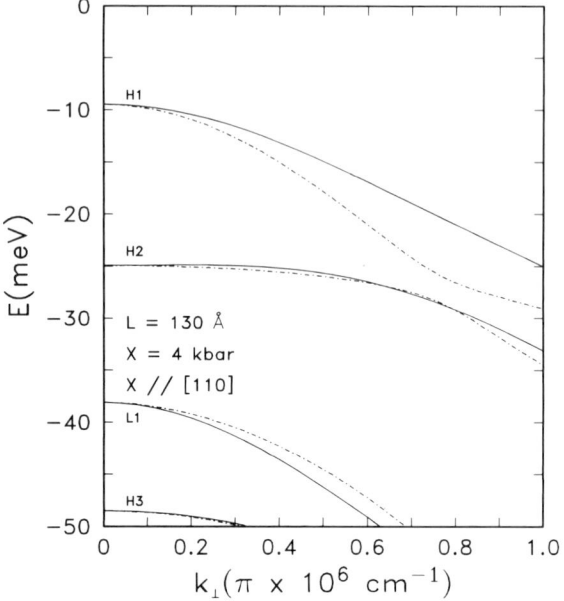

FIG. 15. Same as Fig. 13(a), but for a 4 kbar stress applied parallel to [110].

important, and the stress has a relatively smaller effect on the dispersion relations.

The stress-induced modification of the valence subbands affects the optical properties, shifting the transition energies and changing the line shapes. Although some of the optically observed features can be related to band-to-band transitions (Platero and Altarelli, 1987; Lee and Vassel, 1988), other features require identification in terms of stress-perturbed excitonic states (Broido and Yang, 1988a, 1988b).

The effects of the valence-band mixing can be evidenced in magneto-optical experiments, and some data concerning GaSb–AlSb will be discussed in Section IV.D.2.

IV. Moderately Strained Systems

In this part, we discuss the optical properties of samples where the built-in strains are kept smaller than 1 to 2%. For such lattice mismatches, the critical thicknesses are larger than 50 Å, so that superlattices with rather thick wells and barriers can be grown, where the previously outlined treatment of the electronic states is well suited. Moreover, the details of the interface formation between the two constitutive materials have no noticeable effects on the SLS properties as long as the thicknesses of the layers are not too small. Whereas a large number of strained III–V systems has been investigated, we focus on some examples that are illustrative of the variety of optical properties that can be achieved. $In_xGa_{1-x}As$–GaAs and $In_xGa_{1-x}As$–$In_yAl_{1-y}As$ are examined in Sections IV.A and IV.B in some detail before a summary of results is obtained for $In_xAl_{1-x}As$–GaAs, GaSb–AlSb, and $In_xGa_{1-x}As$–$In_yGa_{1-y}As$ systems in Sections IV.C to IV.E.

A. $In_xGa_{1-x}As$–GaAs

SLS using this system have been grown (Goldstein et al., 1982; Fritz et al., 1982; Roth et al., 1986; Sato and Horikoshi, 1988) either by molecular beam epitaxy (MBE) or metal-organic chemical-vapor deposition (MOCVD) on GaAs substrates and have been the object of numerous structural (Picraux et al., 1983; Quillec et al., 1984) and optical studies (Marzin et al., 1985; Anderson et al., 1985, and references therein; Roth et al., 1986; Menendez et al., 1987; Sato and Horikoshi, 1988). Two approaches have been used depending on the desired overall superlattice thickness and its design parameters. The important quantity is the mean lattice parameter of the superlattice, which is mismatched with respect to the GaAs substrate. Plastic relaxation may occur (between the superlattice and substrate) if the superlattice is too thick. To avoid the presence of the associated defects inside the superlattice, its thickness has to be kept under the critical thickness. Another approach is to grow a thick intermediate $In_yGa_{1-y}As$ buffer layer, where most of the stresses are plastically relaxed, and whose lattice parameter is matched to the mechanical equilibrium lattice parameter of the superlattice. Since it is difficult to avoid defect propagation in the superlattice, the latter method should only be used to obtain thick superlattices, whereas the former method results in better-quality layers with a limited number of periods. The choice of one method or the other also affects the strain distribution inside the superlattice. In the free-standing superlattice, the layers are alternately in biaxial tension (GaAs) and compression ($In_xGa_{1-x}As$), whereas when the superlattice is strained as a whole on GaAs, only $In_xGa_{1-x}As$ layers are strained and experience a biaxial compression. In this latter situation, it is easier to determine the valence-band configuration (band discontinuities), because the strain depends only on the In mole fraction of the alloy, so that a series of samples with varied quantum well thicknesses and equal strains can be readily obtained.

1. Transmission and Photoluminescence Excitation

In the following, we present results obtained on a series of $In_{.15}Ga_{.85}As$–GaAs superlattices containing 10 periods, grown by MBE at 520°C on a 0.5 μm GaAs buffer layer. One period consists of one 200 Å GaAs layer and one $In_{.15}Ga_{.85}As$ layer whose thickness is varied from sample to sample. The sample characteristics are deduced from the growth parameters and checked by a detailed analysis of the x-ray double-diffraction profiles. As is shown by Quillec *et al.* (1984) this technique is a powerful tool for characterizing the SLS: In addition to the period of the superlattice, the sublayer thicknesses and compositions can generally be extracted by a theoretical fit of the experimental profiles in the framework of the kinematic approximation. Table I summarizes these samples parameters.

Liquid N_2 temperature transmission spectra for this series of samples are shown in Fig. 16. These spectra are characteristic of good-quality two-dimensional structures: Pronounced excitonic structures emerge from a steplike continuum, at least for samples with L_t smaller than 150 Å. Figure 17, where the spectrum obtained on a sample with an L_t of 100 Å is compared with a spectrum obtained on a similar $GaAs-Al_{0.3}Ga_{0.7}As$ structure, shows clearly the strain effects. Although the confinement energies should be of the same order in the two cases (similar effective masses), the first excitonic transition is shifted to much higher energy with respect to the well (unstrained) bulk bandgap in the SLS than in the $GaAs-Al_{0.3}Ga_{0.7}As$ sample. The energy separation between the first two transitions is also much larger in Fig. 17b than in 17a. These two effects are due respectively to the increase of the bandgap of $In_{0.15}Ga_{0.85}As$, which experiences a biaxial compression, and to the splitting of its valence band.

In all samples, the individual $In_{0.15}Ga_{0.85}As$ layer thickness is of the order of or below the critical thickness, whereas the whole superlattice thickness, which is mismatched with respect to the GaAs substrate, is above the critical thickness. The superlattice should be partially plastically relaxed with respect to the substrate, according to the thermodynamic equilibrium theory (Section II). This has been indeed observed in these samples using x-ray topography

TABLE I

$In_xGa_{1-x}As-GaAs$ Samples Characteristics as Deduced from X-Ray Analysis. L_t and L_b Are the $In_xGa_{1-x}As$ and GaAs Sublayer Thicknesses, Respectively

Sample $n°$	$x(\%)$	$L_t(Å)$	$L_b(Å)$
1	16.2	51	211
2	15.2	72	208
3	14.4	97	203
4	15	150	200
5	13.8	176	186
6	14.2	291	204

3. OPTICAL STUDIES OF STRAINED III-V HETEROLAYERS 85

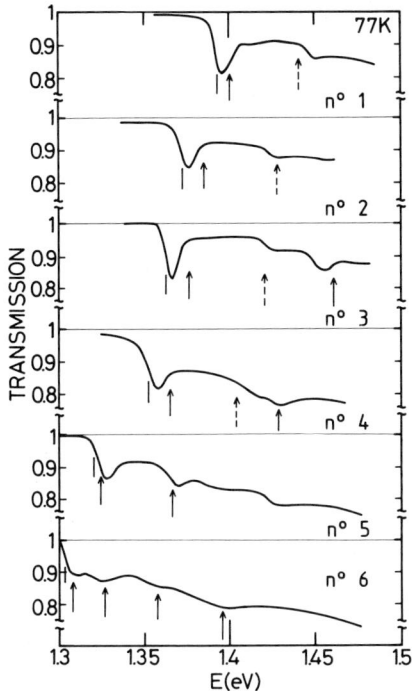

FIG. 16. Transmission spectra obtained at 77K on six $In_{0.15}Ga_{0.85}As$–GaAs SLS for increasing well width (characteristics of the samples are listed in Table I). Arrows mark calculated E_n–HH_n ($\rightarrow$) and dashed arrows mark E_n–LH_n ($-\!\!\!-\!\!\!\rightarrow$) transitions (exciton binding energies are not included). Photoluminescence peak energies are shown by vertical bars.

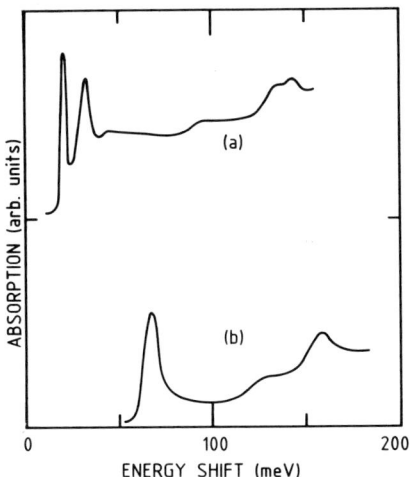

FIG. 17. Low-temperature absorption spectra obtained for $Ga_{0.7}Al_{0.3}As$–GaAs (a) and $In_{0.15}Ga_{0.85}As$–GaAs (b) superlattices are plotted versus the energy shift from the bandgaps of GaAs (a) or unstrained $In_{0.15}Ga_{0.85}As$ (b), so as to illustrate strain-related effects in (b).

on beveled structures (Joncour et al., 1986). Arrays of misfit dislocations are located at the interface between the GaAs buffer layer and the superlattice, but they relax only a small fraction (less than 1%) of the mismatch between the two, for the samples with L_t smaller than 180 Å. For L_t larger than this value, significant plastic relaxation occurs.

We will first discuss the optical properties of the structures where the superlattice is elastically strained on the GaAs substrate.

As in these SLS, the GaAs layers are unstrained, the first valence-band level for the superlattice is HH_1. The fact that the first transition exhibits a strong excitonic feature demonstrates that the first electron E_1 and hole level HH_1 are confined in the same smaller bandgap $In_{0.15}Ga_{0.85}As$ layers. Several valence-band configurations displayed in Fig. 18 are still consistent with this observation. A precise fitting of the observed transitions (and first their assignment) is thus required to establish the actual configuration.

The insert in Fig. 19 shows one possible method (Marzin et al., 1985) for assigning the transitions associated with the excited electron and hole levels. On-edge excitation of the photoluminescence allows testing of the selection rules for these transitions. In standard excitation experiments, the dye laser is focused onto the sample surface, and the exciting beam is almost parallel to the sample growth axis Z, due to the high refraction index of III–V compounds. In on-edge excitation, the laser beam enters the sample through the GaAs substrate edge and propagates nearly in the plane of the sublayers within the sample. This experimental setup, (which is only possible when the substrate is transparent), allows one to obtain excitation spectra with the laser polarization perpendicular to or along the z-axis. The sample for which L_t is 100 Å has been studied in this on-edge configuration, in the energy range of the second and third transitions labeled β and γ, respectively. For P_x or P_y polarization, both heavy- and light-hole transitions are allowed, and the spectrum, displayed in Fig. 19b, is very similar to the standard excitation

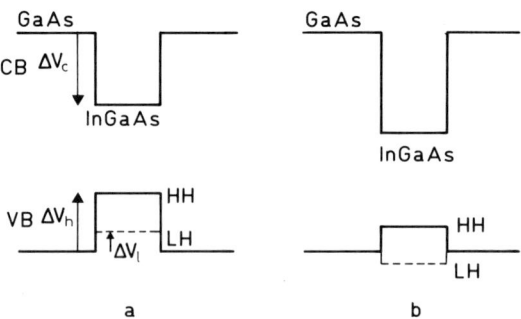

FIG. 18. Possible valence-band configurations in $In_{0.15}Ga_{0.85}As$–GaAs SLS.

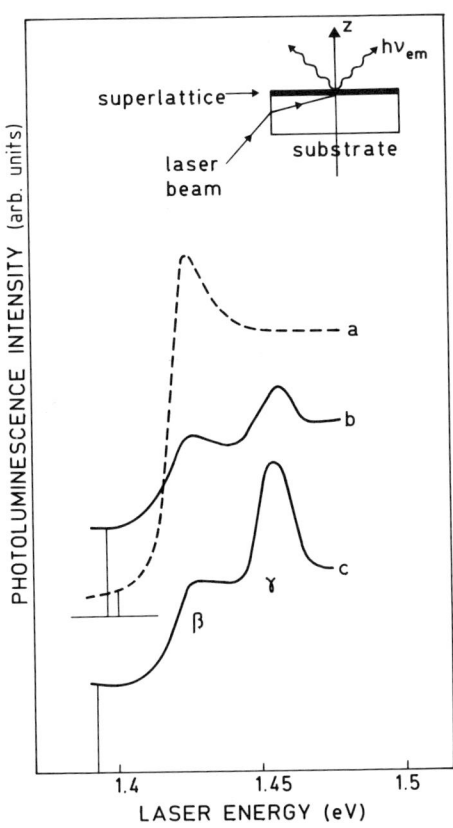

FIG. 19. On-edge photoluminescence excitation spectra obtained for sample 2 for light polarization parallel (a) and perpendicular (b) to the growth axis z. The experimental setup is shown in the inset. (c) Excitation spectrum obtained in the standard configuration, where the laser beam is focused on the sample surface.

spectrum of Fig. 19c, where β and γ transitions are both observed. When the polarization is set along the z-axis (P_z), only the light-hole transitions are allowed, and the spectrum is shown in Fig. 19a. The γ transition is no longer observed, and this transition is then assigned to HH_2-E_2 exciton creation, while β transition is thought to be LH_1-E_1. It has to be pointed out that in these on-edge excitation experiments, the shape of the transitions are rather strongly modified by the residual substrate absorption, as is seen when comparing spectra (b) and (c), so that "excitonic" shape of transition β in Fig. 19a may be spurious.

2. *Band-Offset Determination*

By fitting the energies (and natures) of the observed transitions with the band discontinuity as an adjustable parameter, a 55-meV-deep well for the heavy holes is obtained. This corresponds to the situation of Fig. 18b, where the heavy holes are confined in $In_{0.15}Ga_{0.85}As$, whereas for light holes, GaAs is the well material and $In_{0.15}Ga_{0.85}As$ is the barrier. However, the β transition is observed, since LH_1 wave functions are essentially delocalized in the superlattice. This is due to the small (15 meV) barrier height for the light holes. Other experimental determination led to the same conclusion on the valence-band configuration in samples with the same indium content and strain state (Pan et al., 1988, and references therein).

$In_{0.05}Ga_{0.95}As$–GaAs samples were recently studied using electronic Raman scattering and photoluminescence excitation (Menendez et al., 1987). The observed transitions were assigned by optical pumping. This study clearly indicated that for these samples, the configuration is that of Fig. 18a, that is of type I for both heavy and light holes. Tight-binding self-consistent calculations (Priester et al., 1988) furthermore suggest that, due to the strain effects, the InAs–GaAs system on GaAs should be of mixed type (Fig. 18b), whereas for $In_xGa_{1-x}As$–GaAs for smaller x values, it becomes a type-I system for heavy and light holes (Fig. 18a). This is consistent with both observations. Recent photoluminescence data confirmed that the band offset actually depends on x (Joyce et al., 1988), and they were consistent with these two previous determinations obtained for $In_{0.15}Ga_{0.85}As$–GaAs and $In_{0.05}Ga_{0.95}As$–GaAs.

3. Photoluminescence

Typical photoluminescence spectra for these samples consist of one main line, as shown in Fig. 20, whose energy corresponds to the lowest energy transition observed in transmission. As in GaAs–GaAlAs structures, this line corresponds to the recombination of quasi-two-dimensional free excitons. In the samples mentioned above, typical 10 K half-width at half-maximum is 7 meV. A modified MOCVD growth technique (flow-rate modulation epitaxy) has

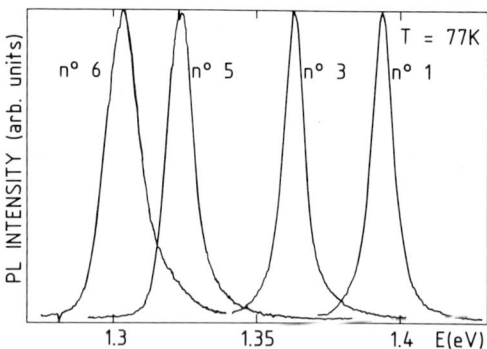

FIG. 20. Typical photoluminescence spectra obtained at 77 K for $In_{0.15}Ga_{0.85}As$–GaAs SLS.

been used recently to grow $In_xGa_{1-x}As$ quantum wells (Sato and Horikoshi, 1988). Impressively narrow low-temperature luminescence lines for $x = 0.06$ (down to 0.6 meV for a 70 Å single-quantum well) have been observed. It should be pointed out here that the effects of well thickness fluctuations in the plane of the layers are amplified in SLS by the accompanying strain fluctuations (which increase with the mismatch). Therefore, the linewidth is expected to increase with x, as was indeed reported by Sato and Horikoshi (1988). On the low-energy side of the photoluminescence, conduction band to neutral acceptor recombination is often observed.

4. *Stress Relaxation and Its Effects on the Optical Properties*

As mentioned above, for all these samples, the superlattices should be far less strained than they actually are. From x-ray analysis, the three samples with the smallest L_t appear to be elastically strained on GaAs, while they are (for increasing L_t) 1.7, 2.7, and 3.4 times thicker than the critical layer thickness. However, arrays of misfit dislocations can be observed either by transmission electron microscope (TEM) or x-ray topography, so that relaxation may have already begun, in agreement with the thermodynamic approach, though the samples are still in a metastable state. On this topic, Fritz *et al.* (1985) discuss the influence of the sensitivity of several measurement techniques for the critical thickness.

For the samples with L_t larger than 150 Å, there is a noticeable plastic relaxation. When the $In_{0.15}Ga_{0.85}As$ and the GaAs layers are 150 Å and 200 Å thick, respectively, the superlattice average lattice mismatch with respect to the substrate is 0.43%. It is 3500 Å thick, which is about six times the critical thickness. From the optical transitions observed in transmission, it can be deduced, assuming that the superlattice relaxes as a whole with respect to the substrate, that only one third of the stresses have been plastically removed (instead of 5/6 at thermodynamic equilibrium).

When such plastic relaxation occurs, we can also observe the following effects:

(i) A degradation in the x-ray double-diffraction profiles while single diffraction profiles still exhibit nice satellites. This is thought to be due to misorientations in the sample to which double diffraction is more sensitive because of the higher beam parallelism.

(ii) A larger shift between the photoluminescence peak and the first transition observed in transmission (see Fig. 16). This first transition is shifted towards lower energy with respect to the completely strained situation. It may even be located (it is the case for $L_t = 300$ Å) at lower energy than the bandgap of the $In_{0.15}Ga_{0.85}As$ strained to match the in-plane lattice parameter of GaAs, despite the confinement energies that increase the transition energy.

(iii) When photoluminescence excitation is performed in the energy region of the GaAs bandgap, stress relaxation affects at least the quality of the interface between the superlattice and the GaAs buffer layer. The diffusion of the carriers created inside the buffer layer toward the superlattice is decreased by numerous nonradiative defects. Typical excitation spectra shown in Fig. 21 switch from that of Fig. 21a when stress relaxation is small to that of Fig. 21b when it is large. The increase (in the former case) of the superlattice photoluminescence when the exciting laser energy is above the GaAs bandgap is completely washed out in the latter. Quantitative analysis is difficult, however, because of the possible re-excitation of the superlattice luminescence by the luminescence of GaAs.

Finally, a discussion of the effects of stress relief on the photoluminescence of quantum wells grown with a higher In composition can be found in the paper by Anderson *et al.* (1984).

5. Conclusion

This strained system allows us to obtain emitting sources or detectors up to 1 μm. Although rapid degradation during room temperature cw operation has often been reported, recent realization show that stable lasers can be

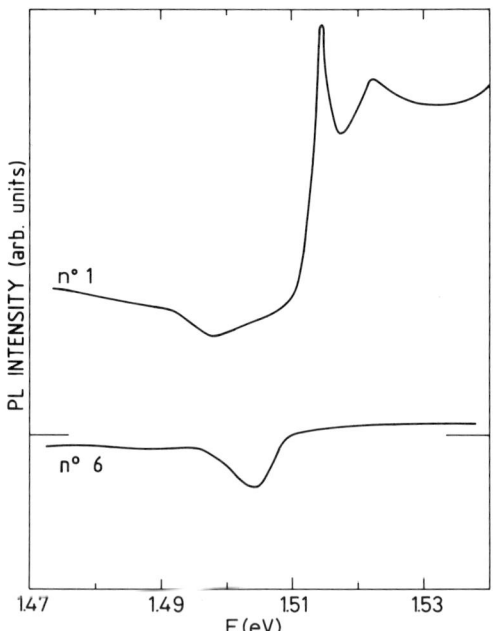

FIG. 21. Photoluminescence excitation spectra obtained near the GaAs bandgap for a small (a) and a large relaxation of $In_{0.15}Ga_{0.85}As$–GaAs SLS.

obtained (see, e.g., Bour et al., 1988, and references therein). More experiments are needed to examine the dependence of the band discontinuities on the indium composition and on the superlattice strain state. For indium compositions around 15%, the band configuration is of mixed type, i.e., the electron heavy-hole system is of type I, whereas the electron light hole one is of type II. Both systems may become of type I for lower indium mole fractions, and the composition limit between these two behaviors has then to be made more precise.

B. $In_xGa_{1-x}As-In_yAl_{1-y}As$

1. *Description of the System*

As stated in the introduction, this system is interesting for its flexibility (Nishi et al., 1986; Gerard et al., 1987). When SLSs made of $In_xGa_{1-x}As$, and $In_yAl_{1-y}As$ are grown on InP and are designed to be lattice-matched to this substrate, several strain configurations can still be obtained, according to the choices of the In content in both alloys. This flexibility is illustrated in Fig. 22. Figure 22a shows the bandgap variation versus the lattice constant for the two alloys. Let d_x and d_y be the $In_xGa_{1-x}As$ and $In_yAl_{1-y}As$ sublayer thicknesses inside the superlattice. Lattice matching to InP requires (assuming for simplicity equal elastic constants in the two materials):

$$d_x(x - 0.53) + d_y(y - 0.52) = 0. \tag{39}$$

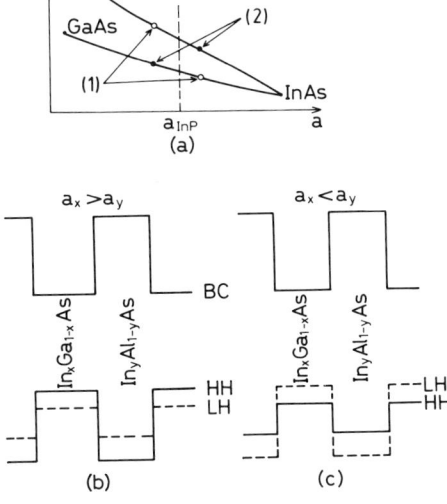

FIG. 22. Two possible strain configurations (a) and the resulting band-extrema line-ups (b) and (c) for InGaAs–InAlAs SLS lattice matched on InP.

In Fig. 22a, two possible choices for x and y are schematized. Choice 1 ($x > 0.53$) results in a smaller bandgap $In_xGa_{1-x}As$ under biaxial compression, and a larger bandgap $In_yAl_{1-y}As$ under tension. The corresponding valence-band configuration is shown in Fig. 22b. It is similar to that obtained in $In_xGa_{1-x}As/GaAs$ apart from the larger valence-band offset between (unstrained) $In_xGa_{1-x}As$ and $In_yAl_{1-y}As$ (Goldstein et al., 1985a). This should lead, for moderate strain situations to a type-I superlattice for both "heavy" and "light" holes. The strain situation is thus the one discussed in Section III.B).

Choice 2, obtained for $x < 0.53$, results in a small bandgap material $In_xGa_{1-x}As$ under tension, and a large one under compression. Concerning the energy separation between HH_1 and LH_1, there is a competition between confinement and strain effects, as discussed in Section III.B(iii).

Technically, one practical way to elaborate such structures using MBE is to use two In cells, each one corresponding to the epitaxy of one superlattice material. It is convenient to add a (nonrestrictive) relation between the two ternary growth rates, so that the simultaneous use of all the type-III species cells allows the growth of an $In_uGa_vAl_{1-u-v}As$ buffer layer lattice matched to InP.

Table II gives the samples characteristics for some such samples. Samples 1 and 2 corresponds to choice 1 and samples 3 and 4 to choice 2.

The extraction of the samples characteristics from x-ray double-diffraction profiles, which should be a priori more difficult than in $In_xGa_{1-x}As$–$GaAs$ structures, is similar in this system, because the diffraction of the buffer layer, grown as mentioned above, gives additional information.

2. Photoluminescence and Transmission

Figure 23a shows the low-temperature photoluminescence and transmission data for samples 1 and 2, in which the well material is under biaxial compression. Again, excitonic transitions are observed, and the lowest energy transition is HH_1-E_1. This first excitonic transition, observed in the

TABLE II
$In_xGa_{1-x}As$–$In_yAl_{1-y}As$ SAMPLES CHARACTERISTICS AS DEDUCED FROM X-RAY ANALYSIS. d_x AND d_y ARE THE $In_xGa_{1-x}As$ AND $In_yAl_{1-y}As$ SUBLAYER THICKNESSES, RESPECTIVELY

Sample n°	x(%)	y(%)	d_x(Å)	d_y(Å)
1	63.5	37.5	97	113
2	63	40	105	95
3	42	59	120	197
4	42	59	30	49

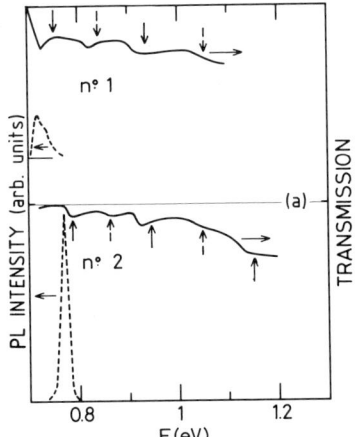 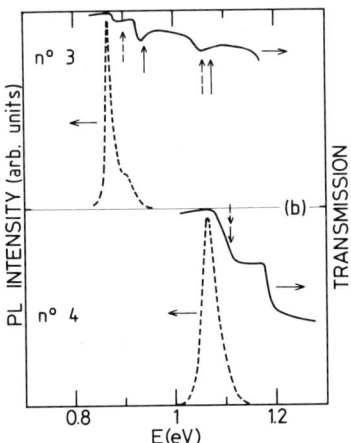

FIG. 23. Photoluminescence and transmission spectra obtained at 8 K for InGaAs–InAlAs SLS, whose characteristics are listed in Table II. In (a), InGaAs in under biaxial tension; it experiences in (b) a constant biaxial tension, with the well width being decreased from n°3 to n°4. Arrows indicate calculated HHn–En (→) and LHn–En (——→) transitions (exciton-binding energies are not included).

transmission spectrum, also corresponds to the photoluminescence line. As in $In_xGa_{1-x}As$–GaAs samples, the second transition, due to LH_1–E_1, is separated from HH_1–E_1 by more than the difference in the confinement energies because of the strain-induced valence-band splitting.

In this system, it was not possible to extract the band discontinuities directly from the optical data on these strained structures. However, one can reasonably assume that the valence-band offset between the two strained materials is rather large, since it is the case for $In_{0.53}Ga_{0.47}As$–$In_{0.52}Al_{0.48}As$ unstrained superlattices. As soon as this offset is larger than the strain effects, the transition energies are found to depend merely on the precise offset value (as in $Al_xGa_{1-x}As$–GaAs). Under these assumptions, the calculated transitions energies, indicated by arrows in Fig. 23, are in reasonable agreement with the experimental data.

Figure 23b shows the optical spectra obtained on samples 3 and 4, where $In_xGa_{1-x}As$ experiences a tensile stress inside the superlattice. For large enough well width (sample 3), the confinement energy difference between HH_1 and LH_1 is smaller than the strain-induced splitting of the $In_xGa_{1-x}As$ valence band that pulls LH_1–E_1 transition to lower energy. The first calculated transition indeed corresponds to LH_1–E_1, and the calculated energy agrees with the experimental spectrum. When the well width is decreased, LH_1 to HH_1 distance decreases. The crossing and recovery of the usual ordering of the levels are not clearly observed here, since well and barrier thicknesses were decreased simultaneously. Due to the resulting coupling from well to well, the two transitions fall at the same energy, and a large oscillator strength is observed at the 1.11 eV bandgap for this sample.

Such data show that strain can be used to tune the ordering of the valence-band levels inside the superlattice. This can also be done in other systems: in GaSb/AlSb or $In_xAl_{1-x}As$–GaAs, where the smaller bandgap materials are under tension. However, the $In_xGa_{1-x}As$–$In_yAl_{1-y}As$ system is well suited for applications requiring defect-free structures, because the superlattice can be designed lattice matched to the InP substrate. Transport measurements have been performed on this system (Hirose *et al.*, 1987), which show the increase of the hole mobility with the compressive strain in the smaller bandgap layers.

C. $In_xAl_{1-x}As$–GaAs ON GaAs

The system was used in order to investigate the tensile-strain situation for the well material, using a binary semiconductor (Sauvage *et al.*, 1986; Kato *et al.*, 1986). Thick $In_xAl_{1-x}As$ buffer layers have to be grown first on the GaAs substrate to obtain this strain state for the superlattice. Photoluminescence excitation spectroscopy can be easily performed in this wavelength range to observe the excited states of the superlattice. Figure 24 shows such optical data obtained on a series of samples grown by MOCVD (Sauvage *et al.*, 1986). The GaAs well thicknesses are the same in all samples, and the biaxial tension experienced by GaAs is increased to illustrate the relative effects of the strains and confinement on the ordering of the valence-band levels. As

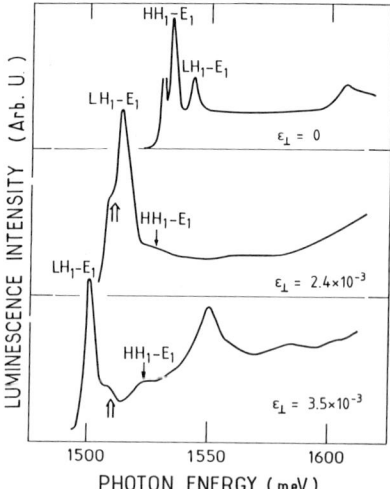

FIG. 24. Low-temperature excitation spectra in three $Al_{1-x}In_xAs$–GaAs SLS; the GaAs well of constant width experiences an increasing biaxial tension (from top to bottom). Arrows at 1.515 eV indicate the signal decay arising from the absorption edge of the GaAs substrate, which removes the contribution of the exciting light reflected at the back of the sample.

discussed in Section IV.B for $In_xGa_{1-x}As–In_yAl_{1-y}As$ samples, for small strain, the first transition observed in photoluminescence and photoluminescence excitation corresponds to HH_1-E_1 exciton (Fig. 24). When increasing the strain in the GaAs well (keeping its width constant), LH_1 and HH_1 levels cross, and the ordering of the two corresponding transitions to E_1 are exchanged. Like $In_xGa_{1-x}As/GaAs$, this system can be used to reduce the bandgap of an epitaxial material on GaAs, but suffers, in our opinion, from the need for an intermediate buffer layer, whose actual strain state is difficult to control and which has to contain many defects.

D. GaSb–AlSb

The first investigation of a strain configuration where the lower bandgap material is under biaxial tension in the superlattice was performed on this system. The lattice mismatch between AlSb and GaSb is 0.67%. The two samples studied by Voisin et al. (1984) were grown on a GaAs substrate. A thick AlSb buffer was grown between this substrate and the superlattice whose characteristics are indicated in Table III. The huge lattice mismatch (7%) between GaAs and AlSb (or GaSb) is plastically accommodated in the thick buffer layer. The grown structure (buffer + superlattice) is thus in self-equilibrium, with its in-plane lattice constant being close to that of AlSb. The strain state for the superlattice was checked for both samples by two methods. The first was the analysis of x-ray double-diffraction profiles, which also showed the high crystalline quality of the GaSb–AlSb superlattice and allowed to correct slightly the individual sublayer thicknesses indicated in Table III. The strain of the GaSb layers were also measured (Jusserand et al., 1985) using Raman scattering from the shift of the GaSb LO-phonon energies with respect to the unstrained-bulk-material value. For both methods, the measured strain state is very close to the one resulting from self-equilibrium of the epitaxial structure.

1. *Luminescence and Transmission*

Figure 25 shows the low-temperature transmission spectra for these two samples, characteristic of a type-I multiquantum well structure. As in the

TABLE III
GaSb (d_1) and AlSb (d_2) Sublayer Thicknesses for Superlattices S1 and S2

Sample	d_1(Å)	d_2(Å)
S1	181	452
S2	84	419

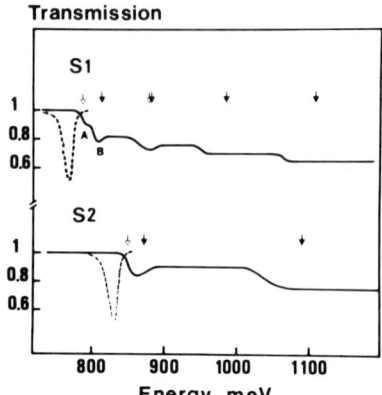

FIG. 25. Low-temperature absorption (solid line) and photoluminescence (dashed line) spectra for two GaSb–AlSb heterostructures. Solid (open) arrows indicate calculated transitions energies involving heavy- (light-) hole states.

InAlAs–GaAs system, the lower gap material GaSb is under in-plane tensile stress. Here again, the bandgap exciton can appear at a lower energy than in the bulk unstrained material. Furthermore, it is here, in both samples, the first light-hole exciton LH_1–E_1, due to the relative magnitudes of the strain and confinement effects.

As far as the photoluminescence is concerned, a systematic Stokes shift (greater than 10 meV) with respect to the absorption edge was reported (Voisin et al., 1984; Ploog et al., 1985; Raisin et al., 1987) in GaSb–AlSb structures (see Fig. 25). Although part of this Stokes shift may be due to interface fluctuations, it is clear that trapping of the excitons on residual neutral acceptors is mainly responsible of this Stokes shift.

The offset value for this system is still the object of a controversy. The energies of the transitions are often rather insensitive to the detailed offset value, but the number of "confined" hole subbands are. The analysis of the number of heavy- and light-hole transitions in the two samples yields a common value of ΔE_{hh} of the order of 40 meV, and $\Delta E_{lh} = 90$ meV. On the other hand, from the study of the resonance of the Raman scattering near the E_1 gap of GaSb, in samples grown on GaSb substrates, and having thus different strain states, a much larger valence-band offset has been found (>300 meV) (Tejedor et al., 1985). The same experiment on samples S1 and S2 did not lead to the same conclusion (Calleja et al., 1986). Again, as in the InGaAs–GaAs system, this discrepancy raises the still open question whether the band offset can strongly depend on the strain state or not.

GaSb–AlSb structures present an additional specific feature: Due to a small Γ–L distance in GaSb (84 meV in unstrained GaSb), and due to the large Γ bandgap differences between AlSb and GaSb (1.5 eV), L-originating states become the lower-energy conduction states for small enough GaSb layer thicknesses. As a result, the structure displays an indirect bandgap, and a sharp decrease is observed in the photoluminescence intensity (Griffiths et al., 1983) together with an increase of the radiative lifetime (Forchel et al., 1986). In the limit of a single-quantum well, the GaSb thickness where the Γ–L crossover is observed depends (like the Γ–L spacing in GaSb) on the strain state. For GaSb strained onto AlSb, the calculated value is of about 60 Å, in agreement with the experimental data.

Finally it has to be mentioned that the in-plane mass reversal was observed (Voisin et al., 1986a) in this system from magneto-optical transmission, and this point is discussed below.

2. Magneto-optical Transmission

The strong mixing of the valence subbands at finite in-plane wave vector and the related complexity of the in-plane dispersion relations in superlattices are now well understood (Altarelli, 1983; Fasolino and Altarelli, 1984; Schulman and Chang, 1985; O'Reilly and Witchlow, 1986; Bastard and Brum, 1986). The part that these effects play in the magneto-optical transitions is far from being as extensively discussed. Indeed, the valence subband mixing breaks the parity selection rules at finite magnetic field B, and a very large number of magneto-optical transitions become allowed, even though only a few of them will have a significant strength. The calculation of the transition oscillator strengths is, in fact, a prerequisite to the comparison of theory and experiments. This has been done recently for the GaAs–AlGaAs systems (Ancilotto et al., 1987), but it is clearly worth attempting this comparison in the case of strained-layer heterostructures, since the combination of strain and confinement can produce a variety of subband spacing and ordering, and thus introduces an additional parameter into this problem. Our calculation of the Landau-levels dispersion and the associated oscillator strengths (Brum et al., 1988) is based on the expansion of the eigenfunctions at finite B on the "basis" of the $B = 0$ envelope functions (Bastard and Brum, 1986; Bastard et al., 1989). This method, though approximate, gives sensible results for the first valence subbands and presents the advantage of making the calculation of the oscillator strength rather simple and easy to interpret.

It is known that the eigenfunctions of the ith state of the quantum-well Hamiltonian in presence of a magnetic field parallel to the growth axis can be written as a spinor of the form

$$\Psi_{i,N,B}(r) = L_y^{-1/2} e^{ik_y y} \sum_v \chi_{i,v,N,B}(z) \phi_{v,N}((x - x_0)/\lambda) |u_v(r)\rangle, \quad (40)$$

where i is the subband index, and N the Landau level index, $\phi_{v,N}((x - x_0)/\lambda)$ is an harmonic oscillator function, with $x_0 = -\lambda^2 k_y$ ($\lambda = (hc/eB)^{1/2}$ is the magnetic length), $|u_\nu(r)\rangle$ is the atomic part of the Block function of the νth host band edge, and $\chi_{i,\nu,N,B}(z)$ is a slowly varying envelope function. Usually, only the Γ_6, Γ_8, and $\Gamma_7 (\nu = 1$ to $8)$ band edges are considered.

On the other hand, the set of the $\chi_{l,\nu}(z)$ eigenfunctions at $k_\perp = 0$ (or $B = 0$) form a basis on which $\chi_{i,N,B}(r)$ can be expanded by writing

$$\chi_{i,\nu,N,B}(z) = \sum_l \alpha_{i,\nu,l,N}(B)\chi_{l,\nu}(z), \tag{41}$$

where, again, i and l refer to the subband labelling and ν to the host band edges. In fact, in most cases, restricting the summation in Eq. (41) to the discrete QW bound states is a sensible approximation. $\alpha_{i,l,\nu,N}(B)$, as well as the Landau level energies, can be obtained from the numerical diagonalisation of the Hamiltonian.

The interband magneto-optical matrix elements can then be calculated as

$$\langle \Psi^e_{i,N,B}(r)|p|\psi^H_{j,M,B}(r)\rangle$$
$$= \sum_{l,m,\nu,\mu} \alpha^{*e}_{i,l,\nu,N}\alpha^H_{j,m,\mu,M}\langle u_\nu|p|u_\mu\rangle\langle\chi^e_{l,\nu}|\chi^H_{m,\mu}\rangle\langle\phi_{\nu,N}|\phi_{\mu,M}\rangle. \tag{42}$$

The main advantage of the present method is that the parity selection rule obeyed by the $B = 0$ envelope functions can be used to simplify the computation, whereas the valence-band mixing is more intuitively discussed in terms of the mixing of the $\chi^H_{m,\mu}(z)$.

To illustrate the necessity of the evaluation of the transition oscillator strength, we show in Fig. 26 the calculation of the allowed transitions toward the first spin doublet of the first conduction subband in the σ^- polarization, for the case of our strained-layer GaSb–AlSb multiquantum-well structure: among these numerous transitions, only three, shown by the bold lines, have an oscillator strength (more specifically, the square of the sum in Eq. (42)) larger than $0.2|\langle S|p_x|X\rangle|^2$. If, for instance, we consider five spin doublets in the conduction band, the equivalent of Fig. 26 is covered by an unreadable network of transitions, while only a restricted number of them are strong enough to be observable.

Figure 27 displays the experimental fan diagram obtained in a ten-period 180–450 Å GaSb–AlSb QW structure, together with its description in a semiempirical model (Voisin et al., 1986). It consists of two distinct fan diagrams extrapolating nearly linearly to about 800 and 830 meV, respectively. These transitions were attributed to the transitions $LH^N_1-E^N_1$ and $HH^N_1-E^N_1$, respectively, from the Landau levels of the LH_1 and HH_1 subbands (assumed to have parabolic in-plane dispersion relations) to those of the first conduction subband. The solid lines and dashed lines in Fig. 27 correspond to a fit using a very heavy in-plane mass $(0.8\, m_0)$ for the LH_1 subband and a

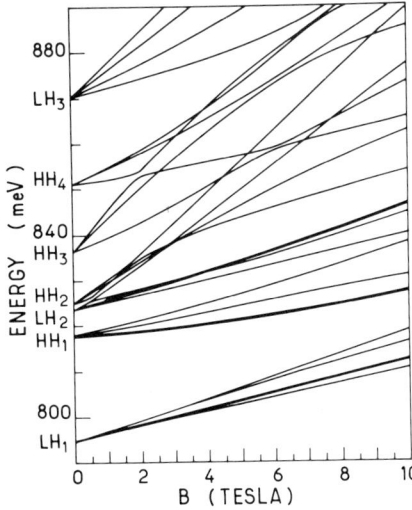

FIG. 26. Calculated interband magneto-optical transitions toward the first spin doublet of the first conduction subband, in the σ-polarisation. The three bold lines show the scarce transitions, which have a significant oscillator strength.

rather light in-plane mass (0.11 m_0) for the HH_1 subband. In addition, Fig. 27 shows two transitions (shown by the dash-dotted lines) having a nonlinear behavior, which extrapolate to the energies of the LH_1–E_1 (795 meV) and HH_1–E_1 (820 meV) excitons. The energy separation between the HH_1–E_1 exciton and the extrapolation of the $HH_1^N \rightarrow E_1^N$ transitions, however, is significantly larger than the possible binding energy of this exciton.

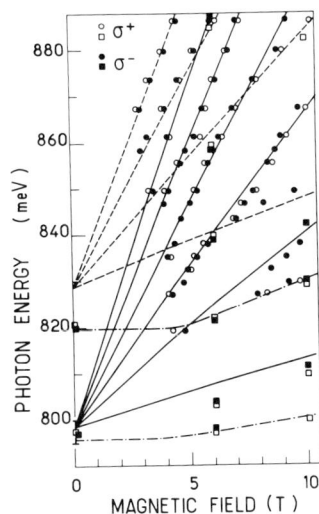

FIG. 27. Experimental fan diagram of the magneto-optical transitions observed in and 180–450 Å GaSb–AlSb multiquantum-well structure, and their interpretation using a semiempirical model.

For the transitions extrapolating to 800 meV, our qualitative interpretation is essentially confirmed by the results of the calculation (Brum et al., 1988) shown in Fig. 28: The transition energies depend about linearly on the magnetic field and involve a very heavy in-plane mass for the "light" hole. Transitions in the σ^+ and σ^- polarizations are split by ~ 6 meV at 10 T, which corresponds essentially to the g-factor $g = -9$ of the conduction band of bulk GaSb. Only a hint of band mixing appears with the emergence of new σ^- transitions at high magnetic field. In the opposite situation, the heavy-hole transitions exhibit a strong band mixing, as the oscillator strength shifts rapidly for the "allowed" $HH_1^M - E_1^N$ transitions at low magnetic field to the "forbidden" $HH_2^M - E_1^N$ transitions at high magnetic field. Except for the lowest transitions, which have a singular behavior, the transitions that can be observed are $HH_2^M - E_1^N$ transitions involving a relatively light hole mass and extrapolate nearly linearly to the energy of the forbidden transition $HH_2 - E_1$ at $B = 0$. This indeed explains our observations.

This calculation, which does not contain any fitting parameter, thus accounts for the main characteristics of our data. However, the fit is relatively poor, essentially because the slope of the transitions is always too large. This remanent discrepancy might be entirely due to the Coulomb interaction. Indeed, calculations of the two-dimensional exciton states in arbitrary magnetic field (MacDonald and Ritchie, 1986; MacDonald and Ritchie, private communication) indicate that excitonic corrections to the Landau level energies are quite important in the whole range of energy and

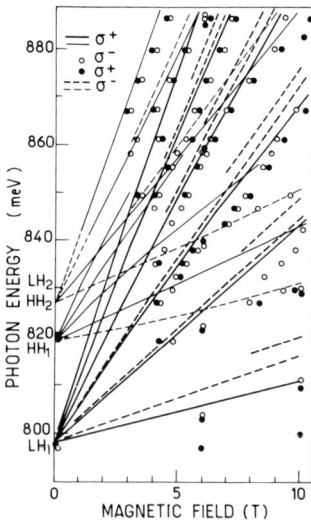

FIG. 28. The same experimental fan diagram as in Fig. 27, and the calculated transitions having an oscillator strength larger than 0.2.

magnetic field of interest. In our case, they would be as large as 10 meV for the lowest transitions at 10 T.

E. $In_xGa_{1-x}As-In_yGa_{1-y}As$ ON InP

$In_xGa_{1-x}As-In_yGa_{1-y}As$ superlattices grown lattice matched on InP have also been studied (Quillec et al., 1985; 1986). The higher bandgap alloy (with the smaller In mole fraction) is under tension, whereas the layers with the higher In content are under compression. The valence-band configuration is similar to the situation in $In_xGa_{1-x}As-GaAs$, i.e., electrons and "heavy" holes tend to be confined in the smaller bandgap material, whereas the "light" holes are confined in the other one. In the optical spectra, the light-hole transitions are no longer observed, supporting this interpretation. Again, the precise offset value is difficult to obtain from transmission spectra, due to the lack of sensitivity of their energies against the band discontinuities. Superlattices with a room-temperature emission of up to 1.9 μm could be obtained in this system.

V. Large-Strain Systems

The study of highly strained systems, as InAs–GaAs, where the lattice mismatch between the two components is 7%, is motivated by the hope that new materials may arise from very large built-in strain effects.

As it can be seen in Fig. 4, when the strains are larger than 2 or 3%, the strain strongly modifies the semiconductor band structure. For example, if InAs is under strong biaxial tension, its bandgap shrinks to very small values, and the higher energy Γ_7 valence band may even cross the Γ_6 conduction band (the Γ's being those of D_{2d}).

Another interest in the high-strain regime is the need for optimizing the growth conditions of the SLS. In this regime, the interactions between the growth process and the strain fields are amplified: Highly strained SLS thus constitute a severe test for the growth conditions, which in turn may allow to improve the quality of less strained structures. In this respect, the study of InAs thin quantum wells embedded in GaAs and strained to this material has yielded precious indications for optimizing the growth of InAs–GaAs short-period superlattices on InP.

One of the specific properties of these systems is that the SLS layers have to be very thin (less than about 10 Å for a 7% mismatch), because of the small critical thicknesses. This implies that the detail of the interface formation between the two materials needs to be carefully considered, specially in the InAs–GaAs system, where the situation is far less ideal than in GaAs–AlAs. Another important point peculiar to the large strains is the intertwining of the growth process and the strain fields. Whereas in the moderate-strain regime the strain relief process involves mainly lattice defects, in the large-strain regime the strains can be reduced by compositional defects.

A. InAs ON GaAs

This is the most highly strained III–V compounds system that has been studied. The growth of InAs on GaAs is a Stranski–Krastanov process (Glas et al., 1987). One to three InAs monolayers (3 to 9 Å) can be deposited two-dimensionally on GaAs, and further InAs deposition results in the formation of three-dimensional islands where dislocation lines are formed when they are large enough. During the MBE growth, the transition from a two- to a three-dimensional process is evidenced by the observation of bulk diffraction spots in the Reflection High Energy Electron Diffraction (RHEED) pattern. When GaAs is subsequently grown in order to form a thin InAs quantum well, two phenomena occur:

(i) If the InAs layer growth was partially three-dimensional, the InAs islands are buried in the final structure. The latter is highly inhomogeneous in composition in the layer planes.

(ii) In segregation takes place during the growth of GaAs on InAs, which smoothes the In composition profile.

The consequences of these two phenomena are examined in the following sections.

1. *Two-Dimensional versus Three-Dimensional Growth*

The existence of large three-dimensional islands can be evidenced in transmission electron microscope (TEM) micrographs (Glas et al., 1987) and from an optical study (Goldstein et al., 1985b; Gerard and Marzin, 1988). These In-rich regions act as radiative defects, when the InAs-deposited thickness is low enough to avoid the presence of dislocations. Figure 29 shows the low-temperature photoluminescence spectra obtained on three samples with thin InAs multiquantum wells embedded in GaAs. The InAs contents are 1.2 (sample 1), 2 (sample 3), and 3 (sample 2) InAs monolayers per well, respectively. During the growth of sample 3, the RHEED pattern indicated three-dimensional growth, and the photoluminescence peak for this sample is shifted by 100 meV to lower energy with respect to that of sample 1 (or 2) where the growth remained two-dimensional. This is thought to be due to the presence of In-rich clusters. This low-energy shift of the photoluminescence line was found to be characteristic of all samples with three-dimensional growth in the InAs layers. The first stages of this three-dimensional growth have been investigated (Houzay et al., 1987) by TEM. These studies showed that the islands are regular in size, this size being

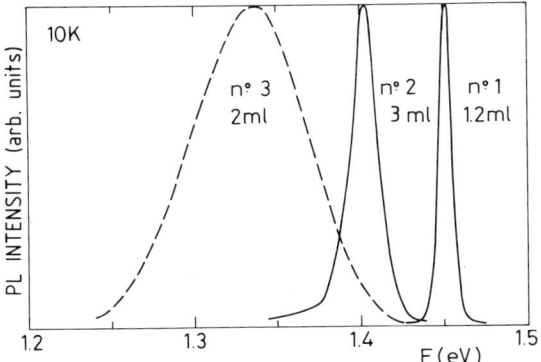

FIG. 29. Photoluminescence spectra obtained at 10 K for narrow InAs quantum wells in GaAs. For samples 1 and 2, the growth of the 1.2- and 3-monolayer-thick well was bi-dimensional, whereas sample 3 (2 monolayers) displays a spectrum typical of samples for which three-dimensional growth occurred. Samples 1 and 3 were grown by MBE, and sample 2 by MEE.

correlated with the number of dislocations they contain. In the very beginning of the three-dimensional growth, where no dislocations are formed, the islands are about 100 Å wide and 30 Å thick (Fig. 30). This correlation in the island sizes results in the moderate full width at half maximum (FWHM) of the associated photoluminescence line (Fig. 29, sample 3).

The thickness of the InAs layer that can be deposited bi-dimensionally depends on the strain, and for a given strain, on the nature of the surface where InAs is deposited and on the growth conditions. Conventional MBE

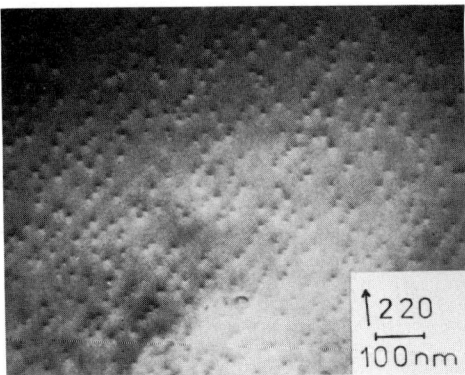

FIG. 30. Scanning transmission electron micrograph obtained for a two-monolayer InAs film grown on GaAs. The growth had been interrupted when three-dimensional growth occurred. Small InAs islands are observed, which are strongly correlated in size. (Courtesy of F. Glas, CNET, Bagneux, France.)

limit is around 2 monolayers, whereas an alternate monolayer deposition of III and V species at low substrate temperature (350°C) allows a two-dimensional (2D) deposition of three InAs monolayers (Gerard and Marzin, 1988). This explains the observed differences between the emission of samples 2 and 3 (Fig. 29).

Much thicker InAs layers (3 to 30 monolayers) have been deposited two-dimensionally on GaAs (Grunthaner et al., 1985; Yen et al., 1986). Indium was deposited in MBE at 420°C by submonolayer increments, and large growth interruptions (a few minutes) followed each indium deposition. This technique led to a two-dimensional growth, thus evidencing the importance of the surface flatness. However, the structures corresponding to the higher InAs thicknesses were clearly metastable, and misfit dislocations were observed at the interfaces. Optical data on these structures are unfortunately not available at the present time.

The microscopic origin of the effects of the growth conditions on the growth mode of the highly strained structures is still not well understood. It is, however, clear that the roughness of the surface may play an important role. Defects such as islands or steps at the surface of the strained layer allow a local partial relaxation. As a result, the growth becomes inhomogeneous, since preferential sites are available for a subsequent incorporation. This may lead to an increase of the roughness and possibly to three-dimensional (3D) growth.

Let us finally mention that the growth of the strained layers is also influenced by the quality of the underlying layers of the superlattice: The memory of the position of compositional inhomogeneities is still present at the growing surface through the strain field they generate. This mechanism tends to pile up the defects in strained superlattices (Goldstein et al., 1985b).

2. Indium Segregation

In situ Auger measurements show clearly that indium tends to segregate during the MBE growth of GaAs on InAs (Houzay et al., 1987; Guille et al., 1987). Although quantitative estimates are difficult, it seems that the last InAs monolayer is affected, and that 80% of the In of this monolayer are segregated. Indium is then incorporated in GaAs at the corresponding rate of 20% per GaAs monolayer deposited. Whereas the InAs on the GaAs interface is abrupt, the GaAs on the InAs one is smooth and about 20 Å thick. Fig. 31 shows the In concentration profiles obtained under these assumptions for thin InAs quantum wells in InAs and InAs–GaAs short-period superlattices. Up until now, no detectable dependence of this segregation on the growth conditions or on the strain has been found.

However, the picture of a segregation mechanism affecting only the last InAs monolayer is likely to be an oversimplification: roughness of the surface may be an important parameter as suggested from the optical data described below.

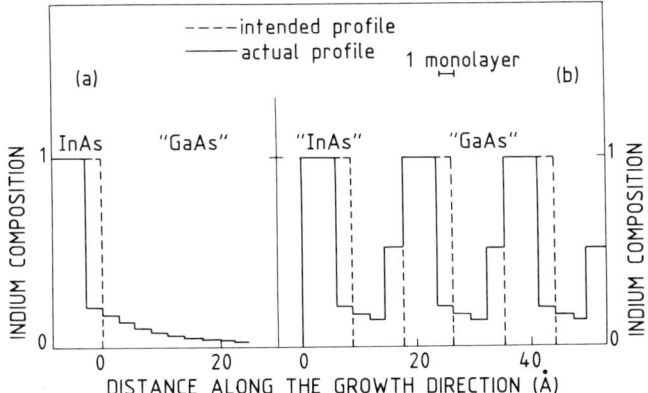

FIG. 31. Calculated indium composition profiles including grading effect of the indium segregation in GaAs: (a) at a GaAs-on-InAs heterojunction, and (b) in an $(InAs)_3(GaAs)_3$ short-period superlattice. Note that this phenomenon does not affect the superlattice periodicity.

3. Optical Properties of Thin InAs Quantum Wells in GaAs

From a basic point of view, these structures are interesting because they constitute the two-dimensional counterpart of isoelectronic impurities in 3D structures. Unfortunately, their properties are dominated by the two phenomena described above and reflect the complexity of the microscopic details.

In order to precisely measure the In content, an accurate method consists in building periodic structures: From the period and mean lattice mismatch to the GaAs substrate deduced from x-ray diffraction profiles, the total amount of Indium per period can be precisely extracted in a way that does not depend on the details of the actual composition profile. However, information on the segregation phenomenon is consequently rather difficult to obtain using this technique.

Figure 32 shows typical low-temperature photoluminescence and transmission spectra obtained here on a multiquantum well structure with thin InAs wells. The luminescence spectrum consists of a single line whose energy decreases with increasing amounts of deposited InAs. The energy of the transitions, the in situ observation of the RHEED pattern, as well as the x-ray profile analysis show that the growth did not switch to a three-dimensional regime. As shown in Fig. 32, this transition can also been observed in transmission (as well as in photoluminescence excitation), where a higher energy transition can also be detected. These two transitions are assigned to $HH_1–E_1$ and $LH_1–E_1$ excitons, respectively.

Figure 33 shows the energy dependence (Tischler et al., 1986; Gerard and Marzin, 1988) of the photoluminescence (PL) peak as a function of the

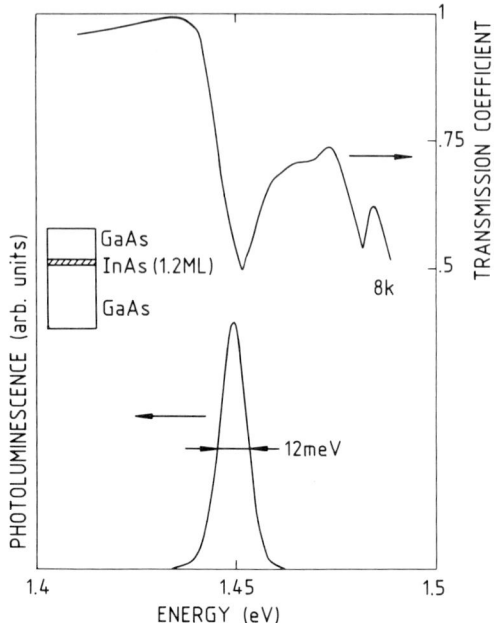

FIG. 32. Typical photoluminescence and transmission spectra obtained at low temperature on a 40-period InAs-GaAs multiquantum-well structure.

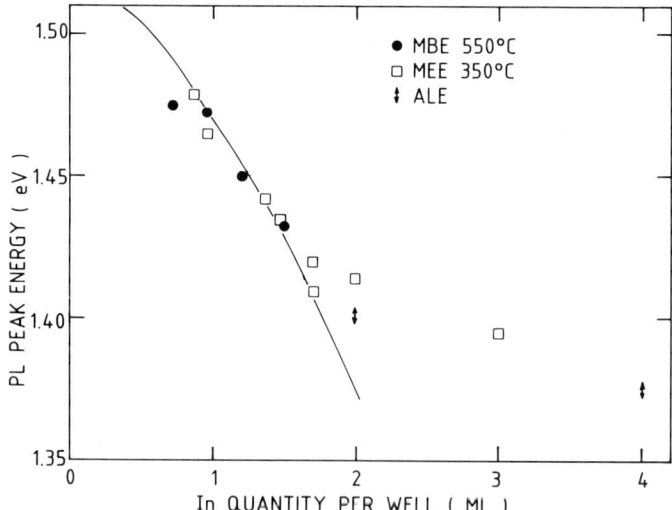

FIG. 33. Low-temperature photoluminescence peak energy as a function of InAs well width. Different growth techniques lead to a consistent set of data (atomic layer epitaxy data are taken from Tischler et al., 1986). Solid line displays a typical theoretical estimate for this variation.

number of InAs monolayers deposited on GaAs. As evidenced in the same figure where the theoretical estimates for these transition energies (derived from tight-binding or envelope functions models) are indicated, one should expect a much steeper energy decrease versus the well thickness in the case of sharp interfaces. When the asymmetric interface grading due to the indium segregation is included in the calculation, a satisfactory agreement with experiments is obtained up to 2 to 2.5 InAs monolayers. On the other hand, the asymmetric grading alone cannot explain the experimental data obtained for three or four monolayers.

Figure 34 shows a high-resolution micrograph obtained on two such samples for 0.9 and 1.7 monolayers. It appears on these micrographs that the interfaces are not equivalent. The second interface, obtained with GaAs on InAs, is broader than the first one, due to the segregation of indium. In-plane inhomogeneities are also detected, and this roughness tends furthermore to increase with the InAs layer thickness. In terms of the mean indium composition per monolayer, this entails a further broadening of the interface. Such an increased spreading of the indium, due to the combined effects of the roughness and of the segregation, should reduce the binding energies of electrons and light holes in the InAs–GaAs quantum well. These effects are thus thought to be responsible for the experimentally smooth dependence of the photoluminescence peak energy versus the InAs-deposited quantity.

B. InAs–GaAs on InP

Since the lattice parameter of InP lies between those of InAs and GaAs, it is possible to grow $(InAs)_m(GaAs)_n$ short-period superlattices approximately lattice matched to InP, and numerous growth studies (Fukui and Saito, 1984; Tamargo et al., 1985; Matsui et al., 1985; Ohno et al., 1985; Razeghi et al., 1987; McDermott et al., 1987; Katsumi et al., 1988; Gerard et al., 1989) have reported on this system.

The GaAs layers are under biaxial tension (+3.1%), whereas the compressive in-plane strain experienced by the InAs layers is significantly smaller (−3.8%) than when grown on GaAs. The growth of thick $(InAs)_m(GaAs)_n$

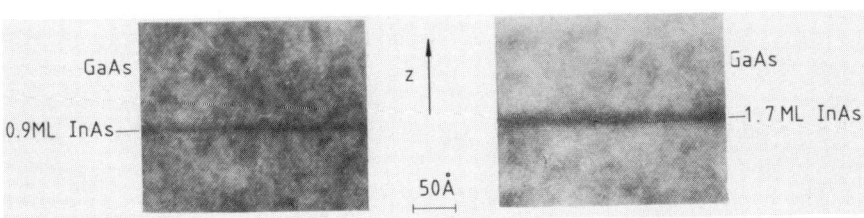

FIG. 34. High-resolution transmission electron micrographs obtained for 0.9- and 1.7-monolayer-thick InAs quantum wells. (Courtesy C. D'Anterroches, CNET Meylan, France).

short-period superlattices (SPS) remains difficult, however, particularly for large values of n and m. During conventional MBE growth, the surface roughness gradually increases, and after the growth of 500 Å ($n = m = 4$) to 900 Å ($n = m = 3$), bulk-type "spotty" RHEED patterns are observed. As in Section V.A.1, the interaction between the strain and the growth mechanism itself may account for this observation. The resulting SPS layers exhibit broad (30 to 50 meV) low-temperature luminescence lines. Confirmation of the rather poor quality of such material can be obtained by studying quantum well structures using $(InAs)_m(GaAs)_n$ as the well material, as is shown below.

An improved growth process results from the use of migration enhanced epitaxy (MEE), and thicker SPS layers have been grown without noticeable degradation of the RHEED pattern. MEE exploits the large migration of metal atoms on the surface under a weak As_4 pressure. This particular MBE technique thus consists in the alternate monolayer deposition of third-column atoms and arsenic on the substrate, which in our studies was done at very low temperature (350°C). MEE, initially introduced in the GaAs–GaAlAs system (Horikoshi et al., 1988), and similar modulated MBE techniques, have been adapted to the growth of highly strained structures, such as InAs quantum wells in GaAs (Gerard and Marzin, 1988), InAs–GaAs SPS (Katsumi et al., 1988; Gerard et al., 1989), or InAs–AlAs SPS (Gonzales et al., 1989). RHEED studies (Katsumi et al., 1988) confirmed the expected improvement of the surface flatness during the MEE growth. This growth technique has allowed a breakthrough in the quality improvement of InAs–GaAs SPS (Gerard et al., 1989).

Structural characterization of a 1200-Å-thick $(InAs)_4(GaAs)_3$ SLS grown on an InGaAs buffer layer lattice matched to InP confirmed the high quality of such layers. Scanning transmission electron micrographs as that shown in Fig. 35 revealed the good in-plane homogeneity of the SLS, and no misfit dislocations could be detected within the structure. A $\theta-2\theta$ diffraction profile of the structure is shown in Fig. 36. Sharp diffraction peaks specific of SPS appear; the width of the more intense diffraction peaks (150" for $N = 13$) is comparable to the experimental resolution ($\Delta\theta = 100$" near InP(004) diffraction order), which attests for the high quality of the structure when compared to the typical values of 0.4° or 1° published for similar MBE samples (Matsui et al., 1985; Ohno et al., 1985). A period of 20.5 Å is determined from the absolute or relative positions of SPS diffraction peaks, which is close to the 20.6 Å period calculated for an elastically strained $(InAs)_4(GaAs)_3$ superlattice.

The Raman scattering spectra (Jusserand and Gerard, 1988) displayed in Fig. 37 present features characteristic of high-quality structures: As in GaAs–AlAs short-period superlattices (Colvard et al., 1980; Jusserand et al., 1984),

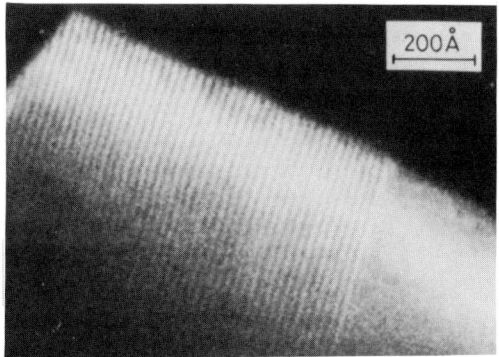

FIG. 35. Scanning transmission electron micrograph obtained on a cleaved edge of a 1200-Å-thick $(InAs)_4(GaAs)_3$ short-period superlattice grown on InP.

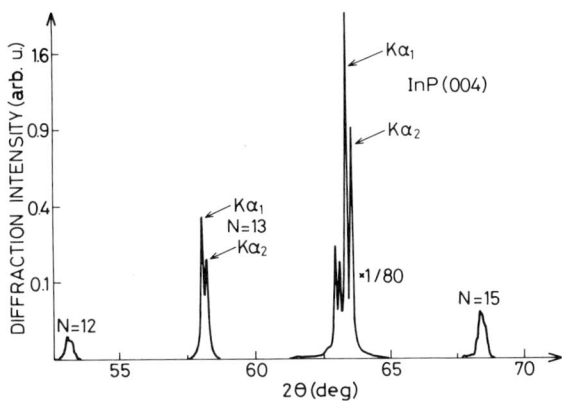

FIG. 36. θ–2θ x-ray diffraction profile obtained on the $(InAs)_4(GaAs)_3$ sample near InP 004 diffraction peak.

one folded acoustic mode and several confined optical modes are observed. From the energy of the acoustic mode, a period of 20.5 Å is determined, in good agreement with x-ray diffraction. The modes labelled 1, 2, and 3 are assigned to confined GaAs-type LO phonon modes, whereas mode 4 is thought to be related to a propagative InAs-type LO phonon mode.

A high degree of perfection has thus been demonstrated, as far as the periodicity is concerned, both from x-ray diffraction and Raman scattering experiments. Both techniques also indicate a strong compositional modulation in this SPS. However, the spacing between the confined optical phonons as well as the analysis of the x-ray satellite intensities indicate deviations from

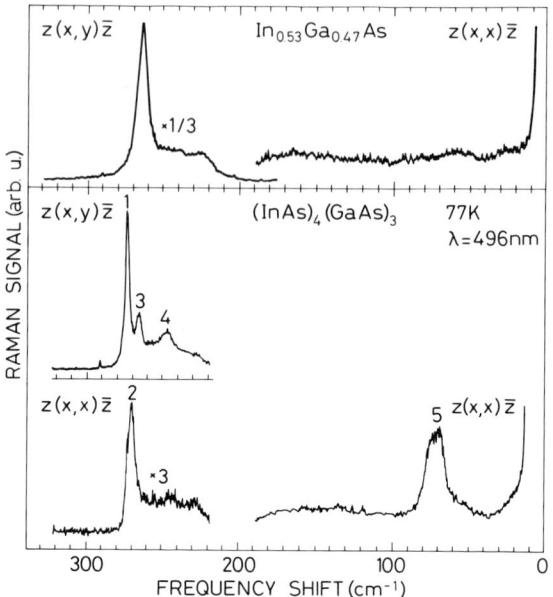

FIG. 37. Raman spectra obtained at 77 K (optical modes) and at room temperature (acoustic modes) on the $(InAs)_4(GaAs)_3$ sample, and on a reference $In_{0.53}Ga_{0.47}As$ MBE-grown alloy.

a perfectly abrupt interface composition profile (Jusserand et al., 1986), though these observations could not be made quantitative up to now. Disorder-activated transverse and longitudinal acoustic modes (DATA and DALA) are still also observed in the Raman spectra of the $(InAs)_4(GaAs)_3$ superlattice, as in the InGaAs random alloys. Indium segregation, which partly intermixes the layers without breaking the periodicity of the structure, is one possible cause for these imperfections.

The low-temperature (PL) spectrum obtained for this sample, shown in Fig. 38, consists of a single intense line at 0.75 eV. One should note, however, some discrepancies in the energy of the transition between previously published data (Fukui and Saito, 1984; Razeghi et al. 1987; McDermott et al., 1987; Voisin et al., 1986a), probably due to the lack of a precise characterization of the samples, and to the sensitivity of this PL position with the average indium composition in the SPS. PL peak FWHM (10 meV) compares favorably with previous reports, with the exception of a 7 meV FWHM observed for an MOCVD-grown $(InAs)_1(GaAs)_1$ monolayer superlattice (Fukui and Saito, 1984).

It is particularly revealing to consider these SPS as a novel material and to insert them in MQW structures to extract information about the material

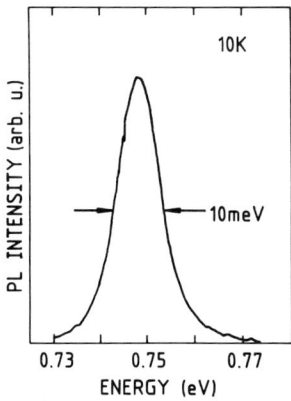

FIG. 38. Photoluminescence spectrum obtained at 10 K on $(InAs)_4(GaAs)_3$.

itself and the growth technique:

(i) The quality of the short-period strained layers tends to decrease with thickness. The study of bulk properties may thus only reveal this degradation, hiding information on the quality of the first few hundred Å.

(ii) More important is the fact that combined photoluminescence and transmission (or photoluminescence excitation) experiments performed on MQW structures using SLS as a well material provide an estimate of the importance of inhomogeneities taking place in the SLS itself. Indeed, the observation (or nonobservation) of step-like features in the optical spectra, the spectral width and Stokes shift of the luminescence with respect to the transmission spectra are important additional pieces of information.

(iii) Precise data concerning details of the SPS electronic band structure can furthermore be extracted from the energies of the optical transitions associated to the MQW.

We thus present next the study of three MQW, designed with seven 105 Å quantum wells (material W) separated by 400 Å MBE-grown InGaAlAs barriers lattice matched to InP (001) substrate. W is either a 9-period $(InAs)_2(GaAs)_2$ SPS-grown by conventional MBE (sample C) or by MEE at 350°C (sample B), or the InGaAs-disordered alloy of same average composition (sample A). High-resolution transmission electron micrographs displayed in Fig. 39 were obtained on sample B and confirmed the successful alternate deposition of InAs and GaAs bilayers, and the homogeneity of this sophisticated well material. A 110 Å overall well thickness is measured, in satisfying agreement with its designed value.

PL and PL excitation spectra obtained at 2 K for these samples are shown in Fig. 40. Whereas the staircaselike shape, characteristic of bi-dimensional systems, and marked excitonic peaks are seen for the reference sample (A),

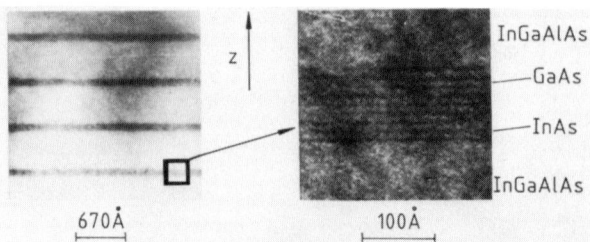

FIG. 39. Low-resolution and high-resolution transmission electron micrographs obtained for multiquantum well structure B: 110-Å-thick wells are built with MEE-grown $(InAs)_2(GaAs)_2$ SPS.

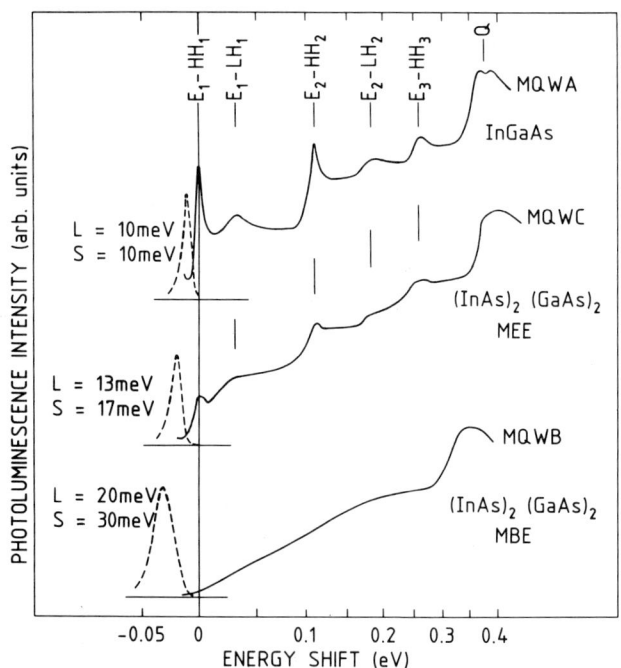

FIG. 40. Low-temperature photoluminescence and photoluminescence excitation spectra for three multiquantum well samples; the well material is either $(InAs)_2(GaAs)_2$ grown by MEE (b) or standard MBE (c), or the InGaAs random alloy of same average composition (a). Spectra have been shifted to make the first structure seen on excitation spectra coincide. L and S denote the spectral width and the Stokes shift of the photoluminescence.

these structures are completely smeared out for an MBE-grown InAs–GaAs well material (sample C). This indicates that large compositional inhomogeneities are still present at the very beginning of the MBE growth of this material, even more clearly than the moderate PL peak width (22 meV) and Stokes shift (30 meV). For an MEE-grown well material (sample B), the PL peak is narrower (13 meV) and the Stokes shift reduced (15 meV). However, these values are still larger than those commonly observed for InGaAs quantum wells. More significant is the observation of staircase-like steplike features and the recovery of excitonic peaks in the excitation spectrum: They confirm a drastic quality improvement for the MEE-grown $(InAs)_2(GaAs)_2$ materials. These experimental results have been confirmed by transmission experiments performed at low temperature on these samples.

The spectra in Fig. 40 have been shifted to line up the first excitonic structures for samples A and B. (We have observed small shifts in the 10 meV range for this bandgap position from sample to sample. These may be only due to small variations of the average In composition for the different well materials. As a result, no detailed information can be obtained on the SLS and alloy bandgaps difference except that this difference is rather small, for example, < 20 meV.) Although different in shape, observed structures strikingly coincide. They can thus be attributed to the same optical transitions for both structures. Three of them involve electron- and heavy-hole subbands (HH_n–E_n n = 1 to 3) and two electron and light holes (LH_n–E_n n = 1, 2).

This study highlights two key electronic properties of $(InAs)_2(GaAs)_2$ superlattices as shown below:

(i) Since well width and barrier height are kept constant for both MQW, the spacing between E_n–HH_n optical transition strongly depends on the light-particle (i.e., electron) effective mass along the growth direction, m_z^* (and not much on the large heavy-hole effective mass). Perpendicular electron effective masses are thus very similar for $(InAs)_2(GaAs)_2$ SPS and InGaAs alloy of same average composition: A clear "pseudo-alloy" behavior is here displayed by these highly strained SPS. E_n–LH_n transitions are not so well resolved, but a similar conclusion seems to hold also for the light-hole dispersion curve along the growth direction.

(ii) $(InAs)_2(GaAs)_2$ SPS as well as biaxially strained InAs or GaAs display a tetragonal symmetry. It is particularly interesting to estimate the resulting valence-band degeneracy splitting δ_{h-l} in these materials. Schematic zone-center dispersion curves have been plotted in Fig. 41 for InAs and GaAs strained on InP. Large δ_{h-l} (−160 and +360 meV, respectively) result from the large experienced strains. One-band-effective mass calculations will thus predict for such small period InAs–GaAs SPS, a light hole-to-conduction fundamental bandgap, and a large valence-band splitting (δ_{h-l} close to 100 meV). Comparing the difference of HH_1–E_1 and LH_1–E_1 transition

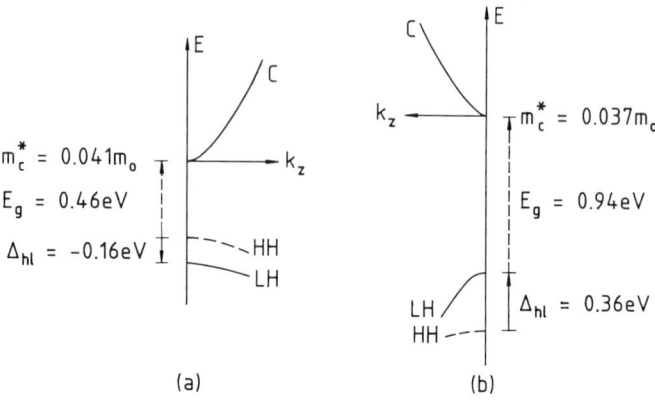

FIG. 41. Schematic zone-center band structure for InAs and GaAs, when they are strained onto InP.

energies in MQW samples A and B allows to reject this hypothesis and leads to the first estimate of this fundamental parameter. When possible experimental errors are taken into account, δ_{h-l} in $(InAs)_2(GaAs)_2$ SPS and in $In_{0.5}Ga_{0.5}As$ should differ from less than 12 meV. A small δ_{h-l} is estimated for $In_{0.5}Ga_{0.5}As$ alloy (18 meV), which is slightly mismatched to InP. Consequently, $(InAs)_2(GaAs)_2$ SPS should have (as $In_{0.5}Ga_{0.5}As$ strained on InP) a light hole-to-conduction fundamental bandgap transition and a rather small δ_{h-l} ($6 < \delta_{h-l} < 30$ meV). Such a small symmetry-induced valence-band splitting had been predicted using a more elaborate theoretical approach (*ab initio* calculations of $(InAs)_1(GaAs)_1$ band structure; Padgen and Paquet, unpublished).

Let us finally mention again that if the simple effective mass model described in Section III.C. fails to describe accurately the valence band of SLS for such large strains and thin layers, it does allow a good description of the conduction band and predicts correctly the pseudoalloy behavior observed for electrons.

VI. Conclusion

The search of heterostructures exhibiting quantum confinement effects has naturally led to the growth of lattice-mismatched materials, since small layer thicknesses are required for both effects. The III–V compounds are ideal candidates for the realization of efficient emitters and detectors operating in the long-wavelength range, since it appears now possible, through the growth of strained heterolayers, to realize almost any structure. Although many puzzling questions (for example, about the stability of strained-layer structures) already received encouraging answers, more studies are required to overcome the remaining problems, encountered in particular in the growth of highly strained layers.

Acknowledgment

It is a pleasure for the authors to thank G. Bastard to whom they are indebted for many discussions about this paper, as well as for his active participation in most of the subjects covered therein.

References

Abstreiter, G., Brugger, H., Wolf, T., Jorke, H., and Herzog, H. J. (1986). *Surf. Science* **174**, 640.
Altarelli, M. (1983). *Phys. Rev.* **B28**, 842.
Altarelli, M. (1985a). In "Semiconductor Heterojunctions and Superlattices," *Proc. 85 Les Houches Winter School*, Springer-Verlag, Berlin, p. 12.
Altarelli, M., Ekenberg, U., and Fasolino, A. (1985b). *Phys. Rev. B* **32**, 5138.
Ancilotto, F., Fasolino, A., and Maan, J. C. (1987). *Superlattices and Microstructures*, **3**, 187.
Anderson, N. G., Laidig W. D., and Lin, F. Y. (1984). *J. Electron. Mater.* **14**, 187.
Anderson, N. D., Laidig, W. D., Lee, G., Lo, Y., and Ozturk, M. (1985). In "Layered Structures, Epitaxy and Interfaces," (J. M. Gibson and L. R. Dawson eds.), *Materials Research Society*, Pittsburgh.
Bastard, G. (1981). *Phys. Rev.* **B24**, 4714.
Bastard, G., and Brum, J. A. (1986). *IEEE J. Quant. Electron.* **OE22**, 1625.
Bastard, G. (1988). In "Wave Mechanics Applied to Semiconductor Heterostructures," *Les Editions de Physique*, Les Ulis.
Bastard, G., Delalande, C., Guldner, Y., and Voisin, P. Optical Characterization of III–V and II–VI Heterostructures (1988) In "Advance in Electronics and Electron Physics." Academic Press, New York.
Bir, G. E., and Pikus, G. E. (1974). In "Symmetry and Strain-Induced Effects in Semiconductors," J. Wiley, New York.
Bour, D. P., Gilbert, D. B., Elbaum, L., and Harvey, M. G. (1988). *Appl. Phys. Lett.* **53**, 2371.
Broido, D. A., and Yang, E. (1988). *Bulletin of the American Society*, March Meeting.
Broido, D. A., and Yang, E. (1988). In "Proceedings of the 19th Conference on the Physics of Semiconductors," Warsaw, to be published.
Brum, J. A., Voisin, P., Voos, M., Chang, L. L., and Esaki, L. (1988) *Surf. Science* **196**, 545.
Calleja, J. M., Meseguer, F., Tejedor, C., Mendez, E. E., Chang, C. A., and Esaki, L. (1986). *Surf. Science* **168**, 558.
Collins, R. T., Viña, L., Wang, W. I., Chang, L. L., and Esaki, L. (1987). *Phys. Rev.* **B36**, 1531.
Colvard, C., Merlin, R., Klein, M. V., and Gossard, A. C. (1980). *Phys. Rev. Lett.* **43**, 298.
Esaki, L., and Tsu, R. (1970). *IBM J. Res. Develop.* **14**, 61.
Fasolino, A., and Altarelli, M. (1984). *Surface Sci.* **142**, 322.
Forchel, A., Cebulla, U., Trankle, G., Kroemer, H., Subanna, S., and Griffiths, G. (1986). *Surface Science* **174**, 143.
Fritz, I. J., Dawson, L. R., Osbourn, G. C., Gourley, P. L., and Biefeld, R. M. (1982). *Int. Phys. Ser.* **65**, 241.

Fritz, I. J., Picraux, S. T., Dawson, L. R. Drummond, T. J. Laidig, W. D., and Anderson, N. G. (1985). *Appl. Phys. Lett.* **46**, 967.
Fukui, T., and Saito, H. (1984). *Japan. J. Appl. Phys.* **23**, L521.
Gérard, J. M., Marzin, J. Y., and Primot, J. (1987). *J. Phys. (Paris)* **C5**, 169.
Gérard, J. M., and Marzin, J. Y. (1988). *Appl. Phys. Lett.* **53**, 568.
Gérard, J. M., Marzin, J. Y., Jusserand, B., Glas, F., and Primot, J. (1989). *Appl. Phys. Lett.* **54**, 30.
Glas, F., Guillé, C., Hénoc, P., and Houzay, F. (1987). *Int. Phys. Conf. Ser.* **87**, 71.
Goldstein, L., Quillec, M., Rao, E. V. K., Henoc, P., Masson, J. M., and Marzin, J. Y. (1982). *J. Phys. (Paris)* **12**, C5, 201.
Goldstein, L., Jean-Louis, A. M., Marzin, J. Y., Allovon, M., Alibert, C., and Gaillard, S. (1985). *Proc. Int. Symp. on GaAs and Related Compounds*, Biarritz, 1984, Int. Phys. Conf. Ser. **74**, 133.
Goldstein, L., Glas, F., Marzin, J. Y., Charasse, M. N., and Le Roux, G. (1985). *Appl. Phys. Lett.* **47**, 1099.
Gonzales, L., Ruiz, A., Mazuelas, A., Armelles, G., Recio, M., and Briones, F. (1989). *Superlattices and microstructures* **5**, 5.
Griffiths, G., Mohammed, K., Subbanna, S., Kroemer, H., and Merz, J. L. (1983). *Appl. Phys. Lett.* **43**, 1059.
Grunthaner, F. J., Yen, M. Y., Fernandez, R., Lee, T. C., Madhukar, A., and Lewis, B. F. (1985). *Appl. Phys. Lett.* **46**, 983.
Guillé, C., Houzay, F., Moison, J. M., and Barthe, F. (1987). *Surf. Science* **189/190**, 1041.
Guillé, C. (1987). Thèse, Paris, unpublished.
Hirose, K., Mizutani, T., and Nishi, K. (1987). *J. Crystal Growth* **81**, 130.
Horikoshi, Y., Kawashima, M., and Yamaguchi, H. (1988). *Japan. J. Appl. Phys.* **27**, 169.
Houzay, F., Guille, C., Moison, J. M., Henoc, P., and Barthe, F. (1987). *J. Cryst. Growth* **81**, 67.
Jagannath, C., Koteles, E. S., Lee, J., Chen, Y. J., Elman, B. S., and Chi, J. Y. (1986). *Phys. Rev.* **B34**, 7027.
Jaros, M., Wong, K. B., and Gell, M. A. (1985). *Phys. Rev.* **B31**, 1205.
Joncour, M. C., Mellet, R., Charasse, M. N., and Burgeat, J. (1986). *J. Cryst. Growth* **75**, 295.
Joyce, M. J., Johnson, M. J., Gal, M., and Usher, B. F. (1988). *Phys. Rev.* **B38**, 10978.
Jusserand, B., Paquet, D., and Regreny, A. (1984). *Phys. Rev.* **B30**, 6245.
Jusserand, B., Voisin, P., Voos, M., Chang, L. L., Mendez, E. E., and Esaki, L. (1985). *Appl. Phys. Lett.* **46**, 678.
Jusserand, B., Alexandre, F., Paquet, D., and Le Roux, G. (1986). *Appl. Phys. Lett.* **47**, 301.
Jusserand, B., and Gérard, J. M. (1988). In "Proceedings of the 19th Int. Conf. Phys. Semiconductors (W. Zawadzki, ed.), Institute of Physics, Polish Academy of Science, 799.
Kane, E. O. (1957). *J. Phys. Chem. Sol.* **1**, 249.
Kasper, E. (1986). *Surface Science* **174**, 630.
Kato, H., Iguchi, N., Chica, S., Nakayama, N., and Sano, N. (1986). *J. Appl. Phys.* **59**, 588.
Katsumi, R., Ohno, H., Ishii, K., Matsuzaki, K., Akatsu, Y., and Hasegawa, H. (1988). *J. Vac. Sci. Technol.* **B6** (2), 593.
Koteles, E. S., Jagannath, C., Lee, J., Chen, Y. J., Elman, B. S., and Chi, J. Y. (1986). In "Proceedings of the 18th Conference on the Physics of Semiconductors," O. Engstrom, ed. World Scientific, Volume 1, p. 625.
Landau, L., and Lifshitz, E. (1967). "Theory of Elasticity," ch. 4. MIR, Moscow.
Landolt-Börnstein (1982). "Numerical Data and Functional Relationships in Science and Technology," O. Madelung, ed. Group III, Vol. 17, Springer-Verlag, Berlin.
Lee, J., and Vassell, M. O. (1988). *Phys. Rev.* **B37**, 8861.
Luttinger, J. M. (1956). *Phys. Rev.* **102**, 1030.

MacDonald, A. H., and Ritchie, D. S. (1986). *Phys. Rev.* **B33**, 8326.
Marzin, J. Y. (1985). In "Semiconductor Heterojunctions and Superlattices," *Proc. 85 Les Houches Winter School*, Springer-Verlag, Berlin, p. 161.
Marzin, J. Y. (1987). Thèse, Paris, unpublished.
Marzin, J. Y., Charasse, M. N., and Sermage, B. (1985). *Phys. Rev.* **B31**, 8298.
Matsui, Y., Hayashi, H., Takahashi, M., Kikuchi, K., and Yoshida, K. (1985). *J. Cryst. Growth* **71**, 280.
Matthews, J. W., and Blakeslee, A. E. (1974). *J. Crystal Growth* **27**, 118.
Matthews, J. W., and Blakeslee, A. E. (1975). *J. Crystal Growth* **29**, 273.
Matthews, J. W., and Blakeslee, A. E. (1976). *J. Crystal Growth* **32**, 265.
McDermott, B. T., El-Masry, N. A., Tischler, M. A., and Bedair, S. M. (1987). *Appl. Phys. Lett.* **51**, 1830.
Menéndez, J., Pinczuk, A., Werder, D. J., Sput, S. K., Miller, R. C., Sivco, D. L., and Cho, A. Y. (1987). *Phys. Rev.* **B36**, 8165.
Nabarro, F. R. N. (1967). "Theory of Crystal Dislocations," Oxford.
Nishi, K., Hirose, K., and Mizutani, T. (1986). *Appl. Phys. Lett.* **49**, 794.
O'Reilly, E. P., and Witchlow, G. P. (1986). *Phys. Rev.* **B34**, 6030.
Ohno, H., Katsumi, R., Takama, T., and Hasegawa, H. (1985). *Japan. J. Appl. Phys.* **24**, L682.
Osbourn, G. C. (1982). *J. Vac. Sci. Technol.* **21**, 459.
Pan, S. H., Shen, H., Hang, Z., Pollak, S. H., Zhuang, W., Xu, Q., Roth, A. P., Masut, R. A., Lacelle, C., and Morris, D. (1988). *Phys. Rev.* **B38**, 3375.
People, R., and Bean, J. C. (1985). *Appl. Phys. Lett.* **47**, 322.
Picraux, S. T., Dawson, L. R., Osbourn, G. C., and Chu, W. K. (1983). *Appl. Phys. Lett.* **43**, 930.
Platero, G., and Altarelli, M. (1987). *Phys. Rev.* **B36**, 6591.
Ploog, K., Ohmori, Y., Okamoto, H., Stolz, W., and Wagner, J. (1985). *Appl. Phys. Lett.* **47**, 384.
Pollak, F. H., and Cardona, M. (1968). *Phys. Rev.* **172**, 816.
Priester, C., Allan, G., and Lannoo, M. (1988). *Phys. Rev.* **B38**, 9870.
Quillec, M., Goldstein, L., Le Roux, G., Burgeat, J., and Primot, J. (1984). *J. Appl. Phys.* **55**, 2904.
Quillec, M., Marzin, J. Y., Benchimol, J. L., Primot, J., and Le Roux, G. (1985). *SPIE*, Vol. 587, Optical Fiber Sources and Detectors, p. 62.
Quillec, M., Marzin, J. Y., Primot, J., Le Roux, G., Benchimol, J. L., and Burgeat, J. (1986). *J. Appl. Phys.* **59**, 2447.
Raisin, C., Lassabatere, L., Alibert, C., Girault, B., Abdel-Fattah, G., and Voisin, P. (1987). *Solid State Comm.* **61**, 17.
Razeghi, M., Maurel, P., Omnes, F., and Nagle, J. (1987). *Appl. Phys. Lett.* **51**, 2218.
Roth, A. P., Sacilotti, M., Masut, R. A., D'Arcy, P. J., Watt, B., Sproule, G. I., and Mitchell, D. F. (1986). *Appl. Phys. Lett.* **48**, 1452.
Sanders, G. D., and Chang, Y. C. (1985). *Phys. Rev.* **B32**, 4282.
Sato, M., and Horikoshi, Y. (1988). *J. J. Appl. Phys.* **27**, L2192.
Sauvage, M., Delalande, C., Voisin, P., Etienne, P., and Delescluse, P. (1986). Proc. of MSSII, Kyoto 1985, *Surf. Science* **174**, 573.
Schirber, J. E., Fritz, I. J., and Dawson, L. R. (1985). *Appl. Phys. Lett.* **46**, 187.
Schulman, J. N., and Chang, Y. C. (1981). *Phys. Rev.* **B24**, 4445.
Schulman, J. N., and Chang, Y. C. (1985). *Phys. Rev.* **B31**, 2056.
Tamargo, M. C., Hull, R., Greene, L. H., Hayes, J. R., and Cho, A. Y. (1985). *Appl. Phys. Lett.* **46**, 569.
Tejedor, C., Calleja, J. M., Meseguer, P., Mendez, E. E., Chang, C. A., and Esaki, L. (1985). *Phys. Rev.* **B32**, 5303.
Tischler, M. A., Anderson, N. G., and Bedair, S. M. (1986). *Appl. Phys. Lett.* **49**, 1199.

Van der Merwe, J. H. (1972). *Surf. Science* **31**, 198.
Voisin, P. (1985). In "Semiconductor Heterojunctions and Superlattices," *Proc. 85 Les Houches Winter School*, Springer-Verlag, Berlin, p. 73.
Voisin, P., Delalande, C., Voos, M., Chang, L. L., Segmuller, A., Chang, C. A., and Esaki, L. (1984) *Phys. Rev.* **B30**, 2276.
Voisin, P., Maan, J. C., Voos, M., Chang, L. L., and Esaki, L. (1986). *Surf. Science* **170**, 651.
Voisin, P., Voos, M., Marzin, J. Y., Tamargo, M. C., Nahory, R. E., and Cho, A. Y. (1986). *Appl. Phys. Lett.* **48**, 1476.
Yen, M. Y., Madhukar, A., Lewis, B. F., Fernandez, R., Eng, L., and Grunthaner, F. J. (1986). *Surf. Science* **174**, 600.
Zipperian, T. E., Dawson, L. R. Drummond, T. J., Schirber, J. E., and Fritz, I. J. (1988). *Appl. Phys. Lett.* **52**, 975.

CHAPTER 4

Structurally Induced States from Strain and Confinement

R. People and S. A. Jackson

AT&T BELL LABORATORIES
MURRAY HILL, NEW JERSEY

I.	INTRODUCTION	119
	A. Strained-Layer Semiconductor Structures	119
	B. Limits of Strained-Layer Epitaxy (Critical Thicknesses)	122
II.	OPTICAL AND ELECTRONIC PROPERTIES OF LATTICES MISMATCH-INDUCED STRAINED LAYERS	134
	A. Effects of Coherency Strain on Fundamental Band Edges	134
	B. Heterojunction-Band Alignment Modifications Induced by Coherent-Layer Strains	143
	C. Mass Modification Due to Coherent Layer Strains	155
III.	ZONE-FOLDING EFFECTS IN ULTRASHORT-PERIOD STRAINED-LAYER SUPERLATTICES.	164
	A. Effective-Mass Estimates of Optical Transition Energies in the (4×4)-ML Si/Ge SLS on Si (001)	165
	B. Quasidirect Gap Structures in Monolayer Si/Ge Elemental Strained-Layer Superlattices	169
IV.	OTHER EFFECTS IN STRAINED-LAYER HETEROSTRUCTURES	169
	A. Piezoelectric Effects in Strained-Layer Heterostructures	169
V.	SUMMARY	171
	ACKNOWLEDGMENTS	171
	REFERENCES	171

I. Introduction

A. STRAINED-LAYER SEMICONDUCTOR STRUCTURES

Interest in epitaxial layers grown coherently on substrates having a different lattice constant has a long history, stretching back to the celebrated work on monolayer growths by Frank and Van der Merwe (1949). The structural stability problem, relating to the coherent growth of epitaxial films containing many monolayers, was first addressed by Van der Merwe (1963) and later by Jesser and Kuhlmann-Wilsdorf (1967). In these models, it was found that there exists a maximum (critical) film height, h_c, above which it costs too much energy to strain additional layers into coherence with the substrate. Instead, a defect appears; in this case, the appropriate defect is called a misfit dislocation, which acts to partially relieve the strain in the overlayer. These and other models for strain relaxation in semiconductor overlayers will be reviewed in the following section on critical layer thicknesses. At this point, a brief synopsis of the pioneering work on the structural properties of strained-semiconductor multilayers will be given; as these studies helped stimulate subsequent interest in such strained-layer heterostructures.

Coherently strained-multilayer structures consist of thin alternating layers of materials, which are lattice *mismatched* in bulk form, but which elastically strain to uniformly match up the lattice constants of the materials in the planes parallel to the heterointerface. The resulting value of the in-plane lattice constant ($a_\parallel$) is intermediate between the unstrained lattice constants of the multilayer constituents. Matthews and Blakeslee (1976) first obtained an expression for ($a_\parallel$) by minimizing the total elastic strain energy of a single pair of layers. If the unstrained lattice parameters of the two materials are denoted by a_A and a_B (assume $a_A < a_B$), then

$$a_\parallel = a_A \left[1 + f_0 \bigg/ \left(1 + \frac{G_A h_A}{G_B h_B} \right) \right], \tag{1}$$

where G_i are the shear moduli, h_i the layer thicknesses, and the *misfit* between the consistent layers is given by $f_0 = (a_B - a_A)/a_A$, respectively. An unstrained set of mismatched layers and the corresponding set of strained layers are shown schematically in Fig. 1. Note that the quantity

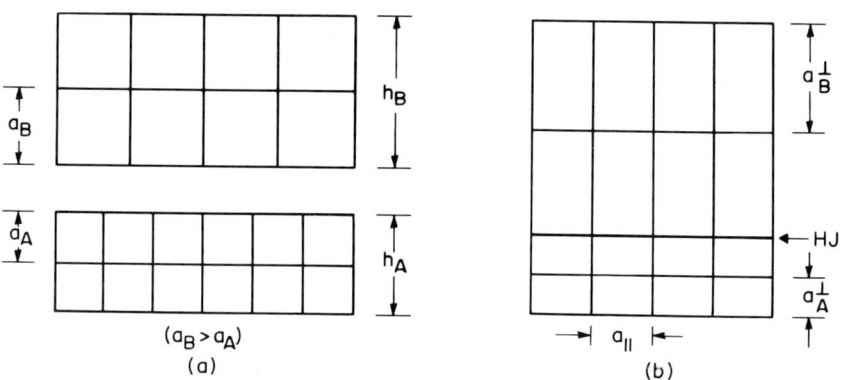

FIG. 1. (a) Schematic of unstrained lattices for two thin bulk materials having lattice constants a_A, a_B and layer thicknesses h_A, $< a_B$. (b) Schematic of the equilibrium configuration of the two thin layers in (a) for epitaxy of B on A.

$f_0/(1 + G_A h_A/G_B h_B)$ is simply the strain in layer A, i.e., $e_\parallel^A$. In equilibrium,

$$e_\parallel^A = -\left(\frac{G_B h_B}{G_A h_A}\right) e_\parallel^B. \tag{2}$$

Note that Eq. (2) implies that the average strain $\langle e_\parallel \rangle$ in the bilayer structure,

$$\langle e_\parallel \rangle \equiv \frac{\Sigma e_i h_i}{\Sigma h_i} \tag{3}$$

is zero under equilibrium conditions. Since the net strain is zero in each period, structurally stable multilayer structures may be fabricated by successive deposition of such bilayer structures with no limit on the number of depositions (periods). Note that multilayer structures in equilibrium may only be realized for either free-standing films or films grown on substrates having lattice parameter $a_s = a_\parallel$.

For arbitrary constituent layer lattice constants and film thickness, it is in general not possible to find bulk substates having lattice parameters equal to the resulting $a_\parallel$. If an application requires an *infinitely* thick multilayer structure, it is possible to fabricate such structures on a relaxed alloy buffer layer (which itself has been grown epitaxially on a readily available bulk substrate). The major disadvantage of such a scheme is the generation of threading dislocations within the relaxed buffer layer, which subsequently thread into the *active* layer, if their motion is not somehow impeded. The use of misfit strain to remove threading dislocations from epitaxial thin films had been suggested by Matthews et al. (1976), and was subsequently demonstrated in the structural studies of Matthews and Blakeslee (1976, 1977), in which a graded layer was inserted between the substrate and the relaxed buffer layer along with a strained multilayer immediately after the buffer layer (as shown in Fig. 2). The strained multilayer acts as a trap for threading dislocations. Osbourn and coworkers (Osbourn et al., 1982) demonstrated that such low dislocation-density multilayers may themselves demonstrate interesting optical properties, whereas Luryi et al. (1984) demonstrated a device-quality Ge p–i–n photodetector on Si (001) substrate using this technique.

Some applications of strained-layer heterostructures, especially those involving high energy densities in the strained active regions, cannot tolerate the high density of threading dislocations as might be found in growth on relaxed alloy buffer layers. Under high-energy-density operations, such dislocations may move and multiply, leading to either a gradual or catastrophic failure of the device. In these instances, it is desirable to have strained-layer heterostructures that are pseudomorphic throughout the entire structure. We shall therefore investigate the criteria for pseudomorphic

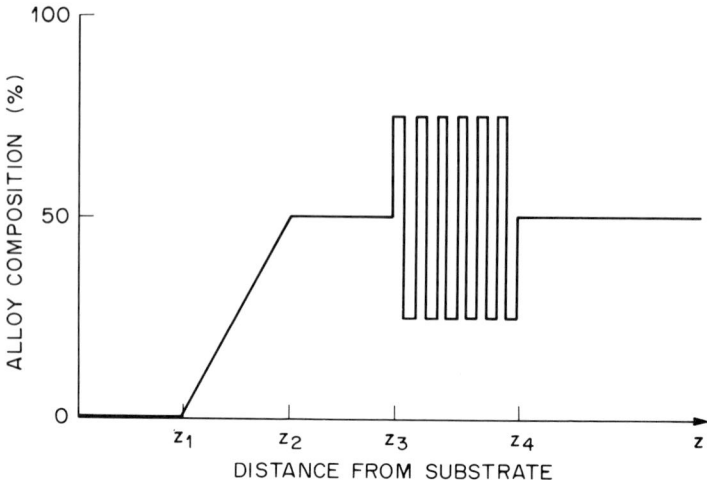

FIG. 2. Use of strained multilayers to block dislocation threading up from a relaxed alloy buffer layer. The region $0 < z < z_1$ corresponds to an alloy composition identical to the bulk substrate; $z_1 < z < z_2$ consists of a graded region; $z_2 < z < z_2$ corresponds to the relaxed buffer containing a high density of threading dislocations; $z_3 < z < z_4$ contains a strained-multilayer structure that blocks dislocations threading from the buffer layer; therefore, the region $z > z_4$ consists of a relaxed alloy layer having greatly reduced threading dislocation density.

growth as presently perceived. It should be noted that even in the case of equilibrium multilayer growth, it has been implicitly assumed that the thickness of each of the individual layers (h_A, h_B) is less than the maximum layer thickness corresponding to the magnitude of the strain ($e_\|^A$, $e_\|^B$) experienced by the individual layer.

B. LIMITS OF STRAINED-LAYER EPITAXY (CRITICAL THICKNESSES)

It is now well known (see, e.g., Vook, 1982) that growth of an overlayer on a bulk substrate may occur in one of three distinct modes: Frank–van der Merwe (FM) growth, Volmer–Weber (VW) growth, or Stranski–Krastanov (SK) growth. In the first of these, growth proceeds one monolayer at a time. The growth is two-dimensional in the sense that no atoms occupy layer i until lay $(i-1)$ is complete. By contrast, VW growth occurs by the nucleation of solid clusters that condense as three-dimensional islands on the substrate. These islands ultimately coalesce to form a continuous film. The SK growth mode is a combination of FM and VW growth modes. A bulk overlayer forms by solid-cluster nucleation after a few monolayers adsorb in layer-by-layer fashion. In the following, we shall briefly review models that attempt to describe the maximum number of monolayers (layer thickness) under which a film grows in the layer-by-layer (FM) growth mode before regions of the overlayer atoms lose registry with the substrate.

1. Single-Layer Critical Thicknesses

To reiterate, if the misfit f_0 between a bulk substrate and a growing epilayer is sufficiently small, the first atomic layers that are deposited will be strained to match the substrate, and a coherent (or pseudomorphic) interface will be formed. However, as the epilayer thickness increases, the homogeneous strain energy ξ_H becomes so large that a thickness is reached wherein it becomes favorable for misfit dislocations to be introduced (i.e., regions in the overlayer where atoms within the overlayer out of registry with the substrate are generated). Such misfit dislocations accommodate a fraction of the misfit, so that a reduced strain remains in the overlayer.

In the work of Van der Merwe (1963), the substrate and epilayer were treated as adjacent elastic media with an external sinusoidal stress imposed at the interface. The interfacial energy ξ_I (per unit area) between film and substrate was calculated. This energy density was assumed to be the minimum available for generation of dislocations, with a periodic array of dislocations being formed when ξ_I exceed the homogeneous strain energy density of the films. The following approximate expression for ξ_I was obtained in the limit of moderate misfits, $f_0 \lesssim 4\%$:

$$\xi_I \simeq 9.5 f_0 (Gb/4\pi^2). \tag{4}$$

Here the shear modulus $G = 2[C_{11} + C_{12} - 2C_{12}^2/C_{11}]$, where C_{ij} are the elastic stiffness constants, and b is the slip distance (i.e., the magnitude of the Burger vector). Note that ξ_I is independent of film thickness. The areal strain energy density, ξ_H, associated with a coherent film (one for which the misfit is totally accommodated by elastic strain, i.e., $e_\parallel = f_0$) is given by

$$\xi_H = 2G \left(\frac{1+v}{1-v} \right) h e_\parallel^2, \tag{5}$$

where v is Poisson's ratio. Note that ξ_H increases linearly from zero with film thickness h. Equating (4) and (5) determines the maximum (critical) layer thickness for coherent growth in the Van der Merwe model. One obtains

$$h_c \simeq \left(\frac{9.5}{8\pi^2} \right) \left(\frac{1+v}{1-v} \right) \left(\frac{b}{f_0} \right) \sim 0.1 (a_0/f_0), \tag{6}$$

where a_0 is the lattice parameter of the bulk substrate, and we have used $b = a_0/\sqrt{2}$. Equation (6) is plotted as the lower dashed curve in Fig. 3, wherein $a_0 = a_0(\text{Si})$ and f_0 is the misfit between a Si substrate and an overlayer of $Si_{1-x}Ge_x$ alloy. The Van der Merwe relation is usually obeyed in

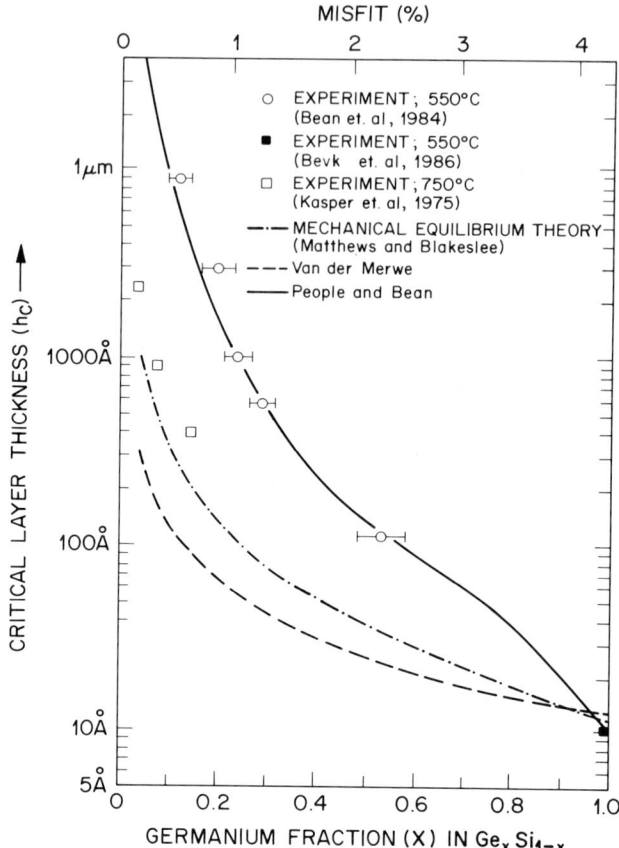

FIG. 3. Various growth temperature-independent models for single-layer critical thicknesses, along with a number of experimental values for growth of Ge_xSi_{1-x} alloys on (001) Si substrates.

epitaxial metals, where the lattice offers little resistance to dislocation motion. In these systems, predicted values of h_c and the thickness dependence of the residual strain following loss of coherence are in rather good agreement with experiment (Kuk et al., 1983).

Matthews and Blakeslee (1974) developed a model for critical layer thickness wherein h_c is determined by that thickness for which a segment of threading dislocation bows and elongates, under the influence of the misfit stress, to form a segment of interfacial dislocation line. Although in their original work the segment of threading dislocation was assumed to be grown in, the analysis applies equally well to threading dislocation segments that

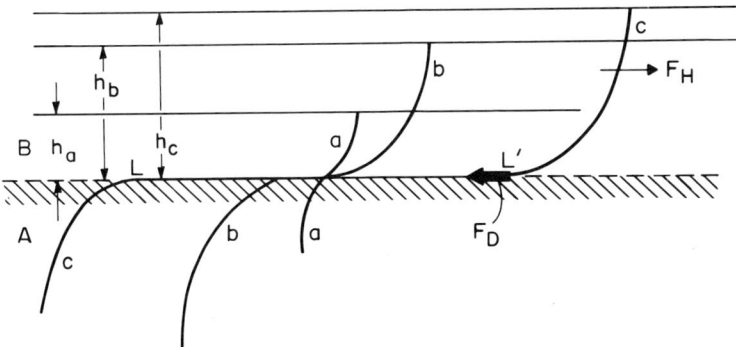

FIG. 4. Transformation of a coherent interface to an incoherent interface via the elongation of a pre-existing segment of threading dislocation. Initially the interface is assumed coherent (a). For film thickness h_b, the interface is critical (i.e., $F_H = F_D$), whereas for film thickness h_c, $F_H > F_D$, allowing the dislocation to elongate in the interface plane, thereby generating the length LL' of misfit dislocation line.

may be associated with critical half-loops. The process is illustrated in Fig. 4 for the case of a grown-in threading dislocation. The misfit strain $(e_\parallel)$ exerts a force F_H on the threading dislocation, where F_H is given by

$$F_H = 2G \left(\frac{1+v}{1-v}\right) bhe_\parallel \cdot \cos \lambda. \quad (7)$$

Here $\cos \lambda = 0.5$ for 60° dislocations. The tension in the segment LL' of the dislocation line, which is generated by the motion of the threading segment, is denoted by F_1.

$$F_1 = \frac{Gb^2}{4\pi} \frac{(1 - v\cos^2 \Theta)}{(1-v)} [1 + \ln(h/b)], \quad (8)$$

where $\cos^2 \Theta = 0.25$ for 60° dislocations. For a single epitaxial layer, the equilibrium critical thickness h_c is determined by equating (7) and (8). One thus obtains

$$h_c = \left(\frac{b}{f_0}\right) \frac{1 - v\cos^2 \Theta}{8\pi(1 + v)\cos \lambda} [1 + \ln(h/b)], \quad (9)$$

where $b = a_0/\sqrt{2}$ and a_0 is the bulk lattice parameter of the epilayer. This result is plotted as the middle dot-dashed curve in Fig. 3. Rather similar predictions for h_c have also been generated by the atomistic simulations of equilibrium growth made by Grabow and Gilmer (1986) and Dodson and Taylor (1986).

Early Ge_xSi_{1-x} growth experiments of Kasper and coworkers (Kasper, 1985; Kasper et al., 1985), having $x \lesssim 0.15$ and a growth temperature of 750°C, gave results in close agreement with Eq. (9). These results are given by the open squares in Fig. 3. Critical thickness measurements on 520°C molecular beam epitaxy (MBE)-grown and coherently strained single-quantum wells of $In_xGa_{1-x}As$ on GaAs by Anderson (1986) and Fritz et al. (1987) also show agreement with Eq. (9). Further, measurements of Fritz et al. (1985) on $In_xGa_{1-x}As/GaAs$ strained-layer superlattice (SLS) grown on relaxed buffer layers matched to the equilibrium (free-standing) SLS lattice constant are well described by the threading-dislocation limited growth model given in Eq. (9). Matthews (1975) has shown that the force balance arguments used to derive Eq. (9) are entirely equivalent to an energy-balance argument if the segments of edge dislocations so generated form a square grid of two perpendicular and noninteracting arrays of spacing $p = b/(f_0 - e_\parallel) \equiv b/\gamma$, where b is the magnitude of the Burger vector and γ is the strain relieved by the segment of line dislocation. Hence, both models previously described are equilibrium in nature.

A number of experimental critical thickness results (Bean et al., 1984; Orders and Usher, 1987) have indicated that in some cases, coherence apparently persists to thicknesses much greater than those predicted by Eq. (9); especially in epilayers having small misfits, $f_0 \lesssim 3\%$. The data of Bean et al. (1984) are shown as open circles in Fig. 3. The threading dislocation mechanism of Matthews and Blakeslee (1974) is not expected to be operational for growth on extremely high-quality Si substrates, due to the paucity of threading dislocations (where initial densities may approach $\sim 1\,cm^{-2}$). Further, electron microscopy studies of MBE-grown Ge_xSi_{1-x} films on Si indicated that initially dislocations in Ge_xSi_{1-x} tended *not* to form regular arrays. These observations led People and Bean (1985, 1986) to propose a nonequilibrium model based on energy balance but wherein the areal strain energy density of the film ξ_H (see Eq. (5)) is balanced against the *self-energy* of an *isolated* dislocation (rather than a square grid). In this model, the energy density of an isolated segment of dislocation line at a distance h from a free surface is taken as

$$\xi_D \simeq \left(\frac{Gb^2}{4\pi\langle w\rangle}\right) \ln(h/b). \tag{10}$$

The parameter $\langle w \rangle$ characterizes the effective "interfacial width" of the isolated dislocation and replaces the separation (p) between elements in a regular array. The width $\langle w \rangle$ of an isolated dislocation is defined as the distance along the glide plane in the glide direction over which the displacement of atoms is greater than half the maximum displacement of $b/2$ (People, 1986a). In general, $\langle w \rangle$ characterizes the effective lateral extent of the dislocation's strain field. Estimates of $\langle w \rangle$ depend on the form of the force-displacement relation assumed, with probable values for semiconductors lying between one and ten atom spacings along $\langle 110 \rangle$, for edge dislocations in (001) planes (see, e.g., Hobart, 1965); however, there are no experimental results to confirm these estimates. Therefore, $\langle w \rangle$ is incorporated into the People–Bean model as a phenomenological parameter that may be determined from a single critical thickness data point. Equating (5) and (10), one obtains h_c as

$$h_c \cong \left(\frac{b^2}{8\pi\langle w \rangle}\right)\left(\frac{1-v}{1+v}\right)\left[\left(\frac{1}{f_0}\right)^2 \cdot \ln(h/b)\right]. \tag{11}$$

To fit the Ge_xSi_{1-x} data for 550°C growth, it is found that $\langle w \rangle \simeq 5(\sqrt{2}/2)\langle a(x) \rangle$, where $\langle a(x) \rangle = 5.54$ Å; the mean (Ge, Si) lattice constant. These results are plotted as the solid curve in Fig. 3. It should be noted that the model proposed by People and Bean is not an equilibrium model (i.e., it is not equivalent to a force balance relation) but rather is an explicit statement on germinate nucleation in a metastable system. This can be seen as follows. The difference in the energy densities ξ_D and ξ_H, i.e., $\delta E_T = \xi_D - \xi_H$, represents the change in the total energy of the epilayer. Note that δE_T constitutes an effective nucleation barrier. This barrier must be reduced to zero by the elastic strain in the growing film before dislocations may spontaneously appear. This condition leads to Eq. (11). Due to the implicit assumption of stochastic nucleation, the People–Bean model does not address the average state of strain in the epilayer once h exceeds h_c. Other shortcomings of this model include (i) the absence of an explicit dependence of h_c on growth temperature, and (ii) the absence of an underlying nucleation mechanism (e.g., homogeneous nucleation at surface steps and the like). Although an explicit theory of germinate (homogeneous) nucleation including growth temperature effects is as yet unavailable, much progress has been made in describing the effects of growth temperature on h_c in the case where the epilayer contains a pre-existing segment of threading dislocation (Matthews et al., 1970; Dodson and Tsao, 1987, 1988; Tsao et al., 1987).

Tsao et al. (1987) proposed that a natural measure of the stability or metastability of a strained heterostructure against a particular plastic flow mechanism is its *excess stress*, this being the difference between the stress arising from misfit strain (Eq. (7) and the stress arising from the tension in the dislocation line (Eq. (8)). The excess stress, considered as the driving force for strain relief, forms the basis of a kinetic model for strain relief. This can be seen as follows. Assume the presence of a pre-existing segment of threading dislocation in a strained epilayer. Under the influence of the misfit strain, this threading segment bows (glides) and eventually generates a segment of misfit dislocation line at the film–substrate interface as shown in Fig. 4. Dislocation motion via glide is inhibited in semiconductors having the diamond structure, due to the Peierls stress (which generates a frictional force on the dislocation). In order for dislocation motion to occur, the force on the dislocation due to misfit strain, F_H, must equal the sum of the tension in the dislocation line, F_1, and the frictional force, F_f, generated by the Peierls stress; i.e.,

$$F_H = F_1 + F_f. \tag{12}$$

Expressions for F_H and F_1 have been given in Eqs. (7) and (8), respectively. Following Haasen (1957), the frictional force is assumed to be generated by the diffusion of a microcrack coincident with the dislocation core resulting in an expression F_f of the form:

$$F_f = \left(\frac{h}{\cos\phi}\right)\left(\frac{k_B T}{b D_0}\right) \exp(U/k_B T) \cdot v. \tag{13}$$

here v is the glide velocity, k_B is Bolzman's constant, and $D_0 \exp(-U/k_B T)$ is the diffusion coefficient of the dislocation core. Further, the general relationship between elastic strain ($e_\parallel$), the misfit accommodated by dislocation lines (γ), and the misfit (f_0) between the bulk lattice parameter of the substrate and the strained layer is

$$f_0 = e_\parallel + \gamma. \tag{14}$$

If the misfit accommodated by dislocation lines is considered plastic strain produced by migrating dislocations, then the relation between glide velocity v and the rate of strain relief ($d\gamma/dt$) is in general given by the expression from Alexander and Haasen (1958)

$$v = \frac{(d\gamma/dt)}{\rho_m \cdot b \cdot \cos\lambda}, \tag{15}$$

where ρ_m is the number of mobile dislocations per unit area that migrate to generate misfit dislocation lines. Substitution of Eqs. (7), (8), and (13)–(15) into Eq. (12), gives a differential equation describing the conversion of elastic strain into plastic strain as a function of time.

If the mobile dislocation density, ρ_m, is assumed to be independent of time, then one obtains the results of Matthews et al. (1970); namely,

$$\left(\frac{d\gamma}{dt}\right) = C_1 \rho_m [\gamma_{eq} - \gamma(t)], \tag{16a}$$

where

$$C_1 = \frac{2Gb^3 \left(\frac{1+\nu}{1-\nu}\right) \cos^2\lambda \cdot \cos\phi \cdot D_0 \exp(U/k_B T)}{k_B T}, \tag{16b}$$

4. STRUCTURALLY INDUCED STATES

and

$$\gamma_{eq} \equiv f_0 - \left(\frac{b}{h}\right)\frac{[1 - \nu\cos^2\Theta]}{4\pi(1 + \nu)}[1 + \ln(h/b)]. \tag{16c}$$

Therefore, in the absence of dislocation multiplication, strain relaxation via plastic flow leads to

$$\gamma(t) = \gamma_{eq}[1 - \exp(-C_1\rho_m t)]. \tag{16d}$$

Equation (16) has been found to describe the relaxation of Ge films grown on (110) GaAs wafers (by GeI_2 decomposition) at $T \approx 350°C$ (Matthews et al., 1970). Here $f_0 \simeq 7 \times 10^{-4}$.

If, on the other hand, one includes dislocation multiplication (i.e., $\rho_m = \rho_m(t)$), then one needs an additional expression relating $\rho_m(t)$ and $\gamma(t)$. Following Dodson and Tsao (1987, 1988) it is assumed that multiplication is described by a phenomenological rate equation of the form:

$$\frac{d\rho_m}{dt} \equiv \rho_m v\delta, \tag{17}$$

where δ is the multiplication constant. Using Eq. (15) to eliminate the glide velocity v in Eq. (17), one obtains

$$\frac{d\rho_m}{dt} = \left(\frac{\delta}{b \cdot \cos\lambda}\right)\left(\frac{d\gamma}{dt}\right). \tag{18}$$

Assuming the existence of a background dislocation density ρ_0 (required for subsequent multiplication), the straightforward integration of (18) yields:

$$\rho_m(t) = \rho_0 + \int_0^{\gamma(t)} \frac{\delta \cdot d\tilde{\gamma}}{(b \cdot \cos\lambda)}. \tag{19}$$

The multiplication constant, δ, may either be constant or a function of the excess stress. The excess stress itself is proportional to $(\gamma_{eq} - \gamma(t))$; as may be obtained via inspection of Eq. (12). It has been observed (see, e.g., Berner and Alexander, 1967) that the dislocation density versus local strain curves in Ge are in good agreement with a multiplication parameter that depends linearly upon the effective (excess) stress. This assumption was therefore adopted by Dodson and Tsao (1987, 1988) and leads to

$$\rho_m(t) = 2\kappa G\left(\frac{1+\nu}{1-\nu}\right)(\gamma_{eq} - \gamma(t))(\gamma(t) + \gamma_0), \tag{20}$$

so that the dynamical equation for strain relaxation via plastic flow and including multiplication is given by:

$$\left(\frac{d\gamma}{dt}\right) \equiv C\mu^2(\gamma_{eq} - \gamma(t))^2[\gamma(t) + \gamma_0], \tag{21a}$$

where

$$C\mu^2 = \left(\frac{4\kappa G^2 b^3 \cos\lambda \cos\phi (1 + \nu/1 - \nu)^2 D_0 \exp(-U/k_B T)}{k_B T}\right). \quad (21b)$$

Note that although Eq. (21a) is nonlinear, a closed-form solution may be obtained due to the fact that (i) it is of first order, and (ii) the right-hand side of Eq. (21a) contains no explicit time dependence. By separating variables, one obtains the following closed-form solution (People, 1988):

$$C\mu^2 t = \frac{-1}{(\gamma_{eq} + \gamma_0)^2}\left\{\ln\left\|\frac{(\gamma_{eq} - \gamma)}{(\gamma + \gamma_0)}\frac{\gamma_0}{\gamma_{eq}}\right\| - \frac{(\gamma + \gamma_0)}{(\gamma_{eq} - \gamma)} + \left(\frac{\gamma_0}{\gamma_{eq}}\right)\right\}. \quad (21c)$$

It has been pointed out by Fritz et al. (1987) that finite experimental resolution can play a significant role in determining apparent critical layer thicknesses. Techniques such as Rutherford backscattering (RBS) and double-crystal x-ray diffraction have estimated strain resolutions $\gamma_R \sim 10^{-3}$, whereas low-temperature Hall-effect measurements in modulation-doped strained-layer heterostructures and photoluminescence imaging of dark-line defects have estimated strain resolutions of $\gamma_R \sim 10^{-5}$. The kinetic model given in Eq. (21) readily lends itself to the investigation of the effects of finite resolution and growth temperature on critical thickness determinations for systems in which relaxation occurs via the plastic flow mechanism. Dodson and Tsao (1988a, b) used this formulism to model the strain-relaxation measurements of Bean et al. (1984) in the 550°C growth of Ge_xSi_{1-x} on (001) Si. Strain relaxation was measured using RBS and x-ray diffraction. In their analysis, the time t on the left-hand side of Eq. (21c) is taken as $t(h) = (10^3 + h(Å)/5)$ sec; for a growth rate of 5 Å/sec and a cool-down time of 10^3 sec. The constants $C\mu^2$ and γ_0 are determined from a fit to the measured strain relief in a 500 Å film of $Ge_{0.5}Si_{0.5}$ on (001) Si. In Fig. 5, we show the critical thicknesses generated by the above analytic solution for various experimental sensitivities to strain relief, γ_R, along with the critical thickness data of Bean et al. (1984) and Bevk et al. (1986). It is noted that the present kinetic model shows good agreement with the RBS and x-ray data assuming an experimental resolution $\gamma_R = 10^{-3}$. However, this kinetic model predicts maximum layer thicknesses quite close to the Matthews–Blakeslee mechanical equilibrium results in the limit $\gamma_R \to 0$, as had been suggested previously by Fritz et al., 1987). Recently Kohama et al. (1988) measured critical layer thicknesses in the Ge_xSi_{1-x} system using electron beam–induced current (EBIC) and transmission electron microscopy (TEM). The wide field ($\sim 100 \mu m$) accessible by the rastered EBIC beam allows for direct observations of misfit dislocations with strain resolution $\gamma_R \approx b/p \sim 4\,Å/100\,\mu m < 10^{-5}$. Their results are given by the cross-hatched region in

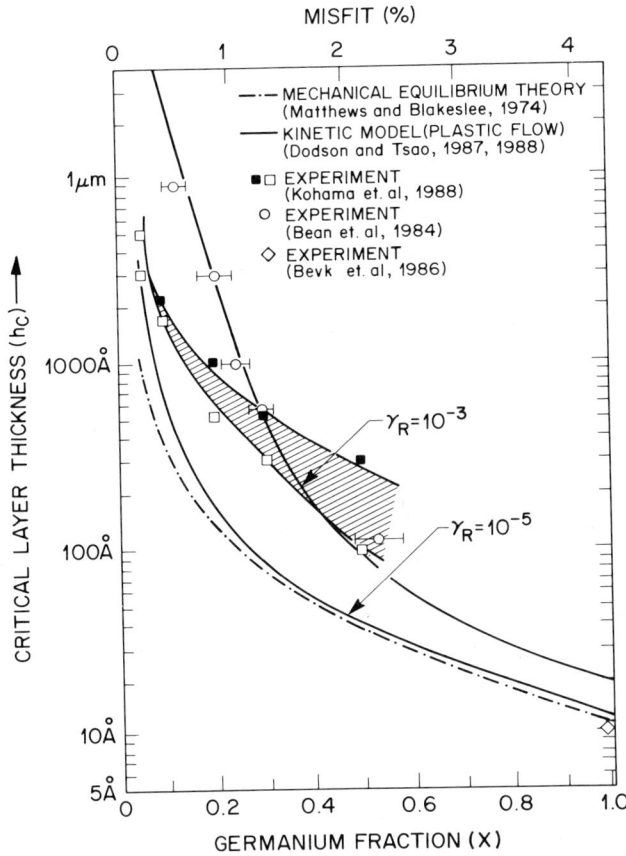

FIG. 5. Critical thickness measurements using RBS and x-ray are given in the data of Bean et al. (1984) and Bevk et al. (1986). High-resolution EBIC measurements (for Ge_xSi_{1-x} films on vicinal Si substrates) by Kohama et al. (1988) lie within the cross-hatched region. Predictions of the kinetic model with plastic flow (Dodson and Tsao, 1988a, b) are shown by solid curves for experimental resolutions $\gamma_R = 10^{-3}$ and 10^{-5}.

Fig. 5, with open squares corresponding to commensurate structures and closed squares corresponding to incommensurate structures. It is noted that the high-resolution EBIC technique gives h_c values lower than those found by RBS and x-ray for Ge contents $x \lesssim 0.3$. However, these high-resolution h_c values remain substantially higher than both the kinetic model having $\gamma_R = 10^{-5}$ and the equilibrium model, demonstrating that the Ge_xSi_{1-x}-strained layers are noticeably metastable. However, this fact is not surprising, because the metastability of Ge_xSi_{1-x} films had been demonstrated previously by Fiory et al. (1985). It should also be noted that the growths of Kohama et al. (1988) were made on tilted Si substrates (oriented $\approx 2°$ off [100] towards [011]). As such vicinal growths give rise to substantial surface steps, it is not yet known if the measured h_c values obtained by using EBIC should not therefore be viewed as lower bonds for h_c in this system.

Other mechanisms (aside from finite resolution) may lead to apparent critical layer thicknesses, which may not be indicative of strain relief via dislocations. In particular, the recent studies of Lievin and Fonstad (1987) on MBE-grown (In, Ga)As and (In, Al)As heterostructures on InP have shown that a strain-induced roughening transition can give rise to an apparent critical layer thicknesses versus lattice-mismatch relation that resembles the equilibrium result given in Eq. (9). A surface-roughening transition would have a profound effect upon surface-sensitive probes of interfacial quality, such as photoluminescence (PL) efficiency and two-dimensional transport.

2. Multilayer Critical Thickness

If coherently strained multilayers are grown on a substrate having lattice parameter a_s, then misfit-free growth is expected if $a_s = a_\parallel$, where $a_\parallel$ is the equilibrium in-plane lattice parameter given in Eq. (1). It is implicitly assumed that the layer thicknesses h_A, h_B are less than the single-layer critical thickness for the given strain they experience. Such equilibrium structures may be repeated *ad infinitum*. If the substrate contains threading dislocations, however, then such threading dislocations may traverse the multilayers and undergo the elongation process shown in Fig. 4, thereby generating interfacial misfit dislocation lines that destroy interfacial coherence (Matthews and Blakeslee, 1974). The equilibrium critical layer thickness for such threading dislocation-limited coherent growth is obtained by the same procedure used to obtain Eq. (9), with the exception that in this multilayer case, it is now required that

$$F_H = 2F_1. \tag{22}$$

The factor of 2 arises due to the fact that in a multilayer structure, each interface may support a segment of line dislocation. It should also be noted that $e_\parallel$ in Eq. (7) now corresponds to the larger one of strains $e_\parallel^A$ and $e_\parallel^B$, whereas in a single layer, $e_\parallel = f_0$. It is perhaps worth emphasizing that the critical thickness h_c so obtained for the multilayer film corresponds to the maximum thickness that an individual layer (not the multilayer as a whole) may be grown.

In those instances where multilayer structures are grown on substrates having lattice constant $a_s \neq a_\parallel$, the structure is no longer in equilibrium, and the strains in the individual layers are no longer related via Eq. (2). The individual layer strains may now be computed as $e_\parallel^i = (a_s - a_0^{(i)})/a_0^{(i)}$, where

$a_0(i)$ is the unstrained bulk lattice parameter for material "i". If loss of coherence occurs via the threading dislocation transformation previously described, then one simply uses these $e_\parallel^i$ values to compute F_H in Eq. (22).

A second scenario for determining multilayer critical thickness has been given by Hull et al. (1986) and People (1986b) and applies to growth of multilayers on high-quality (very low threading dislocation density) substrates such as Si. In this model, one assumes the growth of coherent multilayers on a substrate for which $a_s \neq a_\parallel$. Individual layer thicknesses are assumed to be less than the single-layer critical thickness for the given strain $e_\parallel^{(i)}$ in the layers. Note that to the extent a_s differs from $a_\parallel$, the multilayers, as a whole, will experience an additional in-plane $\bar{e}_\parallel$, which can be expressed in terms of the misfit between the multilayers as a whole and the substrate, namely

$$\bar{e}_\parallel = \left(\frac{a_\parallel - a_s}{a_s}\right). \quad (23)$$

It can be readily shown that a correspondence exists between the total thickness (h_T) and relative strain $(\bar{e}_\parallel)$ for the multilayer as a whole and the thickness h and strain $e_\parallel$ for a single coherently strained layer. For example, if one constituent of the multilayer film (say A) is of the same composition as the substrate, then Eq. (23) reduces to

$$\bar{e}_\parallel = f_0 \left(\frac{G_B h_B}{G_A h_A + G_B h_B}\right). \quad (24)$$

Noting that $e_\parallel(f_0)$ and h describe the state of strain in a single layer, whereas $\bar{e}_\parallel$ and h_T describe the state of strain experienced by the multilayers as a whole relative to the substrate, it follows that the critical thickness for the multilayer film as a whole is obtainable from a knowledge of the single-layer critical thickness for a film of thickness h_T and strain $\bar{e}_\parallel$. Note that $\bar{e}_\parallel$ is simply the spatial average of the strain in a single period of the multilayer, weighed by the respective elastic constants. The result given in Eq. (29) was first obtained by Hull et al. (1986) via calculation of the maximum strain energy that may be relaxed by dislocation formation. Note further that by way of analogy with the single strained layer, exceeding the multilayer critical thickness $(h_T)_c$ implies the generation of misfit dislocations at the interface between the substrate and the multilayers as a whole (with the individual interfaces within the multilayer structure remaining coherent). Loss of coherence in the absence of threading dislocations is therefore expected to proceed in a quite different manner than is suggested by Eq. (22). Experimental studies by Hull et al. (1986) on $Ge_x Si_{1-x}/Si$ strained-layer superlattices (SLS) on Si (001) verified that the relaxation of (Ge, Si) SLS on Si(001) is well described by the present correspondence principle. Further, a number of striking electron micrographs clearly showed that loss of coherence between the (Ge, Si) strained-layer superlattice and the substrate occurred only at the superlattice–substrate interface, with the individual multilayer interfaces remaining coherent.

II. Optical and Electronic Properties of Lattices Mismatch-Induced Strained Layers

The optical and transport properties of bulk semiconductors under external stress had been well established (see, e.g., Bir and Pikus, 1974) during the time of the early structural studies of Matthews and Blakeslee (1974) on $GaAs/GaAs_xP_{1-x}$ strained-layer heterostructures. Interest in the optical and electronic properties of these lattice-mismatch-induced strained layers went largely unnoticed, however, until the work of Osbourn (1982) and Gourley and Biefield (1982). These early studies demonstrated the potential for using coherently layer strains to tailor the bandgap and transport properties of such heterostructures. Progress in the fabrication of pseudomorphic semiconductor structures has led to numerous applications including strained-layer photodetectors, enhanced-mobility field effect transistors, heterojunction bipolar transistors, light-emitting diodes, and photopumped as well as injection lasers that utilize strained layers. The manner in which lattice-mismatch-induced coherency strains effect the bandgap of bulklike strained layers will be reviewed in the present section; within the context of phenomenological deformation potential theory (see, e.g., Kleiner and Roth, 1959). In this formulism, one only requires (i) the bandgap of the unstrained bulk material, (ii) a deformation potential that describes the uniaxial splitting of the valence-band edge, (iii) a deformation potential that describes the uniaxial splitting of multivalley conduction-band edges, and (iv) the bandgap hydrostatic deformation potential, which is required for finite changes in the volume of the unit cell, $\delta V/V \neq 0$.

A. Effects of Coherency Strain on Fundamental Band Edges

In general, the coherency strain, induced by the lattice-mismatched epitaxial growth, can be decomposed into uniaxial and hydrostatic contributions. The uniaxial components give rise to splittings of degenerate levels (such as the Γ_8^v valence-band edge), whereas the hydrostatic components give rise to uniform shifts of the center of gravity, as illustrated in Fig. 6.

1. *Coherency Strain Tensor for $\{001\}$, $\{110\}$, and $\{111\}$ Growth Planes*

Given commensurate heteroepitaxy of lattice-matched epilayers, the entire lattice mismatch is accommodated as a homogeneous tetragonal distortion in

4. STRUCTURALLY INDUCED STATES 135

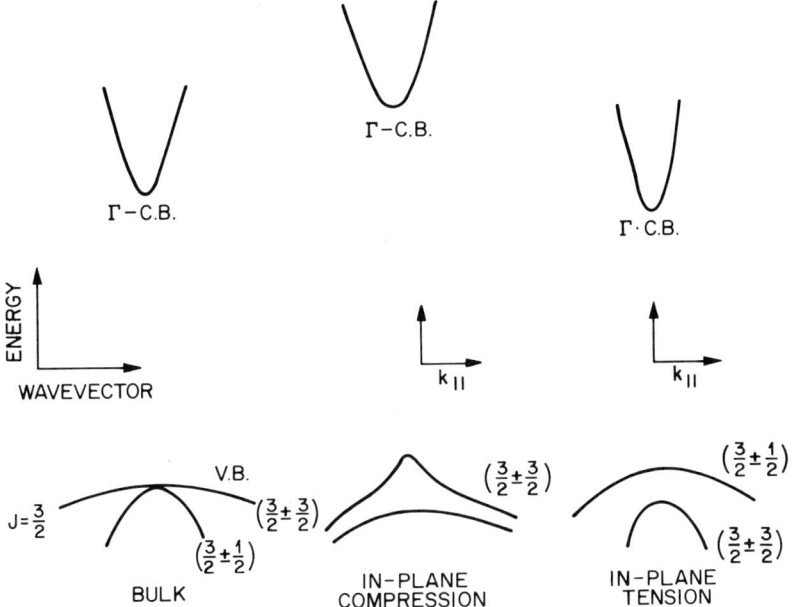

FIG. 6. Schematic of the effects of coherency strain on the $J = 3/2$ valence-band edge and the Γ conduction-band edge of a bulk, direct-gap semiconductor. The wave vector $\mathbf{k}_\parallel$ lies in a direction normal to the growth direction, which is presumed to be (001).

the strained layers only. This fact allows for the calculation of the strain components for the epilayer solely in terms of the bulk lattice parameters and elastic stiffness constants. It is appropriate here to point out the difference between the strain tensor, S_{ij}, and the *conventional* strain components, e_{ij}, namely,

$$S_{ij} = \tfrac{1}{2} e_{ij}(1 + \delta_{ij}), \qquad (25)$$

i.e., the off-diagonal components of S_{ij} differ by 1/2 from the corresponding e_{ij}. This differentiation is important, since the valence-band deformation-potential Hamiltonian is generally defined in terms of the *conventional* strain components (see, e.g., Hensel and Feher, 1963; Hasegawa, 1963). On the other hand, expressions for the condition-band uniaxial splittings are generally defined in terms of the strain tensor (Herring and Vogt, 1956).

The in-plane strain, $e_\parallel$, is defined as

$$e_\parallel = -\left(\frac{b - a_s}{b}\right), \qquad (26)$$

where b and a_s denote the lattice parameters of the unstrained (cubic) epilayer and the substrate, respectively. It has been assumed that the thickness of the strained epilayer is much smaller than that of the substrate, so that we may neglect strains in the substrate. The *conventional* strain components for epitaxy on substrates have surface normals defined by the vector **l** are as follows.

$\mathbf{l} \parallel \{001\}$

$$e_{xx} = e_{yy} = e_\parallel; \quad e_{zz} = -e_\parallel \left(\frac{2C_{12}}{C_{11}}\right), \tag{27a}$$

$$e_{xy} = e_{yz} = e_{zy} = 0. \tag{27b}$$

$\mathbf{l} \parallel \{110\}$

$$e_{xx} = e_{yy} = e_\parallel \left[\frac{2C_{44} - C_{12}}{2C_{44} + C_{11} + C_{12}}\right]; \quad e_{zz} = e_\parallel, \tag{28a}$$

$$e_{xy} = -e_\parallel \left[\frac{2[C_{11} + 2C_{12}]}{2C_{44} + C_{11} + C_{12}}\right], \tag{28b}$$

$$e_{yz} = e_{zx} = 0. \tag{28c}$$

$\mathbf{l} \parallel \{111\}$

$$e_{xx} = e_{yy} = e_{zz} = e_\parallel \left[\frac{4C_{44}}{4C_{44} + 2C_{12} + C_{11}}\right], \tag{29a}$$

$$e_{xy} = e_{yz} = e_{zx} = -e_\parallel \left[\frac{2[C_{11} + 2C_{12}]}{4C_{44} + 2C_{12} + C_{11}}\right]. \tag{29b}$$

2. Valence-Band Uniaxial Splittings

The valence-band edge of Ge, Si, and most III–V semiconductors (e.g., GaAs, InAs, etc.) is threefold degenerate in the absence of spin and transform as an $L = 1$ (p-state) of the cubic group. The inclusion of spin produces six states that are split by the spin–orbit interaction into a fourfold ($J = 3/2$) state having Γ_8^v symmetry, and a twofold ($J = 1/2$) set of states having Γ_7^v symmetry. The splitting of these states (at $\mathbf{k} = 0$), produced by lattice-mismatch-induced coherency strain, is derivable from the strain Hamiltonian of Kleiner and Roth (1959), namely,

$$H_{VB}^{(u)} = \tfrac{2}{3}D_u[(J_x^2 - \tfrac{1}{3}J^2)e_{xx} + (J_y^2 - \tfrac{1}{3}J^2)e_{yy} + (J_z^2 - \tfrac{1}{3}J^2)e_{zz}]$$

$$+ \tfrac{2}{3}D_u'[\{J_xJ_y\}e_{xy} + \{J_xJ_z\}e_{xz} + \{J_yJ_z\}e_{yz}]. \tag{30}$$

Here e_{ij} are the *conventional* strain components, $\{\cdot\}$ implies the anticommutator, and D_u and D'_u are valence-band deformation potentials for a (001) and (111) uniaxial stress, respectively. In general, couplings between the states of the $J = 3/2$ and $J = 1/2$ manifolds are induced by the coherency strain. For substrates oriented along the symmetry directions (001), (011), or (110), Hasegawa (1963) has shown that the (6×6) strain Hamiltonian in Eq. (30) has the following form:

$$H_{VB}^{(u)}(l_1, l_2, l_3) = \begin{array}{c} M_j = \frac{3}{2} \quad \frac{1}{2} \quad -\frac{1}{2} \quad -\frac{3}{2} \quad \frac{1}{2} \quad -\frac{1}{2} \\ \begin{bmatrix} \varepsilon & 0 & B & 0 & 0 & -\sqrt{2}B \\ 0 & -\varepsilon & 0 & B & \sqrt{2}\varepsilon & 0 \\ B & 0 & -\varepsilon & 0 & 0 & -\sqrt{2}\varepsilon \\ 0 & B & 0 & \varepsilon & \sqrt{2}B & 0 \\ 0 & \sqrt{2}\varepsilon & 0 & \sqrt{2}B & -\Delta_0 & 0 \\ -\sqrt{2}B & 0 & -\sqrt{2}B & 0 & 0 & -\Delta_0 \end{bmatrix} \end{array}, \quad (31)$$

where (l_1, l_2, l_3) are the direction cosines of the surface normal, and Δ_0 is the spin–orbit splitting. The matrix elements ε and B for given (l_1, l_2, l_3) is as follows:

$$\mathbf{l} \parallel \{001\}: \varepsilon = \tfrac{2}{3}D_u[e_{zz}^{(001)} - e_{xx}^{(001)}]; \quad B = 0, \quad (32a)$$

$$\mathbf{l} \parallel \{111\}: \varepsilon = D'_u e_{xy}^{(111)}; \quad B = 0, \quad (32b)$$

$$\mathbf{l} \parallel \{110\}: \varepsilon = \tfrac{1}{2}(\xi_0 + \xi'_0); \quad B = \frac{\sqrt{3}}{2}(\xi_0 - \tfrac{1}{3}\xi'_0), \quad (32c)$$

where $\xi_0 = \tfrac{2}{3}D_u[e_{xx}^{(110)} - e_{zz}^{(110)}]$ and $\xi'_0 = D'_u e_{xy}^{(110)}$.

For the cases $\mathbf{l} \parallel \{001\}$ or $\{111\}$, the (6×6) eigenvalue determinant breaks up into three (2×2) matrices. The resulting three eigenvalues at $\mathbf{k} = 0$ (each doubly degenerate) are

$$E_v(\tfrac{3}{2}; \pm\tfrac{3}{2}) = \varepsilon, \quad (33a)$$

$$E_v(\tfrac{3}{2}; \pm\tfrac{1}{2}) = -\tfrac{1}{2}(\varepsilon + \Delta_0) + \tfrac{1}{2}[9\varepsilon^2 + \Delta_0^2 - 2\varepsilon\Delta_0]^{1/2}, \quad (33b)$$

$$E_v(\tfrac{1}{2}; \pm\tfrac{1}{2}) = -\tfrac{1}{2}(\varepsilon + \Delta_0) - \tfrac{1}{2}[9\varepsilon^2 + \Delta_0^2 - 2\varepsilon\Delta_0]^{1/2}. \quad (33c)$$

For the case $\mathbf{l} \parallel \{110\}$, Eq. (31) leads to two equivalent (3×3) determinants, whose solutions are readily obtained numerically. The uniaxial splitting of the $\Gamma_8^v(J = 3/2)$ valence-band edge for pseudomorphic growth of Ge_xSi_{1-x} bulklike alloy layers on Si (001) is shown in Fig. 7 (after People, 1985). Note that the splitting of the Γ_8^v valence-band edge deviates considerably from a

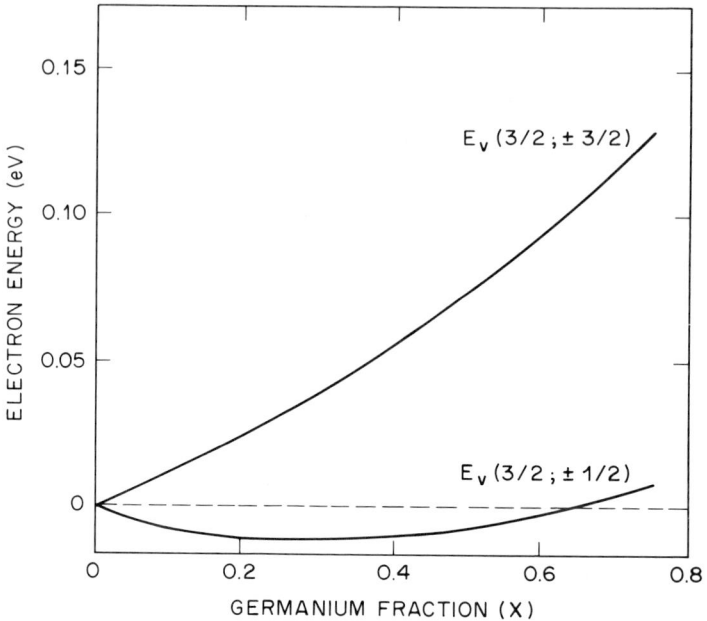

FIG. 7. Uniaxial splittings of the Γ_8^v ($J = 3/2$) valence-band edge of Ge_xSi_{1-x} bulklike alloys on Si (001) substrate. The states are labeled by (J, M_J) descriptors [Reprinted with permission from the American Physical Society, People, R. (1985). *Phys. Rev.* **B32**, 1405.]

linear behavior due to the strong strain-induced interaction between the (3/2; ±1/2) and (1/2; ±1/2) bands.

3. *Multivalley Conduction Bands–Uniaxial Splittings*

In multivalley semiconductors, such as Si and Ge, Herring and Vogt (1956) have shown that the total energy shift $\delta E_c^{(i)}$ of the *i*th conduction minimum, for an arbitrary deformation, can be described by

$$\delta E_c^{(i)} = [\Xi_d \mathbf{1} + \Xi_u \{\hat{a}_i \hat{a}_i\}] : \mathbf{S}, \tag{34}$$

in terms of the conduction-band deformation potentials Ξ_d and Ξ_u (d denotes dilation and u implies uniaxial) and the strain tensor, S_{ij}. Here **1** is the unit tensor, $\hat{a}_i$ is a unit vector parallel to the **k** vector of valley *i*, and $\hat{a}_i \hat{a}_i$ denotes a dyadic product. The shift of the mean (center of gravity) energy of the conduction-band extrema is given by

$$\delta \bar{E}_c = (\Xi_d + \tfrac{1}{3}\Xi_u)\mathbf{1} : \mathbf{S}. \tag{35}$$

The uniaxial splitting of the *i*th valley is given by the difference in Eqs. (34)

and (35), namely,

$$\delta E_{c,u}^{(i)}(\mathbf{l}) = \Xi_u [\{\hat{a}_i \hat{a}_i\} - \tfrac{1}{3}\mathbf{1}] : \mathbf{S}(\mathbf{l}), \tag{36a}$$

for growth on substrate having surface normal along $\mathbf{l}$.

As an example, consider degenerate Δ-conduction minima, for growth of strained epilayers on (001) substrates. In this case, Eq. (36a) gives

$$\delta E_{c,u}^{(\Delta)}(001) = \begin{cases} \tfrac{2}{3}\Xi_u^{(\Delta)} \cdot [e_{zz}^{(001)} - e_{xx}^{(001)}]; & \mathbf{k}_i \parallel \{[001], [00\bar{1}]\} \\ -\tfrac{1}{3}\Xi_u^{(\Delta)} \cdot [e_{zz}^{(001)} - e_{xx}^{(001)}]; & \mathbf{k}_i \parallel \begin{cases} [010], [0\bar{1}0] \\ [100], [\bar{1}00] \end{cases} \end{cases} \tag{36b}$$

Note that $\delta E_{c,u}^{(L)}(001) = 0$, since all L-conduction band extrema are equivalent under an effective (001) stress.

A second case of interest are L-conduction-band minima for growth on (111) substrates. In this case, one obtains:

$$\delta E_{c,u}^{(L)}(111) = \begin{cases} \Xi_u^{(L)} \cdot e_{xy}^{(111)}; & \mathbf{k}_i \parallel [111] \\ -\tfrac{1}{3}\Xi_u^{(L)} \cdot e_{xy}^{(111)}; & \mathbf{k}_i \parallel \{[1\bar{1}1], [11\bar{1}], [1\bar{1}\bar{1}]\} \end{cases} \tag{36c}$$

Note that $\delta E_{c,u}^{(\Delta)}(111) = 0$, since all Δ-conduction-band minima are equivalent under an effective (111) stress.

4. Hydrostatic Contributions to Strained-Layer Bandgaps

Hydrostatic contributions to the strained-layer bandgap will be present if the net change in volume of the unit cell ($\delta V/V$) is nonzero. These hydrostatic terms describe the relative motion of the centers of gravity of a given conduction-band edge with respect to the valence-band edge. The shift of the center of gravity of the conduction band has been previously given in Eq. (35). If we let a_v denote a deformation potential that describes uniform shifts of the valence band, then the bandgap hydrostatic deformation potential is given by $(\Xi_d + \tfrac{1}{3}\Xi_u - a_v)$. The change in bandgap for a conduction-band edge of symmetry γ is then given by

$$\delta E_{\text{HYDRO}}^{(\gamma)}(l_1, l_2, l_3) = \{\Xi_d + \tfrac{1}{3}\Xi_u - a_v\}^{(\gamma)}\mathbf{1} : \mathbf{S}(l_1, l_2, l_3), \tag{37}$$

where $\gamma = \Gamma, \Delta, L$, etc. A representative listing of bandgap hydrostatic deformation potentials for various group-IV and III–V semiconductor compounds has been given by Paul and Warschauer (1963). It should be noted that the quantities $(\Xi_d + \tfrac{1}{3}\Xi_u)$ and a_v are difficult both to calculate and to measure, since they refer to changes in the bands on an *absolute scale* (Verges *et al.*, 1982; Martin and Van de Walle, 1985).

5. Strained Bandgaps for (001), (111), and (110) Growth Surfaces

Given the unstrained bandgap energy, $E_0^{(\gamma)}$, for a particular conduction-band edge (i.e., $\gamma = \Gamma, \Delta, L$, etc), along with the conduction and valence-band uniaxial splittings and hydrostatic contributions, the strained bandgap is given by:

$$\{E_c^{(\gamma)}(\mathbf{k}_i; \mathbf{l}) - E_v^{(j)}(\mathbf{l})\} = E_0^{(\gamma)} + \delta E_{\text{HYDRO}}^{(\gamma)}(\mathbf{l}) + \delta E_{c,u}^{(\gamma)}(\mathbf{k}_i; \mathbf{l}) - E_{v,u}^{(j)}(\mathbf{l}). \quad (38)$$

Here j designates a strain-split valence-band edge, u implies uniaxial, γ denotes the symmetry of the relevant conduction-band edge, and $\mathbf{l}$ is a vector normal to the growth surface. The phenomenological calculation of strained band gaps, as given by Eq. (38) has been applied by People et al. (1987) to Ge_xSi_{1-x} strained-layer heterostructures on (001), (111), and (110) substrates. The results for the lowest Δ and L-bulk band edges are shown in Figs. 8–10.

In Fig. 8, we show the photocurrent measurements of the bandgap of coherently strained Ge_xSi_{1-x} alloy on (001) Si by Lang et al. (1985). The

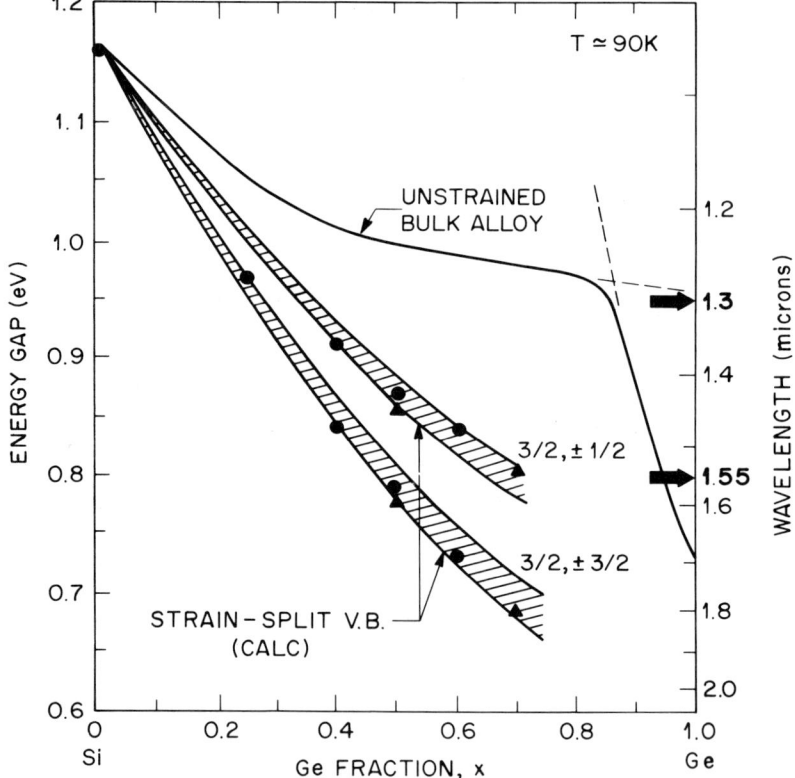

FIG. 8. Measured energy-gap values for coherently strained Ge_xSi_{1-x} alloys on (001) Si. $T = 90$ K, and the data have been corrected for quantum well shifts (circles = 75 Å wells; triangles = 33 Å wells). The unstrained bulk alloy data are from Braunstein et al. (1958). Calculated bandgaps were obtained by using Eq. (38) from text. [Reprinted with permission from the American Institute of Physics, Lang, D.V., People, R., Bean, J.C., and Sergent, A.M. (1985). Appl. Phys. Lett. 47 (12), 1333.]

4. STRUCTURALLY INDUCED STATES

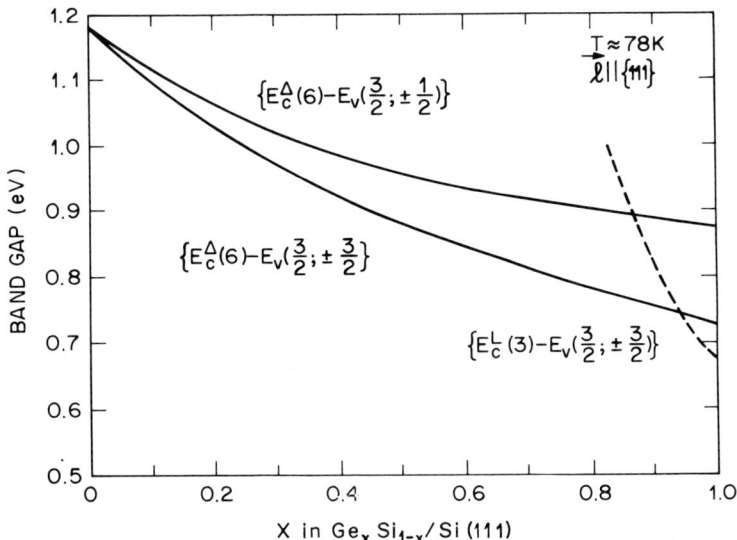

FIG. 9. Lower strained bandgaps derived from the Δ- and L-band edges for growth of pseudomorphic Ge_xSi_{1-x} alloys on Si (111). [Reprinted with permission from World Scientific Publishing Co. Pte. Ltd., People R., Bean, J.C., and Lang, D.V. (1987). *Proc. 18th Int. Conf. Physics Semiconductors*, (O. Engstrom, ed.)

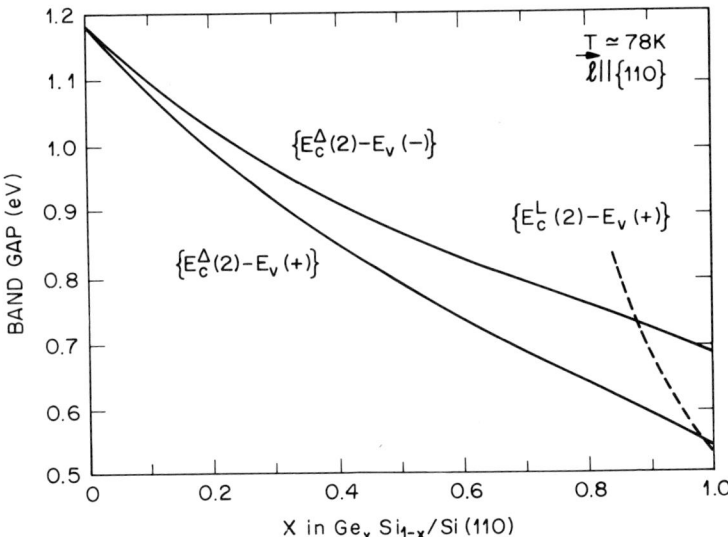

FIG. 10. Lower strained bandgaps derived from the Δ- and L-band edges for growth of pseudomorphic Ge_xSi_{1-x} alloys on Si (110). [Reprinted with permission from World Scientific Publishing Co. Pte. Ltd., People R., Bean, J.C., and Lang, D.V. (1987). *Proc. 18th Int. Conf. Physics Semiconductors*, (O. Engstrom, ed.)]

predictions of Eq. (38) for the two lowest Δ-band edges, derived from the strain split $J = 3/2$ valence band, are also shown (hashed curved) in Fig. 8. It is apparent that phenomenological deformation-potential theory provides an excellent description of the strained-alloy bandgaps in this case. Moving the absorption of Si-based structures into the 1.3 μm range of light-wave communications systems had obvious implications for photodetectors in this wavelength range. The ability to lower the bandgap by using the inherent misfit strain made it possible to achieve 1.3 μm response detectors with better crystal perfection (pseudomorphic throughout) and much lower Ge content than with pure Ge detectors grown on Ge_xSi_{1-x}/Si-relaxed superlattice buffer layers. Better crystalline quality directly translates into lower leakage currents, whereas lower Ge content allows for more robust thermal processing. The characteristics of both $p-i-n$ and avalanche Ge_xSi_{1-x}/Si photodetectors have been investigated by Temkin et al. (1986a, 1986b), Pearsall et al. (1986), and Luryi et al. (1986). Avalanche (Ge, Si) devices having separate absorption (in the Ge_xSi_{1-x}/Si superlattice) and multiplication (in a high-field region of pure Si) have demonstrated the ability of these devices to combine the extremely low excess noise factor of Si with the optical response of Ge.

A second technologically important system which phenomenological deformation potential theory has been shown to describe well is that of InP/In_xGa_{1-x}As. Alloys having an indium content of $x = 0.53$ are lattice-matched to InP substrates. An additional, possibly beneficial, degree of freedom is attained by fabricating coherently strained In_xGa_{1-x}As/InP heterostructures; since device performance is strongly coupled to contingent bandgaps and relative band alignments, parameters that are sensitive to coherent layer strains.

Osbourn (1983) gave estimates of the bandgaps for coherently strained $In_xGa_{1-x}As/In_yGa_{1-y}As$ strained-layer superlattices. The explicit compositional variation of the strained-bulk bandgap of In_xGa_{1-x}As, as applies for pseudomorphic growth on (001) InP, has been given by People (1987) following Eq. (38). These strained In_xGa_{1-x}As bandgap calculations are in good agreement with the optical bandgap measurements of Gershoni et al. (1987). In Fig. 11, we show the measured excitonic (light and heavy hole–to–conduction band) transitions from these authors. Solid and dashed lines are based upon strained bandgaps as calculated by using Eq. (38) and including quantum-confinement effects. Kuo et al. (1985) have investigated strained bulklike layers ($d \sim 1.0$–1.5 μm) of In_xGa_{1-x}As on (001) InP and Ga_xIn_{1-x}P on (001) GaAs for lattice mismatches $f_0 \lesssim 3.5 \times 10^{-3}$. Good agreement is found with deformation-potential estimates of strained bandgaps, ignoring strained-induced mixing of the (3/2; $\pm 1/2$) and (1/2; $\pm 1/2$) valence bands. Mixing effects generally become appreciable for $f_0 \gtrsim 10^{-2}$.

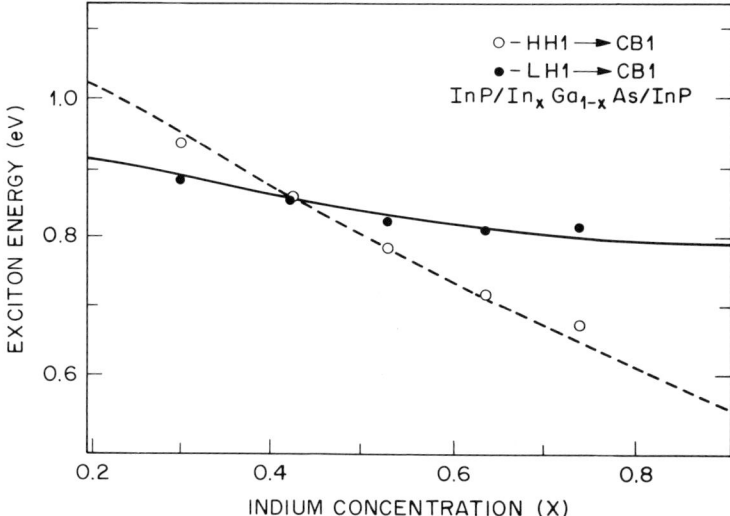

FIG. 11. Light hole and heavy hole–to–conduction band (excitonic) transitions for various coherently strained $In_xGa_{1-x}As$ layers in a quantum-well configuration (having unstrained InP barriers) grown on (001) InP. Open (filled) circles correspond to heavy- (light-) hole transitions, respectively. Solid and dashed curves are calculated by using Eq. (38) of text and including quantum-confinement corrections. (After Gershoni et al., 1987.) [Reprinted with permission from the American Physical Society, Gershoni, D., Vandenberg, J.M., Hamm, R. A., Temkin, H., and Panish, M.G. (1987). Phys. Rev. **B36**, 1320.]

B. HETEROJUNCTION-BAND ALIGNMENT MODIFICATIONS INDUCED BY COHERENT-LAYER STRAINS

The ability to controllably change the band alignments (i.e., band offsets) in heterostructures is highly desirable, as these parameters affect diverse device properties ranging from injection efficiencies of heterojunction bipolar transistors (HJBTs) to real-space transfer effects in selectively doped field-effect transistors (SDFETs). It is now well known that the combined effects of the hydrostatic and uniaxial misfit strain components in pseudomorphic heterostructures can dramatically affect the valence-band offsets (ΔE_v), and hence the band alignments. To date, numerous pseudomorphic heterointerfaces have been investigated. Early studies on the Ge_xSi_{1-x}/Si system exemplified the particularly sensitive nature of ΔE_v to the state of strain in the individual layers of the heterostructure.

Around 1984, several experimental groups had succeeded in growing low-dislocation density, pseudomorphic Si/Si_xGe_{1-x} heterointerfaces and were performing modulation-doping experiments (see, e.g., Störmer et al., 1979), which, while not yielding an explicit value of ΔE_v, still provided qualitative information about the band lineups. Early observations on two-dimensional electrons and holes by People et al. (1984, 1985) in Si/Ge$_{0.2}$Si$_{0.8}$ strained layers grown on Si (001) substrates indicated that $\Delta E_v \gg \Delta E_c$ and further that the band alignment was type I in character (i.e., the narrower Ge$_{0.2}$Si$_{0.8}$ gap falls within the wider Si gap, as shown in Fig. 12). A seeming paradox arose, however, when the experiments of Jorke and Herzog (1985) clearly indicated the band alignment to be type II in character (with the narrower-gap Ge$_x$Si$_{1-x}$ conduction-band edge lying higher in energy than the Si conduction-band edge). A qualitative resolution of this seeming paradox was proposed by Abstreiter et al. (1985), who pointed out that the fundamental difference between the experiments of People et al. and those of Jorke and Herzog was that the Si layers were cubic in the former case but strained in the latter, and hence the type of band alignment obtained may be sensitive to the state of strain in the Si layer. These latter heterostructures had been grown on relaxed Ge$_{0.25}$Si$_{0.75}$ buffer layers, rather than on pure Si.

Even with this hypothesis in hand, it was not yet possible to deduce the (Ge, Si)/Si band alignment due to wide variations in the experimental values for ΔE_v in this system. Kuech et al. (1981) used a chemical vapor-deposit

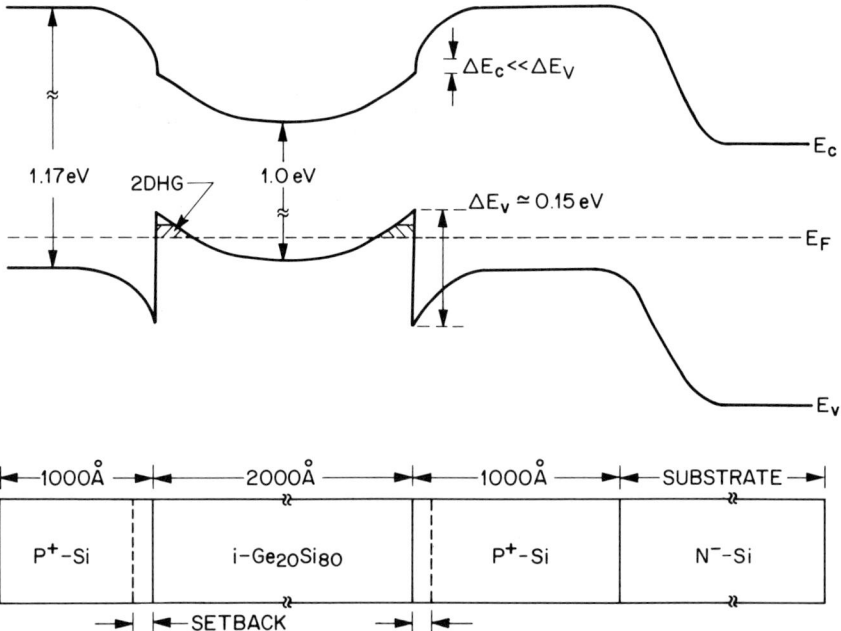

FIG. 12. Schematic of deduced band alignment from modulation-doped $p + \text{Si}/i - \text{Ge}_{0.2}\text{Si}_{0.8}p + \text{Si}$ heterostructure on (001) Si. [Reprinted with permission from the American Institute of Physics, People, R. Bean, J.C., Lang, D.V., Sergent, A.M., Störmer, H.L., Wecht, K.W., Lynch, R.T., and Baldwin, K. (1984). Appl. Phys. Lett. **45**, 1231.]

technique to deposit Ge on (001) Si, and estimated the band discontinuities from reverse-bias capacitance measurements. They obtained $\Delta E_v = 0.39\,\text{eV}$ and $\Delta E_c = 0.05\,\text{eV}$, with a stated experimental uncertainty of $0.04\,\text{eV}$. Margaritondo et al. (1982), using photoemission spectroscopy to observe core-level shifts and valence-band edge energies for thin overlayers of Ge on (111) Si, found $\Delta E_v = 0.2\,\text{eV}$ (with differing sample-preparation techniques producing a change in ΔE_v of less than $\sim 0.2\,\text{eV}$). Further, Mahowald et al. (1985), also using photoemission spectroscopy, determined $\Delta E_v = 0.4 \pm 0.1\,\text{eV}$ for thin overlayers of Si on (111) Ge. In retrospect, the above disperity in ΔE_v most likely emanated from the lack of pseudomorphic growth; it is now well known (Bevk et al., 1986) that the maximum layer thickness for pseudomorphic growth of Ge on Si (and vice versa) is $\approx 10\,\text{Å}$.

A first-principles calculation of the valence-band offset ΔE_v [$\equiv E_v(\text{Ge}) - E_v(\text{Si})$] for pseudomorphic Si/Ge heterointerfaces was performed shortly thereafter by Van de Walle and Martin (1985). These authors performed a full self-consistent, local-density calculation using ab initio pseudopotentials. They found that $\Delta E_v(\text{Ge/Si})$ was dependent upon the substrate lattice parameter (or more appropriately the in-plane lattice parameter $a_\parallel$). For (001)-oriented interfaces, ΔE_v was calculated for (i) $a_\parallel = 5.431\,\text{Å}$ (Si substrates), (ii) $a_\parallel = 5.66\,\text{Å}$ (Ge substrates), and (iii) $a_\parallel = 5.52\,\text{Å}$ ($\text{Ge}_{0.38}\text{Si}_{0.62}$ substrate). It was found that the Ge valence-band edge was consistently higher in energy than the Si valence-band edge, while ΔE_v varied linearly from $0.74\,\text{eV}$ for $a_\parallel = a_0(\text{Si})$ to $0.21\,\text{eV}$ for $a_\parallel = a_0(\text{Ge})$. People and Bean (1986b) combined the valence-band offset calculations of Van de Walle and Martin (1986) with phenomelogical deformation-potential estimates of strained (Si,Ge) bandgaps versus $a_\parallel$, to obtain the anticipated band alignments for the experimental structures of People et al. (1984) and Jorke and Herzog (1985). It was confirmed that one would indeed expect a type-I band alignment with $\Delta E_v \gg \Delta E_c$ for pseudomorphic $\text{Ge}_{0.2}\text{Si}_{0.8}$/Si heterostructures on (001) Si, whereas a type-II (staggered) band alignment was predicted for growth of $\text{Ge}_{0.5}\text{Si}_{0.5}$/Si heterostructures on $\text{Ge}_{0.25}\text{Si}_{0.75}$ substrates (or relaxed buffer layers). These earlier results are shown in Fig. 13, wherein spin–orbit effects on ΔE_v are not included. The inclusion of spin–orbit effects results in a uniform increase of ΔE_v by $0.1\,\text{eV}$ for Si/Ge heterointerfaces (Van de Walle and Martin, 1986). The resulting Δ-bandgap conduction-band discontinuity for $\text{Ge}_x\text{Si}_{1-x}$/Si heterojunctions on Si (001) are shown in Fig. 14.

1. *Estimates of Band Offsets*

Given the bulk band structures of two semiconductor materials A and B, the band-offset problem addresses the question of how these bulk-band structures are aligned relative to one another when a heterointerface A/B is

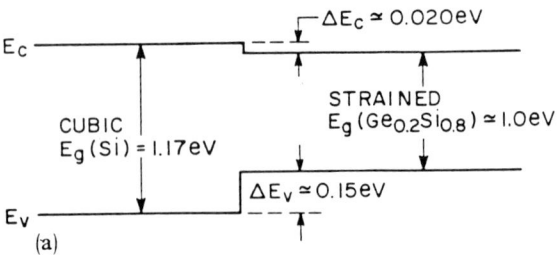

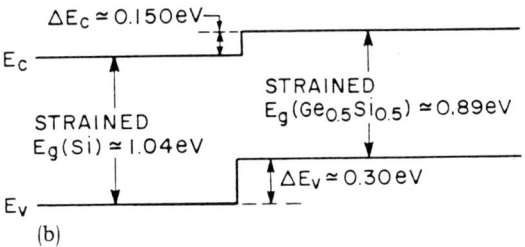

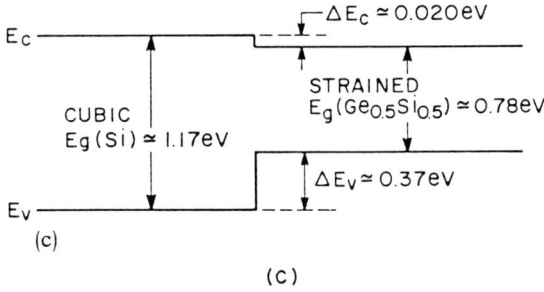

FIG. 13. Band alignments (without spin–orbit effects) for (a) $Ge_{0.2}Si_{0.8}$/Si heterostructures on (001) Si substrates; (b) $Ge_{0.5}Si_{0.5}$/Si heterostructures on an unstrained (001) $Ge_{0.25}Si_{0.75}$ buffer layer; and (c) $Ge_{0.5}Si_{0.5}$/Si heterostructures on (001) Si substrates. [Reprinted with permission from the American Institute of Physics, People R., and Bean, J.C. (1986). *Appl. Phys. Lett.* **48** (8), 538.]

formed. This amounts to determining the lineup of electrostatic potentials. The fundamental problem encountered in deriving band lineups at heterointerfaces is that for a bulk solid, there is no intrinsic energy scale (i.e., no *vacuum zero* is present) to which all energies can be referred (Kleinman, 1981). This ambiguity arises because the long-range nature of the Coulomb interaction does not allow the zero of energy of an infinite (bulk) solid to be

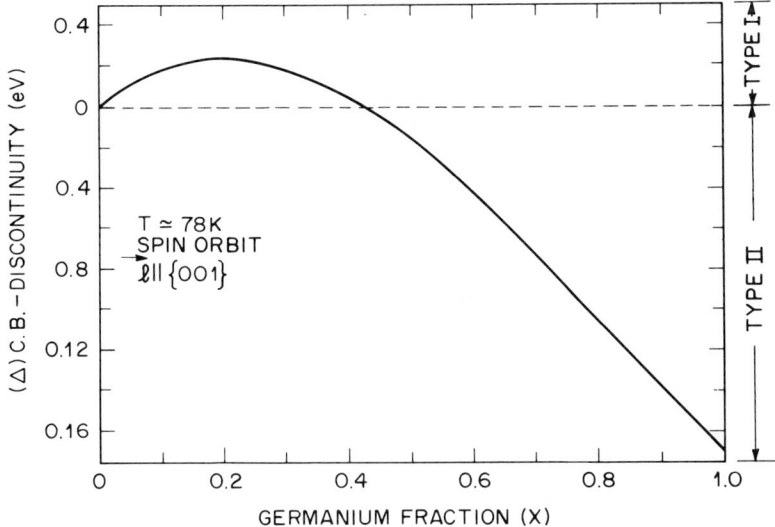

FIG. 14. Δ-band edge conduction-band discontinuity for pseudomorphic Ge_xSi_{1-x}/Si heterostructures on Si (001) and including spin–orbit contributions to ΔE_v. (After People et al., 1987.) [Reprinted with permission from World Scientific Publishing Co. Pte. Ltd., People, R., Bean, J.C., and Lang, D.V. (1987). *Proc. 18th Int. Conf. Physics Semiconductors*, (O. Engstrom ed.).]

well defined. Therefore, there exists no *unique* reference level within each bulk solid with which to compare the potentials for two different solids. To *derive* the potential shift that occurs at a heterojunction, one has to perform some calculation in which both types of materials are present. Such a calculation must account for the redistribution of electronic charge in the neighborhood of the heterointerface. These *charge-transfer* effects set up dipole moments across the interface that given rise to a relative shift of the bulk bands. If a complete solution of the interface problem is desired, then one must carry out self-consistent calculations in which the electrons are allowed to adjust to the specific environment induced by the heterointerface. The first self-consistent interface calculation (SCIC) was that of Baraff et al. (1977), who considered the polar (100) Ge–GaAs heterointerface. These calculations served to elucidate the numerous difficulties associated with calculations involving polar interfaces of zinc-blende structure crystals. Small displacements of charged atoms at polar interfaces can produce additional dipole shifts, whereas the structurally stable interface may require considerable mixing (see, e.g., Kunc and Martin, 1981), leading to dipole shifts that depend strongly on interface stoichiometry. Later such calculations on zinc-blende structures have therefore tended to address the nonpolar (110) interface, using either empirical pseudopotentials (see, e.g., Pickett et al., 1977) or the more recent *ab initio* pseudopotentials of Bachelet *et al.* (1982) as outlined by Van de Walle and Martin (1985, 1987). Results for ΔE_v are presented for a number of lattice-matched (110) heterointerfaces between zinc-blende compounds and for strained Si/Ge heterointerfaces. Unfortunately, the computational complexity of such first-principles calculations is very high, which limits their use as a general tool in the exploration and design of novel heterostructures. Several models have been recently proposed that attempt to establish an *absolute energy level* for each semiconductor. It should be noted that the existence of such an absolute energy scale necessarily implies that one is not dealing with an infinite solid, but rather with a crystal that has been somehow terminated (i.e., a surface has now entered the picture). The means by which this surface (which generally serves as the zero of energy) is defined, underlies and differentiates the various models. A synopsis of a number of the absolute energy level models has been given by Van de Walle and Martin (1987) and will be reiterated here for completeness.

Frensley and Kroemer (1977) constructed a model solid in which they superimposed spherical ions. They chose the mean interstitial potential in the diamond or zinc-blende structure as the electrostatic reference potential for each semiconductor. If the crystal could be viewed as a superposition of spherical charges, this reference potential would correspond to the vacuum potential, provided the charge densities were so localized as to be negligible in the interstitial regions. Valence-band offsets were determined by lining up these potentials, taking into account dipole shifts. The dipole shifts were expressed in terms of charges on the atoms and subsequently in terms of electronegativity differences. It was found that the dipole shifts were small in most cases, indicating that the intrinsic lineup predictions should be close to the true result. Empirical pseudopotentials were used to generate values for the reference potentials. Values for the potential at the interstitial point were found to converge very slowly however, thus requiring a much higher accuracy than is typically required in computing energy eigenvalues. Frensley and Kroemer (1976) acknowledged that their electrostatic potential within the interstitial region of the diamond structure was only flat to ~ 1 eV, using a cutoff energy higher than 18 Ry for Si. Although the Frensley–Kroemer scheme *in principle* offered a very attractive approach for calculating a reference level, the extensive computational requirements render it impractical for generating accurate values for band offsets.

Harrison's (1980) theory of natural bandlineups establishes an absolute energy scale by referring all energies to energy eigenvalues of the free atom. Harrison's model is based on atomic term values, which are assumed to carry over from atom to solid. More recently, Harrison and Tersoff (1986) have developed a rather different theory, wherein a tight-binding calculation is used to estimate the averaged hybrid energy. This energy is then identified as the *neutrality level* that will be *pinned* at a heterojunction. Values for ΔE_v have been calculated for a number of lattice-matched heterointerfaces using this method.

In the approach of Tersoff (1984, 1985) and Tejedor and Flores (1979), screening arguments are used to define a *neutrality level* for each semiconductor. These levels are aligned when an interface is formed, thus determining band offsets. In analogy to metal–metal interfaces, these theories assume that dipoles will be set up that will drive the heterointerface system towards alignment of the *neutrality levels* of the materials (as would be the case at a heterojunction between two metals in which the Fermi levels line up). This model has had much success in predicting band offsets for a large number of lattice-matched nonpolar interfaces (see, e.g., Tersoff, 1987).

Although, in principle, it should be possible to incorporate the effects of coherent layer strains within the afore-mentioned models, no prescription for such has yet been given. Recently, two other model solid calculations based on an *absolute energy reference scale* and allowing for the incorporation of misfit strain have been generated. These include (i) the linear muffin-tin orbitals (LMTO) all-electron calculations of Verges et al. (1982) and Cardona and Christensen (1987), and (ii) the *model solid* calculations of van de Walle and Martin (1987) and Van de Walle (1988).

In the LMTO method, the solid is broken up into atomiclike spheres, and all potentials are referenced to a level that is chosen so that the Hartree potential of a single atomic sphere is zero at infinity. The solid can be terminated at any sphere while leaving the electronic charge distribution in the sphere equal to that in the bulk. In order to calculate valence-band offsets, one first calculates the energies of the upper Γ_8^v valence bands for the individual materials using the LMTO method, with all energies referred to the aforementioned reference level, which, excluding surface dipoles, should represent the potential at infinity. (It should be noted that the LMTO calculations are fully relativistic and hence include the spin–orbit interaction.) When bringing the two materials together to form a heterojunction, it is assumed that a potential difference will appear that will be screened by the electronic polarizability. It is argued that the potential energy associated with this screening effect can be expressed as a band energy divided by the zero-frequency intrinsic dielectric response function, $\varepsilon(q)$. This energy gap is called the dielectric midpoint energy (E_D) and has an associated hydrostatic deformation potential a_D. Since the dielectric function $\varepsilon(q, \omega)$ is obtained via a Brilloun-zone integration, which itself can be replaced by sampling over a small number of special points, so-called Baldereshi points (Baldereshi, 1973); the dielectric midgap energy is calculated as the average LMTO band energy at the first (24 symmetry-equivalent) Baldereshi points. The valence-band offset for lattice-matched nonpolar interface therefore consists of the difference in two energy terms, the first term being the calculated LMTO energy of the upper Γ_8^v valence band, whereas the second term is the calculated average of the LMTO upper valence band and lowest conduction band at the first Baldereshi point $[\mathbf{K}_B = 2\pi/a_0(0.622, 0.295, 0)]$ divided by $\varepsilon(q)$; i.e.,

$$\Delta E_v^{A/B} \equiv E_v^B - E_v^A - (E_D^B - E_D^A)(\bar{\varepsilon} - 1)/\bar{\varepsilon}, \quad (39)$$

where $\bar{\varepsilon}$ is an effective dielectric constant $\simeq 3.5$, independent of material.

When hydrostatic strain is present in one of the layers, say A, one must add to Eq. (39) a term of the form:

$$\Delta_H^{A/B} = (-a_v^A + a_D^A(\bar{\varepsilon} - 1)/\bar{\varepsilon}))[\tfrac{2}{3}(e_\parallel)(1 - C_{12}/C_{11})]. \quad (40)$$

In the above, a_v (a_D) denote the LMTO-calculated hydrostatic deformation potentials for the Γ_8^v valence band (dielectric midgap energy, E_D), and $e_\parallel$ denotes the in-plane strain experienced by the strained-material component. The LMTO a_v and the a_D have the same sign and are comparable in magnitude; therefore, hydrostatic contributions to the LMTO valence-band offsets are expected to be small (Cardona and Christensen, 1987).

The LMTO treatment of the effects on the band offset due to the uniaxial component of the misfit strain tensor is more complicated, and an exact treatment is not available. In particular, for a (001) heterointerface, the first Baldereshi points of the strained-material component are not all equivalent but rather split into three groups (of eight each). The question then arises as to which of these split sets of **k**-space points should be used to calculate the dielectric midgap energy, E_D. Calculations of the splitting associated with these sets of points have not been made due to complications that arise when shear strains are introduced into the LMTO method (Cardona and Christensen, 1987). Instead, a simple model is given to estimate the effects of shear strains on the matching across the interface, i.e., on E_D. In this model, E_D will depend on the direction of **k**, and one must perform the *appropriate* average over **k** in order to obtain that E_D be matched across an interface of given orientation. For the (001) interface, Cardona and Christensen (1987) obtained

$$\langle \Delta E_D \rangle_{(001)} = (\tfrac{4}{5})(\Delta a_D)[-(\tfrac{1}{3})(e_\parallel)(1 + 2C_{12}/C_{11})]. \quad (41)$$

In the above, Δa_D denotes the change in the hydrostatic deformation potential for the dielectric midgap energy (of a given material) before and after screening effects are included. Equations (39)–(41) have been applied to the Ge/Si heterostructure on Ge ($a_\parallel = 5.66$ Å), where it is found that the average valence-band offset $\{E_{v,av}(\text{Ge}) - E_{v,av}(\text{Si})\} = 0.22$ eV. This value is to be compared with the fully self-consistent calculated value of 0.51 eV (Van de Walle and Martin, 1986). Better agreement with the SCIC value for $\Delta E_{v,av}$ is obtained for InAs/GaAs heterointerfaces with $a_\parallel = 6.08$ Å. In this case, the LMTO calculation gives $E_{v,av}(\text{InAs}) - E_{v,av}(\text{GaAs}) = 0.18$ eV, whereas the SCIC value is 0.21 eV (Van de Walle, 1986).

In the *model-solid* theory of Van de Walle and Martin (1987), a procedure is developed for obtaining the lineup of electrostatic potential across the heterointerface by again using only information from the bulk constituents. This model retains the spirit of the SCIC and was the result of an attempt to extract the essential features of the SCIC results, but without having to perform the local-density calculations on a supercell. The infinite (bulk) solid is replaced by a semi-infinite solid having an ideally terminated surface (i.e., one without a surface dipole layer). The semi-infinite model solid is constructed by taking a superposition of neutral atomic spheres. The potential outside each of these neutral spheres goes exponentially to (an absolute) zero; which is taken as the zero of energy for the model solid. Note that the presence of a surface in this model does not induce any shift in the average potential, since no dipole layer can be set up using neutral spheres. The valence-band offset calculation is performed in two parts: (i) The average potential, $\bar{V}$, in each semi-infinite solid, which is a superposition of atomic charge densities, is calculated; (ii) the band structure for each (cubic) bulk solid is calculated (ignoring spin), and the valence-band maximum $E_{v,av}^{(0)}$ added to the afore-mentioned average potential. The average energy of the top of the valence band, $E_{v,av}^{(0)}$, is thus determined on an absolute scale for each material component of the heterojunction in its cubic state. It should be kept in mind that taken separately, the results for the bulk valence-band edge position with respect to the average potential and the average potential itself contain no information, since they depend on the choice of pseudopotential and of the angular momentum used in the nonlocal part of the potential (Bachelet et al., 1982).

In calculating the average potential, the total potential is taken as the sum of the ionic, Hartree, and exchange-correlation potentials, i.e.,

$$V^l \equiv V^{\text{ion},l} + V^H + V^{xc}. \tag{42}$$

where the superscript l denotes the angular momentum component used in the nonlocal pseudopotential. The atomic calculations are then performed on a pseudoatom having atomic configuration $s^x p^y$, where x, y are obtained from tight-binding theory (see, e.g., Chadi, 1987); after which the resulting charge densities are superimposed to generate the average potential. Since the first two terms in Eq. (42) are linear in charge density, they can therefore be expressed as a superposition of atomic potentials. Their average in the solid is therefore

$$\bar{V}^{\text{ion},l} + \bar{V}^H = \sum_i \left(\frac{1}{\Omega}\right) \int (V_i^{\text{ion},l} + V_i^H) d\tau, \tag{43}$$

where Ω denotes the volume of the unit cell and i runs over all atoms in the unit cell. Equation (43) has important consequences with respect to strained layers, since it implies that the average values for these terms follow a simple Ω^{-1} scaling law. The exchange-correlation potential is not linear in density and therefore cannot be expressed as a superposition of atomic potentials. It is argued, however (Van de Walle and Martin, 1987), that due to the local nature of V^{xc}, its value will not depend on how the solid is terminated, and hence values for this quantity may be obtained from calculations for the bulk solid, and added in afterwards. Since V^{xc} is proportional to the 1/3 power of density, it follows that in strained layers, the bulk values should scale as $\Omega^{-1/3}$. The calculation of the energy of the valence-band maximum for the cubic bulk solid proceeds in standard fashion as outlined, e.g., by Chelikowsky and Cohen (1976).

Since (as shown in Section II.A.2) there is an interaction between strain and spin–orbit splittings, the most straightforward means for incorporating strain into the *model-solid* calculation of the valence-band offset is as follows. First compute the afore-mentioned average position of the valence-band maximums on an absolute scale (i.e., ignoring spin–orbit effects) as shown in Fig. 15. Given $E_{v,av}^{(0)}$, the energy of the average bulk valence-band edge for the *cubic* solid, uniaxial strain, hydrostatic strain and spin–orbit effects may be added in afterwards. The motions of the strained valence-band edges relative to $E_{v,av}^{(0)}$ (and including spin–orbit splittings) are given by Pollak and Cardona (1968)

$$\delta E_v(\tfrac{3}{2}; \pm\tfrac{3}{2}) = \tfrac{1}{3}\Delta_0 + \varepsilon, \tag{44a}$$

$$\delta E_v(\tfrac{3}{2}; \pm\tfrac{1}{2}) = -\tfrac{1}{6}\Delta_0 - \tfrac{1}{2}\varepsilon + \tfrac{1}{2}[\Delta_0^2 - 2\varepsilon\Delta_0 + 9\varepsilon^2]^{1/2}, \tag{44b}$$

and

$$E_v(\tfrac{1}{2}; \pm\tfrac{1}{2}) = -\tfrac{1}{6}\Delta_0 - \tfrac{1}{2}\varepsilon - \tfrac{1}{2}[\Delta_0^2 - 2\varepsilon\Delta_0 + 9\varepsilon^2]^{1/2}, \tag{44c}$$

where Δ_0 is the spin–orbit splitting and ε has been given for $\mathbf{l} \parallel \{011\}$ or $\{111\}$ in Eqs. (32a) and (32b) (Van de Walle, 1988).

The effect of hydrostatic strain on $E_{v,av}^{(0)}$ is described by the deformation potential a_v,

$$a_v = \frac{dE_{v,av}^{(0)}}{d(\ln\Omega)}, \tag{45}$$

which may be calculated within the context of the model solid. Therefore, the average energy of the valence bands within a strained solid is given by

$$E_{v,av} \equiv E_{v,av}^{(0)} + a_v\left(\frac{\Delta\Omega}{\Omega}\right), \tag{46}$$

where the fractional volume change $\Delta\Omega/\Omega \equiv \text{Tr}(\mathbf{S})$; as given in Section II.A.1. Note that the quantities $E_{v,av}^{(0)}$ and a_v (along with Eqs. (44)) are the fundamental parameters required to calculate the valence-band offsets within the *model-solid* theory. Results for $E_{v,av}^{(0)}$ and a_v have been tabulated by Van der Walle (1988) for a number of elemental, III–V, and II–VI semiconductors

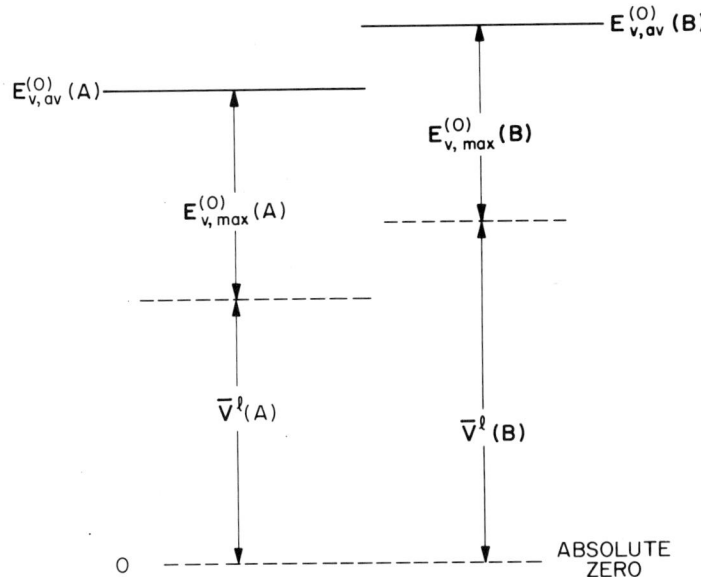

FIG. 15. Schematic of the various energies in the model-solid calculation, relative to an *absolute zero* $\bar{V}^1$, represents the average ion core, Hartree, and exchange-correlation energy, obtained by a superposition of the pseudoatom results. $E_{v,\text{max}}^{(0)}$ denotes the results of a band-structure calculation of the average energy of the valence-band maximum (i.e., Γ_{25}^v for Si, and Ge or Γ_{15}^v for zinc-blende solids). $E_{v,\text{av}}^{(0)}$ is the sum of these terms and represents the average energy of the valence-band maximum for the cubic solid, referenced to an *absolute zero*.

and are listed in Table I. As an example of the application of these data, we consider the pseudomorphic Si/Ge interface on (001)Si. In this case the Si is cubic; using $E_{v,\text{av}}^{(0)}(\text{Si}) = \sim -7.03\,\text{eV}$ and $\Delta_0(\text{Si})/3 = 0.014\,\text{eV}$, we get $(E_{v,\text{av}}^{(0)} + \Delta_0/3) \simeq -7.016\,\text{eV}$. For the strained Ge: $e_{xx} = e_{yy} = -4.2 \times 10^{-2}$, $e_{zz} = 3.155 \times 10^{-2}$ so that using $a_v = +1.24\,\text{eV}$ and $\Delta\Omega/\Omega = -5.24 \times 10^{-2}$, we get $[E_{v,\text{av}}^{(0)} + a_v(\Delta\Omega/\Omega)] = -6.415\,\text{eV}$. Further using $D_u^{(001)} = 3.81\,\text{eV}$ and $\Delta_0 = 0.3\,\text{eV}$, we obtain $(\Delta_0/3 + \varepsilon) = 0.287\,\text{eV}$ so that the upper $(3/2 \pm 3/2)$ valence-band edge of strained Ge lies at $(-6.415 + 0.287)\,\text{eV} = -6.128\,\text{eV}$. The valence-band offset $\Delta E_v \equiv E_v(\text{Ge}) - E_v(\text{Si}) = -6.128 - (-7.016)\,\text{eV} = 0.89\,\text{eV}$. This result compares well with the full SCIC result of 0.84 eV. These values agree well with the experimentally determined value for ΔE_v of $(0.81 \pm 0.06)\,\text{eV}$, given by Ni *et al.* (1987) using *in situ* x-ray photoelectron spectroscopy (XPS) on pseudomorphic (GeSi)/Si heterostructures.

A second interesting example is provided by the pseudomorphic InP/In$_x$Ga$_{1-x}$As interface on (001) InP. The system In$_x$Ga$_{1-x}$As/InP on (001) InP has recently been investigated by Gershoni *et al.* (1987, 1988), using photoluminescence and photocurrent spectroscopy on strained-layer superlattices grown by gas-source MBE. We consider three cases: (i) *Lattice*

TABLE I

Lattice Constant a_0(Å), Spin–Orbit Splitting Δ_0, Average Valence-Band Energy $E_{v,av}^{(0)}$ for Cubic Material, and Hydrostatic Deformation Potential a_v, as Computed in Model-Solid Theory. (After Van de Walle, 1988.) Inverse Mass Parameters, γ_i. (After P. Lawaetz, 1971)

Material	a_0(Å)	Δ_0(eV)	$E_{v,av}^{(0)}$(eV)	a_v(eV)	γ_1	γ_2	γ_3
Si	5.431	0.04	−7.03	2.46	4.22	0.39	1.44
Ge	5.658	0.30	−6.35	1.34	13.35	4.25	5.69
GaAs	5.654	0.34	−6.92	1.16	7.65	2.41	3.28
AlAs	5.661	0.28	−7.49	2.47	4.04	0.78	1.57
InAs	6.058	0.38	−6.67	1.00	19.67	8.37	9.29
GaP	5.451	0.08	−7.40	1.70	4.20	0.98	1.66
AlP	5.462	—	−8.09	3.15	3.47	0.06	1.15
InP	5.869	0.11	−7.04	1.27	6.28	2.08	2.76
GaSb	6.095	0.82	−6.25	0.79	11.80	4.03	5.26
AlSb	6.140	0.65	−6.66	1.38	4.15	1.01	1.75
InSb	6.48	0.81	−6.09	0.36	35.08	15.64	16.91
ZnSe	5.667	0.43	−8.37	1.65	3.77	1.24	1.67
ZnS	5.40	0.07	−9.15	2.31	2.54	0.75	1.09
CdTe	6.48	0.93	−7.07	0.55	5.29	1.89	2.46
HgTe	6.464	1.05	−6.88	−0.13	−18.68	−10.19	−9.56

matched $In_{0.53}Ga_{0.47}As$ on InP. Here the model-solid result for $\Delta E_v \equiv E_v$ ($In_{0.53}Ga_{0.47}As) - E_v(InP)$ is 0.34 eV. This is in excellent agreement with the experimental values for ΔE_v by Lang et al. (1987) of (0.35 ± 0.01) eV, and (0.36 ± 0.01) eV as obtained by Forrest et al. (1984). (ii) Strained InAs on (001) InP. The model-solid result for $\Delta E_v = E_v(InAs) - E_v(InP)$ is 0.55 eV. Given an anticipated uncertainty of $\lesssim \pm 0.1$ eV, the model solid result compares well with the experimental result of 0.47 eV as obtained by Gershoni et al. (1988). (iii) Strained GaAs on (001) InP. The model-solid calculation gives $\Delta E_v \equiv E_v(GaAs) - E_v(InP) = 0.422$ eV for pseudomorphic GaAs on (001) InP. The photocurrent data of Gershoni et al. (1988) is best fit by a value for ΔE_v of 0.56 eV. Although the discrepancy between experiment and theory is <0.2 eV in this case, this difference is critical to the conduction-band alignment, since the bandgap of the coherently strained GaAs is 0.9 eV at 300 K. Hence the model solid predicts a type-I alignment with small $\Delta E_c = E_c(InP) - E_c(GaAs) = [1.35 - (0.90 + 0.422)]$ eV $\simeq 0.03$ eV ~ 0. Given the experimental ΔE_v of 0.56 eV, one obtains $\Delta E_c = E_c(InP) - E_c(GaAs) = [1.35 - (0.90 + 0.56)] = -0.120$ eV; i.e., a type-II alignment with the wide-gap InP conduction-band edge lying lower in energy. The disparity between experiment and theory might (aside from numerical accuracy) be related to (i) the fact that no common atoms are present on either side of the InP/GaAs heterointerface in conjunction with the polar nature of the (001) interface, (ii) differences in electronegativity of various probable interfacial species, e.g., InAs, GaAs, InP, InGaAsP, and/or (iii) details of the normal growth sequence used in gas-source MBE; all factors that may affect both the calculated and measured value of ΔE_v. It should also be mentioned that a linear interpolation of SCIC results for InAs/GaAs (Van de Walle, 1986) in conjunction with measured ΔE_v values for the lattice-matched InGaAs/InP interface and transitivity yields $E_c(\text{InP}) - E_c(\text{GaAs})$ $\sim 0.2\,\text{eV}$ (People, 1987). In particular, this latter result totally ignores any interfacial chemistry involving the InP and the strained InGaAs layers.

C. Mass Modifications Due to Coherent Layer Strains

The dispersion relation $E(\mathbf{k})$ for free carriers in a strained semiconductor is altered from the cubic case, since the strain Hamiltonian, H_ε, and the wave-vector-dependent operator $H_\mathbf{k}$, must be treated on an equal footing. Strain effects are most pronounced within the degenerate valence-band edge states. Indeed, externally applied uniaxial stress was used by Hensel and Feher (1963) to remove the fourfold degeneracy of the upper $J = 3/2$ valence-band edge of Si. The resulting decoupled bands allowed for unambiguous determination of valence-band uniaxial deformation potentials (D_u, D'_u) and the magnitude and sign of the inverse mass parameters. In the following, we shall concentrate mainly on the effects of coherent layer strains on the valence-band mass parameters.

The $H(\mathbf{k})$ terms arise naturally in $\mathbf{k} \cdot \mathbf{p}$ perturbation theory, given the Bloch form of carrier wave function. Dresselhaus et al. (1955) used this approach to derive the $E(\mathbf{k})$ relation for the valence bands in group-IV (elemental) semiconductors, whereas Kane (1957) included efforts due to couplings to the lowest conduction states (and hence obtained results applicable to narrow-gap III–V semiconductors). It is usually sufficient, however, to know only the general form of $H(\mathbf{k})$, since the various constants (inverse mass parameters, momentum matrix elements, etc.) are generally determined either from experimental data or calculation. The general form of $H(\mathbf{k})$ may be readily determined by using the *theory of invariants*, developed by J. M. Luttinger (1956). This approach allows one to write down an explicit operator formulation for the Hamiltonian $H(\mathbf{k})$ based upon the angular momentum operators J_i, in analogy with Eq. (30) for the uniaxial strain Hamiltonian.

The Luttinger Hamiltonian for the valence-band states is given by:

$$H_{\text{VB}}(\mathbf{k}) = -\frac{\hbar^2}{2m_0}\{\gamma_1 K^2 - 2\gamma_2[(J_x^2 - \tfrac{1}{3}J^2)k_x^2 + \text{cp}] \\ - 4\gamma_3[\{J_x, J_y\}\{k_x, k_y\} + \text{cp}]\}, \qquad (47)$$

where m_0 is the free-electron mass, $K^2 = k_x^2 + k_y^2 + k_z^2$, γ_i are the inverse-mass parameters (see, e.g., Lawaetz, 1971), $\{\,,\,\}$ denotes the anticommutator and cp implies cyclic permutations. The matrix elements of $H(\mathbf{k})$ within the (J, M_J) wave-function space have been given by Luttinger and Kohn (1955) and Bir and Pikus (1974).

1. *Hole Dispersion near* $\mathbf{k} = 0 - (6 \times 6)$ *Luttinger Model*

The total strain and wave-vector-dependent Hamiltonian describing the dispersion relation for holes in bulk semiconductors under uniaxial stress is given by

$$H_V = H_{VB}^{(u)}(\mathbf{l}) + H_{VB}(\mathbf{k}) + H_{SO}, \qquad (48)$$

where H_{SO} is the spin–orbit interaction (see Luttinger and Kohn, (1955), $\mathbf{l}$ denotes the unit vector normal to the substrate, and the strain Hamiltonian $H_{VB}^{(u)}(\mathbf{l})$ within the (J, M_J) representation has been given in Eq. (30). Although it would appear that the general (6×6) secular matrix generated by Eq. (48) might be readily obtained by combining the strain matrix elements (as calculated by Hasegawa, 1963, in Eq. (31)) with the explicit $\mathbf{k}$-dependent matrix elements, as given by both Luttinger and Kohn (1955) and Bir and Picus (1974), such an approach immediately fails to generate the proper E versus $\mathbf{k}$ relationship to order k^2. The cause of the breakdown of this approach has been traced to a difference in the form of the six *zeroth-order* valence-band basis functions used to diagonalize the spin–orbit interaction. People and Sputz (1990) have treated strain and $\mathbf{k} \cdot \mathbf{p}$ interactions on an equal footing, using the time-reversal symmetry-invariant basis functions of Luttinger and Kohn (1955) throughout. The resulting (6×6) secular matrix for (001)- or (111)-oriented substrates is given by:

$$\begin{array}{cccccc}
(\tfrac{3}{2}; \tfrac{3}{2}) & (\tfrac{3}{2}; \tfrac{1}{2}) & (\tfrac{3}{2}; -\tfrac{1}{2}) & (\tfrac{3}{2}; -\tfrac{3}{2}) & (\tfrac{1}{2}; \tfrac{1}{2}) & (\tfrac{1}{2}; -\tfrac{1}{2})
\end{array}$$

$$\begin{bmatrix}
H & \alpha & \beta & 0 & \dfrac{i}{\sqrt{2}}\alpha & -i\sqrt{2}\beta \\[1em]
\alpha^* & L & 0 & \beta & i\left(\dfrac{D}{\sqrt{2}} - \sqrt{2}\varepsilon\right) & i\sqrt{\tfrac{3}{2}}\alpha \\[1em]
\beta^* & 0 & L & -\alpha & -i\sqrt{\tfrac{3}{2}}\alpha^* & i\left(\dfrac{D}{\sqrt{2}} - \sqrt{2}\varepsilon\right) \\[1em]
0 & \beta^* & -\alpha^* & H & -i\sqrt{2}\beta^* & \dfrac{-i}{\sqrt{2}}\alpha^* \\[1em]
\dfrac{-i}{\sqrt{2}}\alpha^* & -i\left(\dfrac{D}{\sqrt{2}} - \sqrt{2}\varepsilon\right) & i\sqrt{\tfrac{3}{2}}\alpha & i\sqrt{2}\beta & S & 0 \\[1em]
i\sqrt{2}\beta^* & -i\sqrt{\tfrac{3}{2}}\alpha^* & -i\left(\dfrac{D}{\sqrt{2}} - \sqrt{2}\varepsilon\right) & \dfrac{i}{\sqrt{2}}\alpha & 0 & S
\end{bmatrix}$$

where ε denotes the strain energy of Eq. (32), and

$$H = -\frac{\hbar^2}{2m_0}[(k_x^2 + k_y^2)(\gamma_1 + \gamma_2) + k_z^2(\gamma_1 - 2\gamma_2)] + \varepsilon \equiv H_{hh} + \varepsilon,$$

$$L = -\frac{\hbar^2}{2m_0}[(k_x^2 + k_y^2)(\gamma_1 - \gamma_2) + k_z^2(\gamma_1 + 2\gamma_2)] - \varepsilon \equiv H_{lh} - \varepsilon,$$

$$\alpha = i\frac{\sqrt{3}}{m_0}\hbar^2[k_z(k_x - ik_y)\gamma_3],$$

$$\beta = -\frac{\sqrt{3}}{2}\frac{\hbar^2}{m_0}[(k_x^2 - k_y^2)\gamma_2 - 2ik_xk_y\gamma_3],$$

$$D = (H_{lh} - H_{hh}),$$

and

$$S = \tfrac{1}{2}(L + H) - \Delta_0. \tag{49}$$

Note that the diagonal components of $H_{VB}(\mathbf{k})$, namely H_{hh}, H_{lh}, and $\{(H_{hh} + H_{lh})/2 - \Delta_0\}$, give the dispersion of the heavy-hole (3/2; $\pm 3/2$), light-hole (3/2; $\pm 1/2$), and spin–orbit split-off (1/2; $\pm 1/2$) hole bands to order k^2 in the absence of strain. Note further that the dispersion of the heavy-hole and light-hole bands are anisotropic, having differing mass parameters for dispersion in-plane (i.e., $K_\parallel^2 = k_x^2 + k_y^2$) and along the k_z-axis (which is the assumed axis of quantization). Keep in mind that M_J is a good quantum number in the presence of strain for growth on (001)- or (111)-oriented substrates.

The eigenvalues of the matrix in Eq. (49) are doubly degenerate. The *general secular equation* obtained by People and Sputz (1990) is a cubic form, given by:

$$0 = \left\{-\tilde{S}\tilde{H}\tilde{L} + \tilde{H}\left(\frac{D}{\sqrt{2}} - \sqrt{2}\varepsilon\right)^2 + |\alpha|^2(\tfrac{3}{2}\tilde{H} + \tfrac{1}{2}\tilde{L} + \tilde{S} - D + 2\varepsilon)\right.$$
$$\left. + |\beta|^2(2\tilde{L} + \tilde{S} + 2D - 4\varepsilon) + 3\sqrt{3}\,\text{Re}[(\alpha)^2\beta^*]\right\}, \tag{50}$$

where $\tilde{H} = H - E$, etc., with E denoting the eigenvalue. Note that Eq. (50) shows that the k-dependent matrix elements α and β only enter as modulus squared and hence contribute to the k-dependence of order k^4 or greater (i.e., they relate only to nonparabolicity). Hence the strain-dependent hole-dispersion relations to order k^2 may be readily obtained by setting $\alpha = \beta = 0$ in Eq. (50).

The *secular equation* valid for small k values is therefore:

$$0 = \tilde{H}\left\{-\tilde{SL} + \left(\frac{D}{\sqrt{2}} - \sqrt{2}\varepsilon\right)^2\right\}. \quad (51)$$

It is noted that the $(\frac{3}{2}; \pm\frac{3}{2})$ heavy-hole solution decouples in order k^2, whereas the lattice-mismatch strain couples the light-hole $(\frac{3}{2}, \pm\frac{1}{2})$ and split-off $(\frac{1}{2} \pm \frac{1}{2})$ hole bands. The three (doubly degenerate) givenvalues are:

$$E_v(\tfrac{3}{2} \pm \tfrac{3}{2}) = H_{hh} + \varepsilon, \quad (52a)$$

and (noting that M_J remains a good quantum number for $l = (001), (111)$),

$$E_{1/2}(\pm) = -\frac{\Delta_0}{2}(1+x) + \tfrac{1}{4}(3H_{lh} + H_{hh})$$

$$\pm \frac{\Delta_0}{2}[1 - 2x + 9x^2]^{1/2}\left\{1 + \left(\frac{D}{\Delta_0}\right)\frac{(1-9x)}{[1-2x+9x^2]}\right.$$

$$\left. + \frac{9}{4}\left[\frac{D}{\Delta_0}\right]^2 \frac{1}{[1-2x+9x^2]}\right\}^{1/2} \quad (52b)$$

where $x = \varepsilon/\Delta_0$. The $+$ and $-$ solutions correspond to the $(3/2, \pm 1/2)$ and $(1/2; \pm 1/2)$ valence bands, respectively, in the absence of strain.

In the limit where the spin–orbit splitting exceeds the heavy hole-to-light hole splitting in the bulk (i.e., $\Delta_0 > (H_{hh} - H_{lh})$, the following approximate solutions are obtained (to order k^2):

$$E_{1/2}(+) = E_v^{(+)}(0) - \frac{\hbar^2 k_\parallel^2}{2m_0}(\gamma_1 - Z\gamma_2) - \frac{\hbar^2 k_z^2}{2m_0}(\gamma_1 + 2Z\cdot\gamma_2), \quad (53a)$$

$$E_{1/2}(-) = E_v^{(-)}(0) - \frac{\hbar^2 k_\parallel^2}{2m_0}(\gamma_1 - \gamma_2(1-Z)) - \frac{\hbar^2 k_z^2}{2m_0}(\gamma_1 - 2\gamma_2(1-Z)), \quad (53b)$$

where the function Z is given by

$$Z(x) = \tfrac{1}{2}\left(1 + \frac{(1-9x)}{[1-2x+9x^2]^{1/2}}\right), \quad (53c)$$

and $E_v^{(+)}(0)$ and $E_v^{(-)}(0)$ are given by Eqs. (33b) and (33c), respectively. Equations (52a) and (53a) are identical to the results of Hasegawa (1963). The strain-dependent valence-band mass parameter $Z(x)$ has been plotted in Fig. 16. Note that $Z(x) \to 2$ as $x \to -\infty$, whereas $Z(x) \to -1$ as $x \to +\infty$. Coherence strain can therefore greatly modify both light- and split-off-hole masses, depending upon both the sign and the magnitude of the tetragonal distortion.

In two-dimensional (2D) systems, it is the in-plane mass $(m_\parallel^*)$ that arises in lateral transport studies, whereas the z-mass component (m_z^*) gives rise to

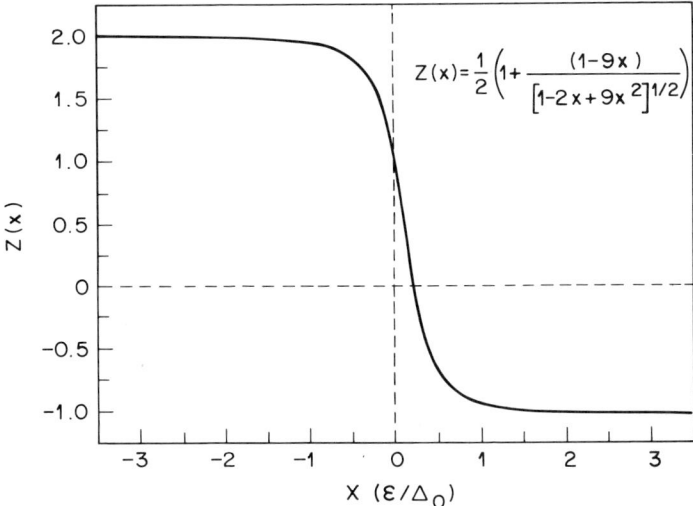

FIG. 16. Strain-dependent valence-band mass parameter $Z(x)$, where $x = \varepsilon/\Delta_0$.

quantum-confinement shifts of the stationary states of the quantum well. The applicability of the decoupled mass formulas of Eq. (53) has been verified for externally strained bulk systems (see, e.g., Hensel and Feher, 1963) as well as for lattice-mismatch-induced strained-layer systems. For the strained $In_{0.2}Ga_{0.8}As/GaAs$ layers studied by Schirber et al. (1985), the upper valence band is $(3/2 \pm 3/2)$ in character, and the calculated in-plane mass $m_{\parallel}^*(\frac{3}{2}\pm\frac{3}{2}) = (\gamma_1 + \gamma_2)^{-1} \cdot m_0$, independent of strain. By linearly interpolating between the InAs and GaAs γ-parameters, one obtains $m_{\parallel}^*(\frac{3}{2}, \pm\frac{3}{2}) = 0.09 m_0$ for strained InGaAs hole transport mass. This mass value is approximately half the measured value of $0.14 m_0$. Similarly the upper valence band of the $Ge_{0.2}Si_{0.8}/Si$ strained layers studied by People et al. (1984) is again $(\frac{3}{2}; \pm\frac{3}{2})$. The linearly interpolated in-plane hole mass calculated by using the Lawaetz γ-parameters for Ge and Si in $m_{\parallel}^*(\frac{3}{2}, \pm\frac{3}{2}) = 0.14 m_0$; to be compared to the measured value of $(0.30 \pm 0.02)m_0$. In both of these strained-layer transport studies, the hole conduction occurred in a 2D gas having similar sheet charge densities (i.e., $p \sim 3.5 \times 10^{11}$ cm^{-2}) giving rise to a Fermi wave vector $k_F \sim 1.5 \times 10^6$ cm^{-1} and Fermi energies of 5.61 meV and 2.62 meV for the pseudomorphic $In_{0.2}Ga_{0.8}As$ on GaAs (001) and $Ge_{0.2}Si_{0.8}$ on Si (001), respectively. Clearly, the strictly parabolic results used to estimate the present transport masses are insufficient to explain observed heavy-hole masses at these wave vectors. The effects of band nonparabolicities, which may cause transport mass enhancements under certain conditions, are therefore investigated in the following sections.

2. Band Nonparabolicities in Bulklike Strained-Layers 6×6 Luttinger Model

To include nonparabolicity effects on transport and quantum-confinement-related masses, one requires the dispersion relation to terms higher than k^2. Over a limited range of band energies, wherein the band energy is small compared to both the strain energy (ε) and the spin–orbit splitting (Δ_0), hole-transport mass modifications may be described by considering $E_v(k)$ to order k^4. In the presence of nonparabolicity, the dispersion relation may, in general, be written as:

$$E_j = \frac{\hbar^2 k^2}{2m_j^*(0)} [1 - \gamma_{NP} k^2], \tag{54}$$

where $m_j^*(0)$ is the effective mass at the band edge and γ_{NP} the nonparabolicity parameter for the jth band.

Hole nonparabolicities within the framework of the (6×6) valence-band Hamiltonian given in Eq. (49) are now considered. If $\mathbf{k} = (0, 0, k_z)$, then $\alpha = \beta = 0$ in Eq. (50), and we again obtain Eq. (51) as the secular equation; however, all solutions are now expanded to order k^6 with terms of order k^4 being retained. Since the heavy-hole solution decouples as in Eq. (52a), we see that the heavy-hole, $E_v(3/2)$, solution is *strictly parabolic* along k_z. Strain admixes the light- and split-off-hole states. Expanding the radical in Eq. (52b), one obtains the following nonparabolicity parameters for the light (+) and split-off (−) hole states within the (6×6) valence-band formulation with k along k_z:

$$\gamma_{1/2}(\pm) = \pm \frac{\hbar^2}{2m_0 \Delta_0} \left(\frac{8\gamma_2^2}{(\gamma_1 + [1 \pm Z']\gamma_2)[1 - 2x + 9x^2]^{3/2}} \right). \tag{55}$$

Here $Z'(x) \equiv 2Z(x) - 1$, with $Z(x)$ given in Eq. (53a), and m_0 is the free electron mass. The associated mass modification is obtained by using

$$m_j^*(E) = m_j^*(0)/[1 - |E|/E_{NP}], \tag{56a}$$

where

$$E_{NP} = (2m_j^*(0) \cdot \gamma_{NP}^{(j)})/\hbar^2. \tag{56b}$$

Although the heavy-hole, $E_v(3/2)$, state is strictly parabolic along k_z, it may be highly nonparabolic along (k_x, k_y). In general, in-plane (transport) mass modifications may be obtained by setting k_z (and hence α) equal to zero in Eq. (50). Heavy, light, and split-off holes remain coupled for $k = k_\parallel = (k_x, k_y)$. However, analytic expressions for band nonparabolicities may still be obtained over a limited range of k (and hence E) values via interative solution of the resulting secular equation. It may be readily shown that for band energies (E) small compared with the strain energy (ε) and the spin-orbit splitting (Δ_0), the heavy-hole in-plane ($k_\parallel$) nonparabolicity parameter is given by

$$\gamma_{NP}(\pm \tfrac{3}{2}) = \tfrac{3}{2}\left(\frac{\gamma_2}{\gamma_1 + \gamma_2}\right)^2 \frac{(1 + 9x)}{\varepsilon} \frac{\hbar^2}{2m_{hh}^*(0)}, \qquad (57)$$

where $m_{hh}^*(0) \equiv m_0/(\gamma_1 + \gamma_2)$. It is noted that the heavy-hole nonparabolicity scales inversely with the strain splitting of the light- and heavy-hole bands ($\sim 2\varepsilon$). A heavy-hole nonparabolicity parameter having such a variation was deduced by Osbourn et al. (1986) from detailed fits of tight-binding calculations on strained $In_xGa_{1-x}As/GaAs$ superlattices. There nonparabolicity parameter (denoted by c) is related to our γ_{NP} via $c = 2m_{hh}^*(0) \cdot \gamma_{NP}/\hbar^2$. The best-fit value of c for the $In_xGa_{1-x}As/GaAs$ superlattice data occurred for $c \approx 1.0$. It is interesting to note that the transport data of Schirber et al. (1985) give a Fermi energy (E_F) of 5.6 meV for the 2D hole gas in a pseudomorphic $In_{0.2}Ga_{0.8}As$ heterostructure. Since $E_F \ll \varepsilon \sim 48$ meV, the assumptions leading to Eq. (57) are satisfied, from which one estimates $c \approx 0.2$. Although the functional form of Eq. (57) is quite appealing, it underestimates the in-plane heavy-hole nonparabolicity. The source of this discrepancy is likely due to band-mixing effects, which play an important role in determining the in-plane dispersion of single and multiple-quantum well structures (see, e.g., Lee and Vassell, 1988). It should be noted that the band nonparabolicities (i.e., mass corrections) along k_z remain unchanged in going from the bulk to quantum well structures (Nelson et al., 1987), and hence Eq. (55) is expected to be valid in both bulk and quantum well strained-layer structures.

3. Kane Model in the Presence of Lattice-Mismatch Strain

To determine conduction band nonparabolicities near $\mathbf{k} = 0$, we introduce two degenerate spin functions having s-like ($L = 0$) orbital symmetry, namely $|S\uparrow\rangle$ and $|S\downarrow\rangle$ for the $\varnothing_c(\tfrac{1}{2}, \tfrac{1}{2})$ and $\varnothing_c(\tfrac{1}{2}, -\tfrac{1}{2})$ conduction band states, respectively. Again assuming $\mathbf{k} = (0, 0, k_{(z)})$, the resulting $8 \times 8 - \mathbf{k} \cdot \mathbf{p}$ Hamiltonian reduces to two equivalent 4×4 blocks. Lattice mismatched growth on (001) substrates gives rise to a doubly degenerate (4×4) determinant. Using the symmetrized basis functions of Luttinger and Kohn (1955), the following secular determinant is obtained for a *first-order* $\mathbf{k} \cdot \mathbf{p}$ treatment; given by People and Sputz (1990) as:

$$0 = \begin{bmatrix} E_G' - \lambda & \frac{1}{\sqrt{3}}kP & -i\sqrt{\frac{2}{3}}kP & 0 \\ \frac{1}{\sqrt{3}}kP & S_0 - \lambda & i\sqrt{2}\varepsilon & 0 \\ i\sqrt{\frac{2}{3}}kP & -i\sqrt{2}\varepsilon & L_0 - \lambda & 0 \\ 0 & 0 & 0 & H_0 - \lambda \end{bmatrix}. \qquad (58a)$$

Here $k = k_z$ and $P = \hbar/m_0 \langle S|p_z|Z \rangle$. Note that in the present case the eigenvalues $\lambda_j = E_j - \hbar^2 k^2/2m_0$ and the diagonal terms S_0, L_0, and H_0 are calculated to first order in $\mathbf{k} \cdot \mathbf{p}$, and hence contain no k-dependences. The bandgap E'_G is given by

$$E'_G \equiv E_G(0) + (\Xi_d + \tfrac{1}{3}\Xi_u - a_v)\mathrm{Tr}(\overset{\leftrightarrow}{\mathbf{e}}). \tag{58b}$$

Here $E_G(0)$ is the bandgap of the unstrained epitaxial layer with the second term giving the hydrostatic contribution to the bandgap of this biaxially strained epilayer.

It is noted that again the $M_j = \pm\tfrac{3}{2}$ heavy hole state remains strictly parabolic. This is to be expected since the higher lying conduction band states (having symmetry Γ^c_8, etc.) have been ignored on the present model. Therefore, the heavy-hole dispersion is given as $\hbar^2 k^2/m_0$. A more realistic expression for the heavy-hole dispersion may be obtained by including terms in $\tilde{H}_0$ to second-order, in which case one obtains Eq. (52a). Conduction, light-hole, and split-off hole states are coupled by the remaining 3×3 determinant, which leads to the secular equation:

$$0 = (E'_G - \lambda)\{(S_0 - \lambda)(L_0 - \lambda) - 2\varepsilon^2\}$$
$$- \tfrac{1}{3}(kP)^2 \{L_0 + 2S_0 + 4\varepsilon - 3\lambda\}. \tag{58c}$$

It should be noted that Eq. (58c) reduces to Kane's (1957) secular equation in the limit $\varepsilon \to 0$. The dispersion relations contained in Eq. (58c) may be obtained either by implicit differentiation or by iteration. Using iteration, one assumes zeroth order solutions of the form: $E_c(0) = E'_G$ and $E_v^{(0)}(\pm)$. The conduction band dispersion relation to order k^4 is:

$$E_c(k) = E'_G + \frac{\hbar^2 k^2}{2m^*_e(\varepsilon)} + E^{cb}_4 k^4 + \cdots, \tag{59a}$$

where

$$\left(\frac{m_0}{m^*_e(\varepsilon)}\right) \equiv \left[1 + \left(\frac{m_0}{m^*_e(0)} - 1\right)\left(\frac{1+y}{1+\tfrac{2}{3}y}\right)\left\{\frac{E_G(0)(E'_G + \tfrac{2}{3}\Delta_0 - \varepsilon)}{[(E'_G)^2 + E'_G(\Delta_0 + \varepsilon) + \Delta_0 \varepsilon - 2\varepsilon^2]}\right\}\right],$$
$$\tag{59b}$$

and

$$E^{cb}_4 = -\left(\frac{\hbar^4}{4m_0^2}\right)\left(\frac{m_0}{m^*_e(\varepsilon)} - 1\right)$$
$$\times \frac{\left[(2E'_G + \Delta_0 + \varepsilon)\left(\frac{m_0}{m^*_e(\varepsilon)} - 1\right) - E^{(0)}_G\left(\frac{1+y}{1+\tfrac{2}{3}y}\right)\left(\frac{m_0}{m^*_e(0)} - 1\right)\right]}{[(E'_G)^2 + E'_G(\Delta_0 + \varepsilon) + \Delta_0 \varepsilon - 2\varepsilon^2]}.$$
$$\tag{59c}$$

Here $y = \Delta_0/E_G(0)$. Note that in the limit $\varepsilon \to 0$, Eq. (59) reduces to a form similar (but not identical) to the expression obtained by Raymond *et al.* (1979).

Similarly to order k^4, the light-hole dispersion relation in the strained bulk is given by:

$$E_{lh}(k) = E_V^{(0)}(+) - \frac{\hbar^2 k^2}{2m_{lh}^*(\varepsilon)} + E_4^{lh} k^4 + \cdots, \tag{60a}$$

$$\left(\frac{m_0}{m_{lh}^*(\varepsilon)}\right) = \left[-1 + \frac{1}{4}\left(\frac{m_0}{m_{lh}^*(0)} + 1\right)\left(\frac{E_G(0)}{E_G' - E_V^{(0)}(+)}\right)\left\{3 + \frac{(1-9x)}{F(x)}\right\}\right], \tag{60b}$$

and

$$E_4^{lh} = \left(\frac{\hbar^4}{4m_0^2}\right)\left(\frac{m_0}{m_{lh}^*(\varepsilon)} + 1\right)$$

$$\times \frac{\left[\frac{3}{2}E_G(0)\left(\frac{m_0}{m_{lh}^*(0)} + 1\right) - (E_G' - E_V^{(0)}(+) - \Delta_0 F(x))\left(\frac{m_0}{m_{lh}^*(\varepsilon)} + 1\right)\right]}{[(E_G' - E_V^{(0)}(+))\Delta_0 F(x)]}, \tag{60c}$$

where $x = \varepsilon/\Delta_0$, $F(x) = [1 - 2x + 9x^2]^{1/2}$, $(m_0/m_{lh}^*(0)) = (\gamma_1 + 2\gamma_2)$, and $E_V^{(0)}(\pm)$ has been defined in Eqs. (33b) and (33c).

The split-off hole dispersion relation to order k^4 is given by

$$E_{SO}(k) = E_V^{(0)}(-) - \frac{\hbar^2 k^2}{2m_{SO}^*(\varepsilon)} + E_4^{SO} k^4 + \cdots, \tag{61a}$$

where

$$\left(\frac{m_0}{m_{SO}^*(\varepsilon)}\right) = \left[-1 + \frac{1}{2}\left(\frac{m_0}{m_{SO}^*(0)} + 1\right)\left(\frac{E_G(0) + \Delta_0}{E_G' - E_V^{(0)}(-)}\right)\left\{3 - \frac{(1-9x)}{F(x)}\right\}\right], \tag{61b}$$

and

$$E_4^{(SO)} = -\left(\frac{\hbar^4}{4m_0^2}\right)\left(\frac{m_0}{m_{SO}^*(\varepsilon)} - 1\right)$$

$$\times \frac{\left[3E_G(0)(1+y)\left(\frac{m_0}{m_{SO}^*(0)} + 1\right) - (E_G' - E_V^{(0)}(-) + \Delta_0 F(x))\left(\frac{m_0}{m_{SO}^*(\varepsilon)} + 1\right)\right]}{[(E_G' - E_V^{(0)}(-))\Delta_0 F(x)]}, \tag{61c}$$

where $(m_0/m_{SO}^*(0)) = \gamma_1$.

It is noted that the introduction of strain into the Kane model leads to light and split-off hole band-edge masses, which may differ appreciably from the (6 × 6) valence band model, wherein the effects of the lower-lying conduction band edges are ignored. Differences are also noted for band nonparabolicities.

III. Zone-Folding Effects in Ultrashort-Period Strained-Layer Superlattices

There is considerable interest at present in ultrashort-period strained-layer superlattices from indirect gap materials, particularly Si, Ge, and (Ge, Si) alloys due to the potential for enhancing both the optical absorption coefficient near the bandgap and possibly radiative recombination rates, in this technologically important materials system. The most intriguing question is that of bandgap conversion—from indirect-gap Si and Ge to a quasidirect gap due to structurally induced foldings of the $E(\mathbf{k})$ relation along the direction in k-space that is complementary to the superlattice growth direction.

Bevk et al. (1986) demonstrated the ordered growth of elemental Ge/Si strained-layer superlattices on Si (001) having periods of atomic-monolayer dimensions. The first evidence for structurally induced modifications of the Si and Ge band structures were reported in the electro-reflectance measurements of Pearsall et al. (1987) on superlattices having periods consisting of four monolayers (ML) of Si and four ML of Ge on a Si (001) substrate. These measurements indicated the occurence of superlattice optical transitions centered near 0.76, 1.25, 1.8, and 2.3 eV. These energies are characteristic of neither the cubic Si nor the strained Ge. People and Jackson (1987) have shown that the near-band-edge transitions energies are well described by using an effective mass envelope function approach wherein square well potentials are generated by using band-extrema energies calculated by using phenomenological deformation-potential theory in conjunction with the valence-band offsets of Van de Walle and Martin (1985), which are calculated by using *ab initio* pseudopotentials within the local-density approximation. Hybertsen et al. (1988) accounted for observed optical transitions in the (4 × 4)-ML Si/Ge SLS by using a first-principles quasiparticle (excitation) energy approach. Quasiparticle and effective-mass-transition energies were generally in good agreement. The superlattice band structure for the (4 × 4)-ML Si/Ge SLS has been calculated by using a tight-binding approach by Brey and Tejedor (1987) and using local-density-functional (LDA) methods by Froyen et al. (1987). Complications that arose due to the inability of LDA methods to correctly estimate bandgaps were avoided in the superlattice band-structure calculations of Hybertsen and Schlüter (1987). The effective-mass estimates of superlattice transition energies will be briefly reviewed in the next section.

A. Effective-Mass Estimates of Optical Transition Energies in the (4 × 4)-ML Si/Ge SLS on Si (001)

The first question that arises in the application of a bulklike approach (such as the effective-mass envelope function approach) to atomic-layer dimension structures, is one of whether such an approach remains valid on these length scales. People and Jackson (1987) have pointed out the concept of zone folding, and carrier confinement remains applicable in these short-period superlattices. Their analysis relied upon results obtained from an earlier full SCIC of the ground-state properties of the (4 × 4)-ML Si/Ge structure, performed by Van de Walle and Martin (1985) using *ab initio* pseudopotentials. The SCIC clearly showed that charge densities and potentials within the (4 × 4)-ML supercell were identical to those of cubic Si and coherently strained Ge, except for a transition region confined to one monolayer on either side of the heterointerface. This result was extremely important in that it established that for the (4 × 4)-ML structure, the respective material components could be characterized by their associated bulk parameters (bandgaps, etc.); however, the potential step (ΔE_v) across the interface still required a fully self-consistent ground-state calculation.

One-dimensional superlattice potentials were then constructed for symmetry-allowed couplings of band-edge (extrema) states in the Si and Ge layers. The k-space character of the various band extrema were determined by folding of the Si and Ge bulk band structures along $\mathbf{k}_\perp = (0, 0, k)$, the superlattice-equivalent momentum direction for growth on Si (001). Note that the cubic Si and strained Ge have similar conduction-band structures for growth on Si substrates (see, e.g., People *et al.*, 1987). In particular, the six lowest Si conduction-band minima may be grouped in two classes for superlattices grown on (001) substrates. The first class consists of the two valleys along $\langle 001 \rangle$ in momentum space. These experience the superlattice potential (i.e., they undergo zone folding) and are denoted by $\Delta_1^\perp$. The second class consists of the four (in-plane) valleys, normal to $\langle 001 \rangle$, which are denoted by $\Delta_1^\parallel$. Similar conditions apply for X-point minima. In the case of the (4 × 4)-ML Si/Ge superlattice, the minizone boundaries occur at $\pm \pi/d$, where $d \approx 2a_0$, and a_0 is the bulk lattice parameter of Ge. Note that $\pm \pi/2a_0 = \pm(1/4)(2\pi/a_0)$, where $2\pi/a_0$ is the magnitude of the wave vector from the zone center to the X-Brillouin-zone boundary. Here X denotes any of the six $\mathbf{k}$-vectors of the form $2\pi/a_0\langle 1, 0, 0 \rangle$. Therefore, the superlattice zone-boundary wave vector is approximately (1/4) of the bulk (three-dimensional) zone. Hence the dispersion relation along $k_\perp$ experiences two-folding. These are shown schematically in Fig. 17 for the Si-like conduction-band structure. Note that the second folding of the bulk dispersion relation gives rise to a new zone-folding-induced quasidirect gap denoted by $\bar{\Gamma}_2(\bar{X})$. The single representations associated with the bands in cubic Si are used for

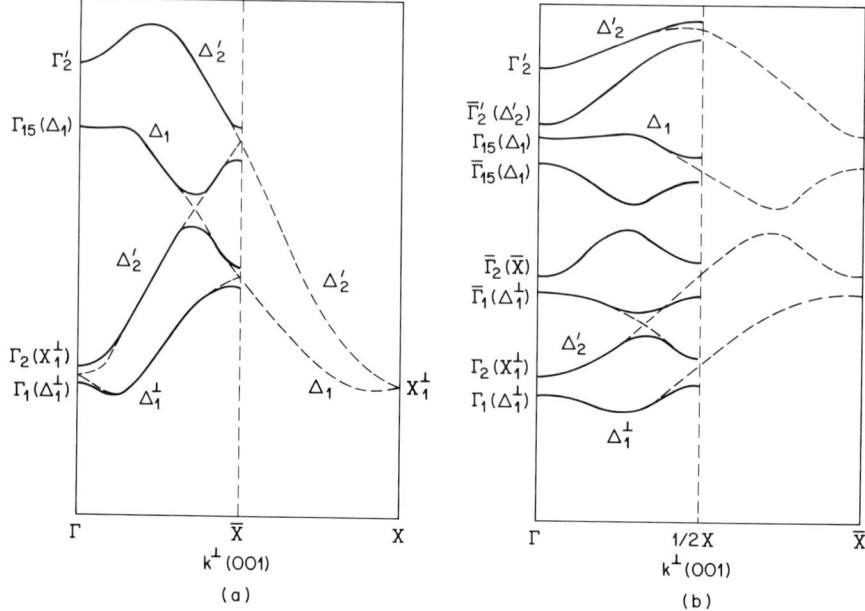

FIG. 17. (a) First folding of the bulk $E(\mathbf{k})$ relation in the (4×4)-ML Si/Ge SLS. (b) Second folding the the bulk $E(\mathbf{k})$.

identification. Note that in Fig. 17a the zone-folded Δ_2'-band (solid curve) would intersect the not yet folded Δ_1-band originating from Γ_{15}. Although such an intersection is allowed under cubic symmetry, it is strictly forbidden under the reduced (C_{2v}) symmetry associated with the Δ-direction for the (4×4) SLS. Indeed Δ_1 and Δ_2' both transform as the totally symmetrical (A_1) representation of C_{2v}. Note further that the zone-fold $\Delta_\perp^\perp$-bands have extrema near, but not at $\mathbf{k} = 0$. The four in-plane $\Delta_\parallel^\parallel$ valleys (not shown) are unaffected by the $\langle 001 \rangle$ superlattice potential. Note that Fig. 17 is only useful for determining the $\mathbf{k}$-space character of the zone-folded bulk bands. The actual energy positions of these extrema must be calculated in the presence of the associated confining superlattice potentials. This is also the case for the actual superlattice dispersion; however, details of the actual dispersion relation are not required if one is interested only in energy positions of band extrema, as is the case at present.

Superlattice-allowed band-edge couplings are determined by requiring that: (i) in-plane crystal momentum ($\mathbf{k}_\parallel$) is conserved, and (ii) wave-function overlap between barrier and well states must be allowed by symmetry. Various symmetry-allowed conduction-band couplings are shown in Fig. 18 for the D_{2d} symmetry associated with the (4×4)-SLS unit cell. Note that in

4. STRUCTURALLY INDUCED STATES

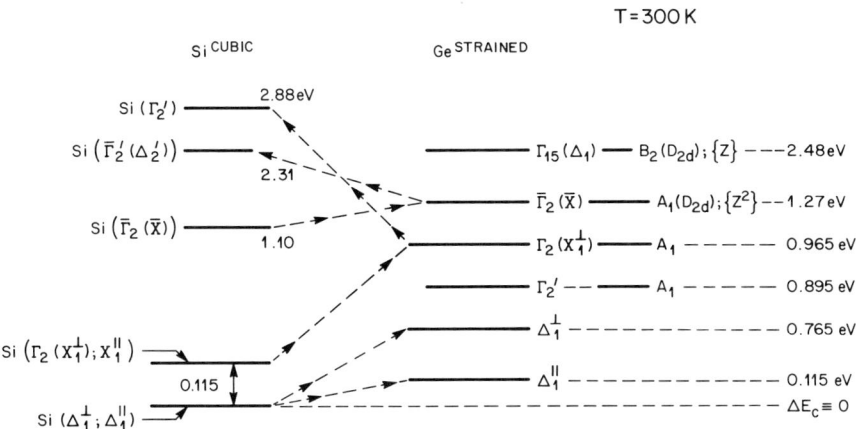

FIG. 18. Various symmetry-allowed conduction-band couplings for the (4 × 4)-ML Ge/Si SLS, under the D_{2d} symmetry of the SLS unit cell.

the strained Ge, the $\Delta_1^\perp$ and $\Delta_1^\parallel$ conduction-band extrema are split by 0.6 eV, whereas these extrema remain degenerate in the cubic Si. The upper valence band in the cubic Si remains fourfold degenerate, having $J = 3/2$ character. The coherence strain splits the $J = 3/2$ valence band of Ge into a set of doublets separated by 0.15 eV, which we denote as $(3/2; \pm 3/2)$ and $(3/2; \pm 1/2)$. SCICs (see Van de Walle and Martin, 1985) place the Ge $(3/2; \pm 3/2)$ upper valence-band edge at 0.84 eV above the Si $(J = 3/2)$ valence-band edge.

Lower-lying $q = 0$ superlattice optical-transition energies are then calculated by using an effective-mass envelope function approximation in the limit of decoupled bands (see, e.g., Bastard and Brum, 1986). The results are summarized in Table II. Note that the lowest-lying optical transitions are centered near 0.76 eV, 1.03 eV, 1.20 eV, 1.89 eV, and 2.31 eV. The transitions near 0.76 and 1.03 eV are strictly indirect within the present model. This conclusion was also reached in the calculation of the full superlattice band structure (Hybertsen and Schlüter, 1987). Zone-folding-induced quasidirect superlattice transitions occur at 1.20 eV and 1.89 eV, with the transition near 2.31 eV being attributed to purely direct transitions involving the nearly degenerate Ge (Γ_2'), Ge $[\Gamma_2(X_1^\perp)]$ well extreme to the Si (Γ_2') barrier extremum.

Since it has been assumed that the in-plane momentum $(\mathbf{k}_\parallel)$ is conserved, the transitions centered at 0.76 eV are indirect and arise entirely from non-zone-folded bulk bands. Intervalley coupling mechanisms, such as surface steps, would relax this $\mathbf{k}_\parallel$ conservation rule, in which case the Ge $(\Delta_1^\parallel)$ state could act as a barrier for the zone-folded Si $(\Delta_1^\perp$ state shown in Fig. 18. The

TABLE II

Lowest Energy q = 0 Superlattice Optical Transitions for the (4 × 4)-Monolayer Si/Ge Strained-layer Superlattice on Si(001), for Light Polarized in the (x, y) Plane. The Material that Acts as the Quantum Well has been Specified in the Last Three Columns

C.B. State V.B. State	$E_e^{(0)}(\Delta_1^{\parallel})$	$E_e^{(0)}(\Delta_1^{\perp})$	$E_e^0\{Si-\Gamma_2(X_1^{\perp}); \Gamma_2(X_1^{\perp})\}$	$E_e^0\{Si(\bar{\Gamma}_2); \bar{\Gamma}_2(\bar{X})\}$	$E_e^0 Ge\{\Gamma_2(X_1^{\perp}); \Gamma_2'\}$ $E_e^0\{Ge(\bar{\Gamma}_2); Si(\bar{\Gamma}_2)\}$
$(\frac{3}{2}; \pm\frac{3}{2})$	0.71 eV	0.97	1.15	1.84	2.26
$(\frac{3}{2}; \pm\frac{1}{2})$	0.82 eV	1.08	1.26	1.95	2.37
Character	Indirect	Z.F. Indirect	Z.F. Quasi-direct	Z.F. Quasi-direct	Direct
C.B. State Decay Length (Å)	24.0 Å	4.0 Å	4.3 Å	9.5 Å	3.5 Å

resulting zone-folded transition would again be centered near 0.76 eV, however it would remain indirect in character (in a manner similar to the 1.03 eV transition). Note further that the zone-folded Si ($X_1^\perp$) extremum is degenerate with the Ge ($\Delta_1^\parallel$) extremum, as shown in Fig. 18. In the absence of $k_\parallel$ conservation, these states would strongly admix, thus giving rise to a quasidirect superlattice transition near 0.8 eV. Other possible effects that might lead to a direct characterlike transition near 0.8 eV include a finite number of superlattice periods and the presence of an electric field in the superlattice region associated with the electro-reflectance modulation technique.

B. Quasidirect Gap Structures in Monolayer Si/Ge Elemental Strained-Layer Superlattices

Miniband formation in indirect gap superlattice materials was considered by Gnutzmann and Clausecker as early as 1974. In their analysis, it was shown how an indirect gap material having its conduction-band extrema at the X-Brillouin-zone edge (e.g., gap) might be transformed into a quasidirect gap material if the superlattice minizone boundary occurred at $(X/2)$, i.e., if the material experienced a single folding. The required superlattice period (d) is related to the bulk lattice parameter via the relation of $q = (X/2)$ or $\pi/d = (\frac{1}{2})(2\pi/a_0)$, so that $d = a_0$. Since $a_0 = 4$ ML, such an effect should be observable in a 4-ML-period superlattice. In the case of Si and biaxially compressed Ge, the conduction-band extrema do not occur exactly at the bulk X-zone boundary, but rather at $\Delta_0 \approx 0.8\,X$. Jackson and People (1986) performed a zone-folding analysis of such an indirect gap semiconductor superlattice. In summary, the generation of a quasidirect gap superlattice is expected to occur for $q = \pi/d = (\frac{1}{5})X$ or $d \approx \frac{5}{2}a_0 \approx 10$ ML. The possibility for observing such a quasidirect gap superlattice in the ($d = 10$ ML)-structures is further enhanced if growth parameters are adjusted to provide a small amount of in-plane tension in the Si-quantum well. This tension would cause the Si ($\Delta_1^\perp$) and Si ($\Delta_1^\parallel$) states to split, with the zone-folded Si ($\Delta_1^\perp$) state lying lower in energy. The magnitude of this splitting is $\sim |\Xi_u^{(\Delta)}(\text{Si}) \cdot (e_{zz} - e_{xx})|$, where $\Xi_u^{(\Delta)}(\text{Si}) \sim 9$ eV.

IV. Other Effects in Strained-Layer Heterostructures

A. Piezoelectric Effects in Strained-Layer Heterostructures

1. *III–V Strained-Layer Structures*

Smith and Mailhiot (1987) have shown that large internal piezoelectric fields are generated when SLS are grown from either III–V or II–VI semiconductors (which possess a nonvanishing piezoelectric constant e_{14}). Lattice-mismatch strain induces polarization fields, P_i, which in turn generate electric fields. In III–V SLS structures, these strain-induced fields approach 10^5 V/cm for lattice-mismatch strains $\sim 1\%$, and hence significantly modify the superlattice electronic structure and optical properties. Note, for example, that the electric fields will change the shape of the quantum well potentials, thereby changing subband energy levels and wave functions. Consequently, one obtains modifications of optical transition energies and oscillator strengths (which reflect wave-function overlaps). The magnitude of these changes have been shown to be second order in the magnitude of the electric fields (second-order Stark effect). The internal electric fields may be externally modulated by either: (i) photo-pumping (since photogenerated carriers screen these internal fields), (ii) external electric fields, and (iii) external strain. As a result of such modulations, one expects to observe large nonlinear optic, electro-optic, and piezo-optic effects.

In zinc-blende-structure semiconductors, off-diagonal strains ε_{jk} induce polarization fields, P_i, whose magnitude is given by

$$P_i = 2e_{14}\varepsilon_{jk}.$$

Note that diagonal strains (ε_{xx}, ε_{yy}, ε_{zz}) do not induce a polarization, since $\varepsilon_\parallel = 0$ in these materials. Therefore, SLS structures grown on (001)-oriented substrates will not give rise to strain-induced polarization fields. For a (111) growth axis, $\mathbf{P}_i$ is along the superlattice direction, whereas for (110) growths, $\mathbf{P}_i$ lies in the superlattice plane.

In (111) SLS structures, Smith and Mailhiot (1987) have shown that very strong features occur in the third-order nonlinear susceptibility $\chi^{(3)}$ (ω; ω, ω, $-\omega$) ~ 0.1 esu near the HH1 $\rightarrow$ CB1 and HH2 $\rightarrow$ CB1 excitonic transitions. It has been further shown that index-of-refraction changes are expected when a (111) SLS structure is modulated with an external field. As a result, a *linear electro-optic* effect is expected in (111) SLS structures. For example in $In_{0.53}Ga_{0.47}As/In_{0.3}Al_{0.7}As$ (111) SLS structures, it is estimated that for an externally applied modulation field of 40 kV/cm, the linear electro-optic coefficient r can obtain values in excess of 5×10^{-8} cm/V. By comparison, in potassium dihydrogen phosphate (KDP), the coefficient r is $\sim 10^{-9}$ cm/V, whereas $|r| \sim 1.4 \times 10^{-10}$ cm/V in bulk GaAs.

2. *Group-IV Strained-Layer Structures*

In zinc-blend structures such as GaAs, the inversion symmetry of the unit cell is broken due to the two different kinds of atoms in the covalent bond. This allows for existence of a linear electro-optic effect due to the nonvanishing of the susceptibility $\chi^{(2)}(\omega)$. In bulk Si and Ge, the existence of a center of inversion causes $\chi^{(2)}$ to vanish identically, since $\chi^{(2)}$ is a third-rank tensor and hence transforms as the cube power of the band coordinate. It has been pointed out by Friedman and Soref (1986) that the monolayer Si/Ge pseudomorphic monolayer structures which should show a linear electro-optic effect.

In particular the (2×2)-ML elemental Si/Ge superlattice on Si (111) substrates gives rise to a pseudo-zinc-blende structure in which the atom arrangement in $\langle 111 \rangle$ planes follows the sequence Si Si Ge Ge Si Si.... Estimates for the magnitude of the linear electro-optic coefficient, r, range from 0.7 to 1.6×10^{-10} cm/V, which is comparable to the value of 1.4×10^{-10} cm/V in bulk GaAs. Such optically active SLS Ge/Si structures might be useful as active elements in rib wave-guide p-i-n electro-optic phase modulators.

V. Summary

In this review, we have attempted to give a flavor of the many facets of strained-layer heteroepitaxy relevant to pseudomorphic growth, characterization, and device applications of such structures. It is our sincere hope that the rudimentary manner in which we have approached this subject will serve to enhance the expertise of present and future workers in this field.

Acknowledgments

We would like to thank our many colleagues and collaborators in this field. D. V. Lang, V. Narayamurti, John C. Bean, and T. P. Pearsall were a source of much inspiration during our early work on (Si, Ge) strained-layer heterostructures. H. L. Störmer introduced us to the wealth of knowledge obtainable in 2D-transport studies and has been the source of many stimulating conversations. John Hensel has provided much insight into strain effects in Si and Ge *p*-type structures. M. Schlüter, M. Hybertsen, Chris Van de Walle, and G. Baraff have provided much insight into *ab initio* pseudo-potential calculations. We would also like to thank J. M. Gibson and R. E. Slusher for encouragement during the course of preparing this manuscript. Many authors have kindly provided preprints of their work, for which we are greatly appreciative. We would also like to acknowledge Carolyn Heaps for the timely preparation of this manuscript.

References

Abstreiter, G., Brugger, H., and Wolf, T. (1985). *Phys. Rev. Lett.* **54**, 2441.
Alexander, H., and Haasen, P. (1968). "Solid State Physics," Vol. 22. Academic, New York, pp. 27–158.

Bachelet, G. B., Hamann, D. R., and Schluter, M. (1982). *Phys. Rev.* **B26**, 4199.
Baldereschi, A. (1973). *Phys. Rev.* **B7**, 5212.
Baraff, G. A., Appelbaum, J. A., and Hamann, D. R. (1977). *Phys. Rev. Lett.* **38**, 237.
Bastard, G., and Brum, J. A. (1986). *IEEE J. Quantum Electron.* **QE-22**, 1625.
Bean, J. C., Feldman, L. C., Fiory, A. T., Nakahara, S., and Robinson, I. K. (1984). *J. Vac. Sci. Technol.* **A2**, 436.
Berner, K., and Alexander, H. (1967). *Acta Met.* **15**, 933.
Bevk, J., Mannaerts, J. P., Feldman, L. C., Davidson, B. A., and Ourmazd, A. (1986). *Appl. Phys. Lett.* **49**, 286.
Bir, G. L., and Pikus, G. E. (1974), "Symmetry and Strain-Induced Effects in Semiconductors," J. Wiley, New York.
Braunstein, R., Moore, A. R., and Herman, F. (1958). *Phys. Rev.* **109**, 695.
Brey, L., and Tejedor, C. (1987). *Phys. Rev. Lett.* **59**, 1022.
Cardona, M., and Christensen, N. E. (1987). *Phys. Rev.* **B35**, 6182.
Chadi, D. (1979). *Phys. Rev.* **B19**, 2074.
Chelikowsky, J. R., and Cohen, M. L. (1976). *Phys. Rev.* **B14**, 556.
Dodson, B. W., and Taylor, P. A. (1986). *Appl. Phys. Lett.* **49**, 642.
Dodson, B. W., and Tsao, J. Y. (1988a). *Appl. Phys. Lett.* **52** (10), 852.
Dodson, B. W., and Tsao, J. Y. (1988b). *Appl. Phys. Lett.* **51** (17), 1325.
Dresselhaus, G., Kip, A. F., and Kittel, C. (1955). *Phys. Rev.* **98**, 368.
Fiory, A. T., Bean, J. C., Hull, R., and Nakahara, S. (1985). *Phys. Rev.* **B31**, 4063.
Forrest, S. R., Schmidt, P. H., Wilson, R. B., and Kaplan, M. L. (1984). *Appl. Phys. Lett.* **45**, 1199.
Frank, F. C., and Van der Merwe, J. H. (1949). *Proc. Roy. Soc. London* **A198**, 216.
Frensley, W. R., and Kroemer, H. (1976). *J. Vac. Sci. Technol.* **13**, 810.
Frensley, W. R., and Kroemer, H. (1977). *Phys. Rev.* **B16**, 2642.
Friedman, L., and Sorif, R. A. (1986). *Electronics Lett.* **22**, 819.
Fritz, I. J. (1987). *Appl. Phys. Lett.* **51** (14), 1080.
Fritz, I. J., Picraux, S. T. Dawson, L. R. Drummond, T. J. Laidig, W. D., and Anderson, N. G. (1985). *Appl. Phys. Lett.* **46**, 967.
Fritz, I. J., Gourley, P. L., and Dawson, L. R. (1987). *Appl. Phys. Lett.* **51** (13), 1004.
Froyen, S., Wood, D. M., and Zunger, A. (1987). *Phys. Rev.* **B36**, 4547.
Gershoni, D., Vandenberg, J. M., Hamm, R. A., Temkin, H., and Panish, M. G. (1987). *Phys. Rev.* **B36**, 1320.
Gershoni, D., Temkin, H., Vandenberg, J. M., Chu, S. N. G., Hamm, R. A., and Panish, M. B. (1988). *Phys. Rev. Lett.* **60**, 448.
Gnutzmann, V. and Clausecker, K. (1974). *Appl. Phys.* **3**, 9.
Gourley, P. L., and Biefeld, R. M. (1982). *J. Vac. Sci. Technol.* **21** (2), 473.
Grabow, M. H., and Gilmer, G. H. (1987). "Initial Stages of Epitaxial Growth," p. 15. Materials Research Society, Pittsburgh.
Harrison, W. A. (1980). "Electronic Structure and the Properties of Solids," p. 253. Freeman, San Francisco.
Harrison, W. A., and Tersoff, J. (1986). *J. Vac. Sci. Technol.* **B4**, 1068.
Hasegawa, H. (1963). *Phys. Rev.* **129**, 1029.
Hensel, J. C., and Feher, G. (1963). *Phys. Rev.* **129**, 1041.
Herring, C., and Vogt, E. (1956). *Phys. Rev.* **101**, 944.
Hobart, R. (1965). *J. Appl. Phys.* **36**, 1944.
Hull, R., Bean, J. C., Cerdeira, F., Fiory, A. T., and Gibson, J. M. (1986). *Appl. Phys. Lett.* **48**, 56.
Hybertsen, M. S., and Schlüter, M. (1987). *Phys. Rev.* **B36**, 9683.
Hybertsen, M. S., Schlüter, M., People, R., Jackson, S. A., Lang, D. V., Pearsall, T. P., Bean, J. C., Vandenberg, J. M., and Bevk, J. (1988). *Phys. Rev.* **B37** (10), 195.

Jackson, S. A., and People, R. (1986). *Mater. Res. Soc. Symp. Proc.* **56**, 365.
Jesser, W. A., and Kuhlmann-Wilsdorf, D. (1967). *Phys. Status Solidi* **19**, 95.
Jorke, H., and Herzog, H. J. (1985). *Proc. 1st Int. Symposium on Silicon Molecular Beam Epitaxy* (J. C. Bean, ed.), p. 352. Electrochemical Society, Pennington, New Jersey.
Kane, E. O. (1957). *J. Phys. Chem. Solids* **1**, 249.
Kasper, E. (1985). *Proc. 2nd Int. Conf. Modulated Semiconductor Structures*, p. 703. Kyoto, Japan.
Kasper, E., Herog, H. J., and Kibbel, H. (1975). *Appl. Phys.* **8**, 199.
Kleiner, W. H., and Roth, L. M. (1959). *Phys. Rev. Lett.* **2**, 334.
Kleinman, L. (1981). *Phys. Rev.* **B24**, 7412.
Kohama, Y., Fukuda, Y., and Seki, M. (1988). *Appl. Phys. Lett.* **52** (5), 380.
Kuech, T. F., Maenpaa, M., and Lau, S. S. (1981). *Appl. Phys. Lett.* **39**, 245.
Kuk, Y., Feldman, L. C., and Silverman, P. J. (1983). *Phys. Rev. Lett.* **50**, 511.
Kunc, K., and Martin, R. M. (1981). *Phys. Rev.* **B24**, 3445.
Kuo, C. P., Vong, S. K., Cohen, R. M., and Stringfellow, G. B. (1985). *J. Appl. Phys.* **57** (12), 5428.
Lawaetz, P. (1971). *Phys. Rev.* **B4**, 3460.
Lang, D. V., People, R., Bean, J. C., and Sergent, A. M. (1985). *Appl. Phys. Lett.* **47** (12), 1333.
Lang, D. V., Panish, M. B., Capasso, F., Allan, J., Hamm, R. A., Sergent, A. M., and Tsang, W. T. (1987). *Appl. Phys. Lett.* **50** (12), 736.
Lee, J., and Vassell, M. O. (1988). *Superlattices and Microstructures* **4** (3), 311.
Lievin, J.-L., and Fonstad, C. G. (1987). *Appl. Phys. Lett.* **51**, 1173.
Luryi, S., Kastalsky, A., and Bean, J. C. (1984). *IEEE Trans. on Electron Devices*, **ED-31**, 1135.
Luryi, S., Pearsall, T. P., Temkin, H., and Bean, J. C. (1986). *Electron. Device Lett.* **EDL-7**, 104.
Luttinger, J. M. (1956). *Phys. Rev.* **102**, 1030.
Mahowald, P. H., List, R. S., Spicer, W. E., Woicik, J., and Pianetta, P. (1985). *J. Vac. Sci. Technol.* **B3**, 1252.
Margaritondo, G., Katnani, A. D., Stoffel, N. G., Daniels, R. R., and Zhao, T.-X. (1982). *Solid State Commun.* **43**, 163.
Martin, R. M., and Van de Walle, C. G. (1985). *Bull. Am. Phys. Soc.* **30**, 226.
Matthews, J. W. (1975). *J. Vac. Sci. Technol.* **12** (1), 126.
Matthews, J. W., and Blakeslee, A. E. (1974). *J. Cryst. Growth* **27**, 118.
Matthews, J. W., and Blakeslee, A. E. (1976). *J. Cryst. Growth* **32**, 265.
Matthews, J. W., and Blakeslee, A. E. (1977). *J. Vac. Sci. Technol.* **14** (4), 989.
Matthews, J. W., Mader, S., and Light, T. B. (1970). *J. Appl. Phys.* **41** (9), 3800.
Matthews, J. W., Blakeslee, A. E., and Mader, S. (1976). *Thin Solid Films* **33**, 253.
Nelson, D. F., Miller, R. C., and Kleinman, D. A. (1987). *Phys. Rev.* **B35**, 7770.
Ni, W.-X., Knall, J., and Hansson, G. V. (1987). *Phys. Rev.* **B36**, 7744.
Osbourn, G. C., Biefeld, R. M., and Gourley, P. L. (1982). *Appl. Phys. Lett.* **41** (2), 172.
Osbourn, G. C. (1982). *J. Appl. Phys.* **53**, 1586.
Osbourn, G. C. (1983). *Phys. Rev.* **B27**, 5126.
Osbourn, G. C., Schirber, J. E., Drummond, T. J., Dawson, L. R. Doyle, B. L., and Fritz, I. J. (1986). *Appl. Phys. Lett.* **49** (12), 731.
Paul, W., and Warschauer, D. M. (1963). *"Solids Under Pressure,"* p. 226 McGraw-Hill, New York.
Pearsall, T. P., Temkin, H., Bean, J. C., and Luryi, S. (1986). *Electron. Device Lett.*, **EDL-7**, 330.
Pearsall, T. P., Bevk, J., Feldman, L. C., Bonar, J. M. Mannaerts, J. P., and Ourmazd, A. (1987). *Phys. Rev. Lett.* **58**, 729.
People, R. (1985). *Phys. Rev.* **B32**, 1405.
People, R. (1986a). *IEEE J. Quantum Electron.* **QE-22**, 1696.
People, R. (1986b). *J. Appl. Phys.* **59** (9), 3297.
People, R. (1987). *Appl. Phys. Lett.* **50** (22), 1604.

People, R., and Bean, J. C. (1985). *Appl. Phys. Lett.* **47**, 322.
People, R., and Bean, J. C. (1986a). *Appl. Phys. Lett.* **49** (E), 229.
People, R., and Bean, J. C. (1986b). *Appl. Phys. Lett.* **48** (8), 538.
People, R., and Jackson, S. A. (1987). *Phys. Rev.* **B36**, 1310.
People, R., and Sputz, S. K. (1990). To be published, *Phys. Rev. B*.
People, R., Bean, J. C., Lang, D. V. Sergent, A. M., Störmer, H. L., Wecht, K. W., Lynch, R. T., and Baldwin, K. (1984). *Appl. Phys. Lett.* **45**, 1231.
People, R., Bean, J. C., and Lang, D. V. (1985). *J. Vac. Sci. Technol.* **A3** (3), 846.
People, R., Bean, J. C., and Lang, D. V. (1987). *Proc. 18th Int. Conf. Physics Semiconductors*, (O. Engström, ed.), p. 767. World Scientific, Singapore.
Pickett, W. E., Louie, S. G., and Cohen, M. L. (1977). *Phys. Rev. Lett.* **39**, 109.
Pollak, F. H., and Cardona, M. (1968). *Phys. Rev.* **172**, 816.
Raymond, A., Robert, J. L., and Bernard, C. (1979). *J. Phys.* **C12**, 2289.
Schirber, J. E., Fritz, I. J., and Dawson, L. R. (1985). *Appl. Phys. Lett.* **46** (2), 187.
Smith, D. L., and Mailhiot, C. (1987). *Phys. Rev. Lett.* **58**, 1264.
Störmer, H. L., Dingle, R., Gossard, A. C., Wiegman, W., and Sturge, M. D. (1979). *Sol. State Commun.* **29**, 705.
Tejedor, C., and Flores, F. (1979). *J. Phys.* **C11**, L19.
Temkin, H., Pearsall, T. P., Bean, J. C., Logan, R. A., and Luryi, S. (1986a). *Appl. Phys. Lett.* **48**, (15) 963.
Temkin, H., Antreasyan, A., Olsson, N. A., Pearsall, T. P., and Bean, J. C. (1986b). *Appl. Phys. Lett.* **49** (13), 809.
Tersoff, J. (1984). *Phys. Rev.* **B30**, 4874.
Tersoff, J. (1985). *Phys. Rev.* **B32**, 6968.
Tersoff, J. (1987). "Heterojunctions: A Modern View of Band Discontinuities and Applications," (G. Margaritondo and F. Capasso, eds.), p. 3. North-Holland, Amsterdam.
Tsao, J. Y., Dodson, B. W., Picraux, S. T., and Cornelison, D. M. (1987). *Phys. Rev. Lett.* **59**, 2455.
Van de Walle, C. G. (1986). Ph.D. thesis, Stanford University (unpublished).
Van de Walle, C. G. (1988). *Mat. Res. Soc. Symp. Proc.* **102**, 565.
Van de Walle, C. G., and Martin, R. M. (1985). *J. Vac. Sci. Technol.* **B3**, 1256.
Van de Walle, C. G., and Martin, R. M. (1986). *Phys. Rev.* **B34**, 5621.
Van de Walle, C. G., and Martin, R. M. (1987). *Phys. Rev.* **B35**, 8154.
Van der Merwe, J. H. (1963). *J. Appl. Phys.* **34**, 123.
Vergés, J. A., Glotzel, D., Cardona, M., and Andersen, O. K. (1982). *Phys. Status Solidi* **B113**, 519.
Vook, R. W. (1982). *Int. Met. Rev.* **27**, 209.

CHAPTER 5

Microscopic Phenomena in Ordered Superlattices

*M. Jaros**

IBM THOMAS J. WATSON RESEARCH CENTER
YORKTOWN HEIGHTS, NEW YORK

I.	INTRODUCTION	175
II.	PHENOMENOLOGY OF THE BREAKDOWN OF THE EFFECTIVE-MASS APPROXIMATION IN SEMICONDUCTOR SUPERLATTICES	179
	A. Semiclassical Particle-in-a-Box Picture of Confinement . . .	179
	B. Breakdown Mechanisms	182
	C. Microscopic Formulation	187
III.	MICROSCOPIC THEORY OF ORDERED SUPERLATTICES	189
	A. Computational Techniques.	189
	B. Heterojunction Band Lineups.	194
	C. Structural Stability and Ordering Effects.	198
	D. Nonlinearity in Semiconductor Superlattices.	201
IV.	QUANTITATIVE ASSESSMENT OF MICROSCOPIC PHENOMENA . . .	203
	A. Introduction	203
	B. $GaAs$–$Ga_{1-x}Al_xAs$	204
	C. Si–$Si_{1-x}Ge_x$	221
	D. $Hg_{1-x}Cd_xTe/Hg_{1-y}Cd_yTe$ Superlattices	240
	E. Concluding Remarks.	253
	ACKNOWLEDGMENTS	254
	REFERENCES	254

I. Introduction

The concept of electronic structure engineering originated in the 1950s. It was realised soon after the discovery of binary compound semiconductors that suitable solid mixtures of such materials could be used to optimise the magnitude of the forbidden-gap and other important material parameters. The advent of low-loss optical fiber gave a fresh impetus to this effort and

---

*Permanent address: Physics Department, The University, Newcastle upon Tyne, United Kingdom.

resulted in a new technology (Pearsall, 1982) based on quaternary alloys (e.g., $Ga_{1-x}In_xAs_{1-y}P_y$) and high-quality InP substrates. Analogous efforts have been made to provide alloys for optical applications in the far-infrared and visible bands.

Concurrently, with the development of quaternary alloys, it became possible to grow quantum well structures, i.e., alternating layers of semiconductors of practically arbitrary thickness. In an alloy, the electronic structure is changed by a change in alloy composition. There is a simple approximately linear relationship between the bandgap and alloy composition. A similar relation exists between the bandgap and lattice constant. However, when a layer of GaAs is sandwiched between layers of, say, $Ga_{0.7}Al_{0.3}As$, the larger gap material (the alloy) acts as a potential barrier for electrons as well as for holes in GaAs. Then the electronic levels in the GaAs well, and consequently the magnitude of the forbidden gap, depend not only on the Al concentration but also on the thickness of the GaAs layer. Clearly, the bandgap engineering is taken one step further.

The electronic levels resulting from electron confinement in GaAs can be estimated from the one-dimensional particle-in-a-box model familiar from elementary texts. In this model (better known as the effective-mass approximation), the two semiconductors forming a microstructure such as a GaAs well are treated as if they were identical from a microscopic point of view. The only difference between them is an electrostatic potential barrier at the interface. This assumes that the periodic part of the electron wave function, which reflects the position and strength of the microscopic atomic potentials residing at the lattice sites in both GaAs and $Ga_{0.7}Al_{0.3}As$, does not change significantly as it crosses the interface between the two layers. We can describe this also by saying that an electron at the bottom of the conduction band of the GaAs layer approaching the interface undergoes a scattering event during which the magnitude of its bulk crystal momentum remains approximately the same. The effect of the interfaces is simply that standing waves are formed when the electron wavelength is a multiple of the width of the GaAs well. The effective-mass approximation proved extremely valuable in modelling of a wide variety of device structures, in the framework of well-established concepts accessible to the general reader (Jaros, 1989).

Since GaAs and $Ga_{0.7}Al_{0.3}As$ are very similar materials, it is not very surprising that such an approximation is useful. However, there are a number of structures in which the difference in the microscopic crystal potential at the interface is significant. Then an electron colliding with the atoms at the adjacent layer is scattered into quantum states of different bulk momenta, and the simple semiclassical picture of a particle in a box is no longer adequate. This concept is in fact familiar from the work on GaP doped with nitrogen (see, for example, Bergh and Dean, 1976) used to make light-emitting diodes. The potential of N is deeper than that of P so that an electron at the bottom of the conduction band of GaP approaching a N impurity experiences a scattering event during which a substantial mixing of bulk momenta takes place. We say that N in GaP behaves as a "deep" impurity. Strictly speaking, the difference of N and P potentials in GaP contains a number of strong short-wavelength Fourier components that are responsible for the mixing (Jaros, 1982). One such component mixes conduction-band states from the Brillouin-zone edge at X and those at the centre (at Γ) so that the N-impurity electron wave function contains a significant admixture of bulk Γ states. This means that an optical transition from the N level near the conduction-band edge to the top of the valence band at Γ is greatly enhanced. In the absence of the momentum mixing, the transition would be from X to Γ, i.e., forbidden. This would be the case if we doped GaP with a "shallow" dopant such as S or Se, whose electronic structure can be described by effective-mass theory akin to the particle-in-a-box concept and whose wave function is derived from the bottom of the conduction band of GaP at the X valley.

This kind of band-structure effect can be achieved in superlatices in many different ways. We can use the analogy with our impurity problem to understand the conceptual difference between the situation where the particle-in-a-box picture is appropriate and that where this band-structure effect or simply momentum mixing plays a part. For example, we can imagine that we form a GaAs/AlAs superlattice by starting with a bulk GaAs and by replacing Ga atoms with Al atoms at the relevant lattice sites. We know that an isolated Al impurity in bulk GaAs is a shallow center, i.e., that the difference between Ga and Al atomic potentials is small. Hence, the scattering event experienced by electrons in GaAs encountering Al atoms/impurities that occupy the same sites (the lattice constants of GaAs and AlAs are very much the same) normally leads to no significant momentum mixing. Let us contrast this situation with that of a Si/Ge superlattice. An electron in Si approaching the Ge layer sees a potential representing the difference between Si and Ge atoms as well as the potential difference due to the displacement of Ge from what would be its position in the Si lattice. This is because the lattice constants of Si and Ge are quite different. Furthermore, unlike the case of an isolated impurity, the Ge atoms form regularly spaced layers. The scattering is also affected by this order, which enhances constructive interference at certain wavelengths. All these factors affect the degree and the form of mixing of bulk momentum components. As in the case of GaP:N, the mixing of Γ states into the conduction-band wave function relaxes the selection rule for optical transitions across the gap, and the superlattice may behave as a quasidirect-gap material although its bulk constituents Si and Ge are indirect-gap materials.

This class of problems is the subject of this chapter: Our main task here is to show (1) when and to what extent the momentum-mixing effect takes place and how its magnitude depends on the chemical composition of the bulk constituents, lattice-mismatch, and ordering lengths, (2) what constitutes a correct way of modelling this effect, and (3) key systems and experimental situations where such effects have been demonstrated.

Since the chief motive for developing superlattices is to find an alternative to existing materials, for example alloys, it would be useful to obtain a simple means of comparing them. In the effective-mass picture, such a comparison is quite straightforward. In particular, it is possible to draw general guidelines concerning energy gaps versus composition, which depend only on bulk parameters and which specify the differences between a superlattice and an alloy of comparable composition. When momentum mixing takes place, the effective-mass picture of energy levels may, with some adjustments, remain approximately valid, i.e., to the extent to which it represents a correct dimensional analysis. However, phenomena that depend on the wave function, such as transition probabilities, often cannot be accounted for even qualitatively. This means that in such cases no clear-cut answer can be given to some of the most basic questions without a detailed calculation. Take, for instance, the concept of band offset. This is normally defined as the difference in energy of, say, conduction-band minima of the constituents taken at a given point in the bulk Brillouin zone (e.g., Γ, X). If, however, an electron colliding with atoms on the other side of the interface suffers excitations into states of different bulk momenta, the strength of the confining potential at one particular point in momentum space is not necessarily a reliable measure of the strength of all such excitations. This is particularly relevant in highly strained structures and when the atomic potentials of the constituents are very different. Furthermore, when we seek insights from large-scale calculations, we must bear in mind that they too may require an interpretation in terms of a model whose concepts often lie outside the formalism in question (for example, the first-principles local-density and quasiparticle calculations).

In Section II, we shall identify the causes of the breakdown of the effective-mass approximation. This will provide a basis for identifying the structures and materials where momentum mixing characteristic of novel ordered structures should occur.

In Section III, we develop a microscopic formulation of the problem and outline the general concepts and issues that dominate computational considerations, modelling, and design of microstructures. We shall show that the differences in atomic potentials and lattice spacing, and the effect of ordering over a critical length, combine to transform the wave functions of bulk bands and to create a new material species. We shall discuss the relative role of these contributions and compare superlattices with other materials, e.g., with alloys of analogous composition, and address the conceptual questions of heterojunction band offset, structural stability, ordering, and nonlinear response in superlattices.

In Section IV, we shall focus on quantitative assessment of some archetypal problems and structures where momentum mixing has been demonstrated by experimental data and/or by calculations. This is not an attempt to review the literature on the subject. It should serve as a means of illustrating the general issues outlined in Section III and as a detailed exposition of competing views.

We begin with the simplest example of momentum mixing that occurs when bulk states of different momenta (e.g., of Γ and X character) cross (i.e., become near-degenerate in energy) as a result of changes of alloy composition or pressure applied to GaAs-Ga$_{1-x}$Al$_x$As superlattices. The momentum mixing is best observed in luminescence as an enhancement of indirect $X-\Gamma$ cross-interface transitions.

Si/Ge superlattices represent an example where momentum mixing is dominated by both the difference in lattice spacing (strain) and by the order length (period) of the superlattice system. This leads to an enhancement of optical transitions across the gap.

Finally, HgTe/CdTe superlattices are used as an example of a system where the difference between atomic potentials (of Cd and Hg) is significant. This gives rise to a characteristic broadening of confined states in momentum space and links the position and localisation of the interface state closely to the interface condition.

The effect of momentum mixing is a unifying concept that enables us to appreciate the common origin of structural, compositional, and ordering-induced phenomena in terms of differences in microscopic crystal potentials, and cuts across the usual compartmentalisation of the field of semiconductor microstructures.

I. Phenomenology of the Breakdown of the Effective-Mass Approximation in Semiconductor Superlattices

A. SEMICLASSICAL PARTICLE-IN-A-BOX PICTURE OF CONFINEMENT

The simplest example of confinement in semiconductor nanostructures is a single-GaAs-quantum well (see review by Ando *et al.*, 1982).

Consider a slab of GaAs of thickness l sandwiched between two thick (semi-infinite) slabs of Ga$_{1-x}$Al$_x$As. Since the lattice constant of these semiconductors is very much the same, one can begin by assuming that an electron moving at the bottom of the conduction band of GaAs experiences at the interface a potential barrier whose height equals the difference between the energies of the bulk conduction-band minima of the constituent semiconductors at the centre of the Brillouin zone (Γ). The band structure of GaAs is shown in Fig. 1. To obtain this potential barrier—the band offset—we must correctly line up the band structures of GaAs and $Ga_{1-x}Al_xAs$. So long as we choose $x < 0.4$, the alloy is a direct-gap material (Fig. 2) so that the electron crystal momentum in the direction (z) perpendicular to the interface remains unchanged. However, even if we choose $x > 0.4$, it is only the energy difference between bulk states of the same linear momentum (k vector) that matters when it comes to constructing the potential barrier. We must also assume that the interface is perfect and free of impurities and that the alloy can be modelled as a virtual crystal.

The barrier height does depend on the way the difference between the bulk bandgaps of GaAs and $Ga_{1-x}Al_xAs$ is divided up. For the present purposes, we shall assume that this division is known. In fact, it turns out that this too can be predicted quite accurately from bulk properties. The problem of finding the magnitude of band offsets is addressed in Section III.B.

Following this prescription, we arrive at the familiar steplike potential characterising the potential difference experienced by electrons (holes)

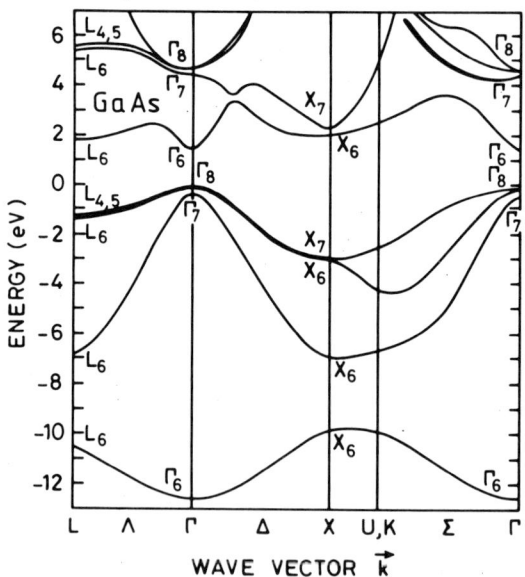

FIG. 1. The band structure of GaAs. The top of the valence band is at $E = 0$.

5. MICROSCOPIC PHENOMENA IN ORDERED SUPERLATTICES

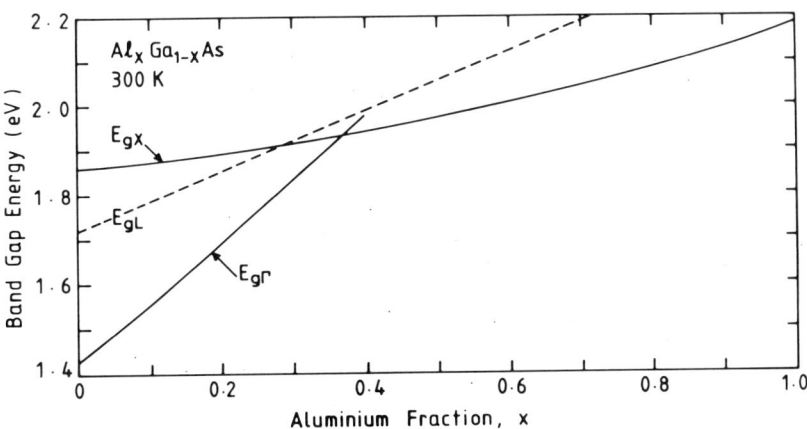

FIG. 2. The variation of the direct and indirect (X and L) gaps in $Ga_{1-x}Al_xAs$ as a function of x.

moving across the interface at the conduction- (valence-) band edge. The effect of this potential upon the electronic structure reduces to that of a particle in a one-dimensional rectangular well and can be solved by elementary means, provided the well width, the band offset, and the effective masses of the corresponding bulk bands are known. In Fig. 3, there is a sketch of quantum states, confined in GaAs, with energies above the bottom of the well. We can imagine that the particle makes a near-total reflection at the

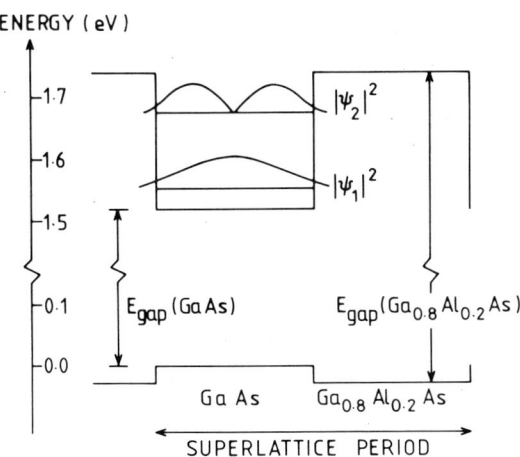

FIG. 3 The steplike potentials obtained by lining up the bulk band structures of GaAs and $Ga_{0.8}Al_{0.2}As$. The confined states in the conduction-band well are also indicated.

interface without experiencing a significant part of the band structure. As in the effective-mass model of shallow impurity levels in semiconductors (Kohn, 1957), the total wave function of such a confined level (j) is a product of the Bloch function $\phi(n, \mathbf{k})$ from the corresponding (e.g., GaAs) bulk-band-edge band n and wave vector $\mathbf{k}$, and a slowly varying envelope function $F_j(z)$ sketched in Fig. 3 (Dingle, 1975).

The quantum numbers n, k, and j can be used to characterise transition rates involving confined states. For instance, the optical transition probability I for a jump across the gap of the quantum well structure is proportional to $|\langle F_j \phi(n, \mathbf{k}) | \hat{p} | F_i \phi(m, \mathbf{k}) \rangle|^2$, where $\hat{p}$ is the momentum operator. Since F_j and F_i are slowly varying functions of z, and the transition between the bulk band-edge valence and conduction states is allowed, the transition probability depends only on the overlap in real space of F_j and F_i, which gives a simple selection rule $I \simeq \delta_{j,i}$. The confined states are pinned to the corresponding bulk band edge, and this makes it easy to determine their response to external fields (e.g., pressure).

When a number of alternating layers of reduced thickness are grown to create a superlattice, the states from adjacent wells can overlap, and the quantum well levels are broadened into minibands. The calculation of the corresponding energies and transition rates can be carried out along the lines outlined for the case of an isolated well. The solution of the Schrödinger equation for a one-dimensional periodic potential of the form shown in Fig. 3 can be obtained by matching the wave functions at the interfaces and by invoking the translational symmetry of the potential. For example, one can use the familiar Kronig–Penney model or a more general transfer-matrix technique based on similar concepts but making it possible to account for the actual symmetry properties of the bulk band structure, particularly at the valence-band edge. An excellent account of the way such calculations should be set up including the boundary conditions has been given, for example, by Bastard (1981), Altarelli (1983), Folland (1983), and others. This class of calculations is also discussed elsewhere in this book and will be omitted here. It might be worth pointing out that unlike in the case of point defects (shallow donors and acceptors), it is not possible to obtain a rigorous derivation of the effective-mass-approximation (EMA) equations for the heterostructure problem (Zhu and Kroemer, 1983; Morrow, 1987a, b; Trzeciakowski, 1988, and references therein). This is essentially because of the steplike discontinuity of mass at the interface. However, in most cases of practical interest, corrections can be made to make this conceptual obstacle irrelevant.

B. BREAKDOWN MECHANISMS

The simple outline given above of what is often called the *envelope-function effective-mass model* also conveys the limits of applicability of the EMA.

First of all, for **k** to be a good quantum number, the change at the interface of the periodic rapidly warying part of the Bloch function $\phi(n, \mathbf{k})$ must be negligibly small. This imposes a severe restriction on the kind of materials that can be studied. For example, it means that the band structures of one of the constituents must be obtainable from that of the other by first-order perturbation theory, at least at the band edge, i.e., $E(\text{Ga}_{1-x}\text{Al}_x\text{As}) = \langle \phi(\text{GaAs}) | V(\text{Al}) - V(\text{Ga}) | \phi(\text{GaAs}) \rangle + E(\text{GaAs})$. In the case of GaAs-Ga$_{1-x}Al_x$As ($x < 0.4$), this relation is in fact approximately valid. This is because the difference between the microscopic atomic potentials of Ga and Al is small enough.

Indeed, it is well known from the work on localised defects, where the problems of mixing of bulk Bloch waves of different momentum is quite familiar, that an isolated Al impurity in GaAs does not introduce any localised states in the gap (see, for example, Jaros, 1982). The collision of a conduction electron with such an isoelectronic centre can also be described in terms of the EMA. Therefore, the degree of coupling of Bloch waves of different bulk momentum by such an impurity potential can be used as a measure of the strength of the potential difference between the host and impurity atoms. Similarly, the changes in the electronic structure of GaAs introduced by the existence of the confining barriers can be viewed as if we constructed the superlattice out of a bulk GaAs by replacing in certain layers Ga atoms with Al. The relationship between the potential difference and momentum mixing is in both cases quite analogous. In the following section, we shall formalise this way of looking at the problem and express it in a mathematical form.

Now consider in this manner the case of, say, CdTe/HgTe superlattices. Again, we have here a well-lattice-matched system. However, whereas the potential difference between Ga and Al is of order 0.01 Ry; in the case of Hg and Cd, it is of order 0.1 Ry. In particular, it has been shown that Hg behaves as a "deep" impurity in CdTe. Although it does not introduce any localised states in the forbidden gap of CdTe, its resonance levels in the continuum are complicated wave packets with many different n and k components. This tells us that an electron in the conduction band of CdTe colliding with Hg impurity is scattered into states of different momenta. Therefore, the first-order perturbation theory breaks down and a more sophisticated procedure is needed to describe the electronic structure and the relevant transition rates concerning CdTe/HgTe superlattices.

The strength of the difference between the microscopic potentials of the constituent atomic species is only one possible source of momentum mixing in heterojunction nanostructures and consequently of the breakdown of the effective-mass model. For example, we can think of materials whose lattice constants (or even the lattice-point-group symmetry) are different. In general, the lattice-constant differences of order 0.1% and larger, and any differences in the lattice symmetry, are significant. Then an electron moving across the interface sees atoms in "wrong" positions, and in addition to the difference in the atomic (bulk) potentials, we must account for the "displacement," i.e., for the difference in the position of the atoms in the lattice on the other side of the interface. Even a small displacement represents a strong scattering potential that mixes waves of different momentum. This is the physical process underlying the effect of strain in, say, Si/Ge superlattices.

Another class of momentum mixing arises when an electron approaching the interface faces a quantum state of similar energy but different bulk momentum. Then the continuity of the wave function requires that the mixing of the two states be carefully considered. For example, we can think of a GaAs/AlAs superlattice where the well and barrier widths are chosen so that the lowest conduction-band level of Γ character in GaAs coincides with or lies close to the lowest X-like state in AlAs. It transpires that the magnitude of this mixing, and consequently of the observable transitions, is very sensitive to the details of the Bloch functions near the relevant band extrema and their amplitude at the interface. A similar crossing of states of different bulk momenta occurs in the valence band where it involves the interaction between the heavy- and light-hole bands.

Let us summarise: The EMA breaks down whenever the difference between the microscopic potentials in adjacent layers leads to mixing of bulk states of different momenta. This occurs when the layers possess (1) different point-group symmetry, (2) atomic potentials whose strength difference is of order 0.1 Ry or more, (3) lattice constants whose difference is of order 0.1% or more, and, (4) in the energy interval of interest bulk states of different momenta quasidegenerate in energy.

It is noteworthy that the breakdown mechanisms 1–4 (hereafter referred to as BM) listed above stem from the difference in bulk properties of the constituents and are independent of the physical dimensions and the degree of order in the microstructure itself. We must therefore ask whether there are structures that satisfy the criteria BM for the validity of the EMA only so long as their characteristic dimensions or order parameter attain a certain critical value.

Consider first the physical dimensions. The relevant parameters are the layer (well and barrier) widths. When any of these become comparable to the bulk lattice constant, the Heisenberg uncertainty principle tells us that the momentum wave function must extend over a significant subspace of wave vectors. In superlattices, this corresponds roughly to the miniband width, whereas in isolated wells, the simple cosine or sine form of the wave function is replaced by what resembles more a wave packet. However, this does not necessarily lead to a serious breakdown of the EMA, provided that the dominant part of the momentum wave function remains well localised and that the BM criteria for the validity of the EMA are still satisfied. This is because under such circumstances the first-order perturbation theory retains its validity. Indeed, calculations have shown that energy levels of very narrow wells (e.g., 10 Å thick) obtained in full-scale calculations agree well with the predictions of the EMA (e.g., Gell et al., 1986). A quantitative example of this situation is presented in Section IV.B.6 for GaAs/AlAs superlattices. We shall see there that qualitative deviations from the EMA occur only when the bulk levels of Γ and X character cross.

Of course, there are quantitative corrections to the EMA of order 1–10 meV for medium thick (40–50 Å) wells (Wong et al., 1986), which increase as we move further up the energy scale towards minibands located at and above the top of the semiclassical barrier, and even larger ones for ultrathin wells (Gell et al., 1986). A detailed comparison of the EMA with larger calculations is difficult to make because of the uncertainties in the EMA parameters as well as in the accuracy of complex numerical methods. This is quite understandable given the range of energies (of order 10 meV) in question. However, there is no doubt that the trend in energy levels versus well and barrier widths and chemical composition are well accounted for by the EMA. This is so to a surprising degree even in systems that do not quite satisfy the BM conditions. For example, modified Kronig–Penney calculations yield a correct estimate of energy levels of four-monolayer Si_4Ge_4 superlattices discussed in Section IV.C.2 (People and Jackson, 1987). This means that although the momentum-mixing effect is significant enough to warrant full-scale calculations to compute the correct momentum wave function, it is not strong enough to lead to a complete breakdown of the particle-in-a-box picture. Indeed, a complete breakdown of the EMA would lead to a much larger increase of the oscillator strength of the transitions across the forbidden gap of this superlattice than that observed in optical experiments.

In fact, it is worth noting that none of the systems for which sufficient experimental data are available exhibits such a complete breakdown. This must be partly because of the difficulties of achieving high-quality alternating layers of two dissimilar materials and/or preparing ultrathin layers. We argued above that superlattices whose layers consist of less than four atomic planes (i.e., whose width is less than the bulk zinc-blende or diamond-lattice constant) represent a special category. It is well known from elementary band theory (e.g., Bassani and Pastori-Parravicini, 1975) that in order to recover the band structure of, say, Si away from the zone centre (Γ), it is required by symmetry that we retain at least the next-nearest-neighbour shell of atoms. Hence such structures are better viewed as novel ordered compounds even if in their bulk form they are quite similar. We should then abandon the concept of band-offset and bulk parameters (e.g., effective masses) normally used to describe superlattices. However, we shall see (Section IV.B.6) that when the constituents have very similar atomic potentials, the conventional band offset leads to at least qualitatively correct results even for $(GaAs)_1$ $(AlAs)_1$. Unfortunately, there are not enough reliable data for such fine superlattice structures to allow a detailed comparison between theory and experiment.

Next we come to consider the role of ordering upon the validity of the EMA. Since the EMA parameters represent average physical quantities over the width of the individual layer, the breakdown of the EMA can only occur when lack or change of order violate one of the BM conditions. This may happen when (1) the virtual crystal approximation for alloys breaks down, and (2) a strongly ordered structure replaces a random or nearly-random alloy. The two are really just different degrees of the same phenomenon. By (1), we normally mean small displacements of atoms from their virtual crystal positions in an alloy or some other small deviation from purely random distribution. The manifestations of this breakdown are broadening of optical lines and reduction of mobility (see Pearsall, 1982; also a review of semiconductor alloy systems by Jaros, 1985). Although (1) and (2) may lead to observable shifts of levels, they are significant only if they lead to violation of the BM. We shall return to this problem in Sections III.C and IV.C.

The effect of confinement has been observed in high-quality samples of GaAs in wells as wide as 400 Å. When the wells and barriers become narrower, the states from adjacent wells overlap and form minibands. The critical dimensions for this to occur depend on the band offset and on the nature of the band lineup. In the ground state of GaAs/AlAs, we need well and barrier width of less than 50 Å (Ivanov and Pollmann, 1979). In Si/Ge, no significant broadening is found unless the width is less than 15 Å (Morrison *et al.*, 1987; Morrison and Jaros, 1988). The effect of periodic order then increases in importance with decreasing layer width, and the superlattice period becomes as important as the width of individual layers. This is because the periodic boundary condition dominates the selection of waves whose wavelength is suitable for forming the superlattice wave function. However, this has no serious implications for the validity of the EMA other than those discussed in connection with the role of well widths. The critical widths for the miniband formation can be estimated from the EMA quite well.

The above figures for band-structure formation imply that the physical dimensions of structures (the superlattice period) suitable for band-structure engineering and exploitation of momentum mixing in semiconductors must be of order 0–100 Å. For larger periods, the width of minibands of most known materials is too small. Epitaxial layers whose width is less than one bulk lattice constants (5 Å) are potentially the best candidates for electronic-structure engineering because they are in essence novel compounds.

However, in spite of tremendous progress of recent years, the extent to which this potential can be exploited has not been established because of the uncertainties concerning growth and processing of such fine structures.

C. MICROSCOPIC FORMULATION

In the course of our effort to identify the conditions under which the particle-in-a-box picture works, we also developed a view of the electronic structure of semiconductor nanostructures that will enable us to formulate a tractable microscopic theory. In particular, we constructed, say, a GaAs/AlAs superlattice starting from a bulk GaAs crystal and generated the AlAs layers by replacing Ga atoms by Al atoms at the relevant sites. This way of approaching the problem enabled us to appreciate differences in the electronic structure in terms of differences in atomic potentials. It will transpire that this is also a useful way of setting up a computational scheme that does not suffer from the limitations of the EMA (Jaros, 1985).

Let us begin by choosing a suitable bulk-crystal Hamiltonian H_0 for, say, GaAs such that the corresponding band-structure energies $E(n, \mathbf{k})$ and wave function $\phi(n, \mathbf{k})$ associated with the wave vectors $\mathbf{k}$ and bands n can be generated in good agreement with experiment, i.e.,

$$H_0 \phi(n, \mathbf{k}) = E(n, \mathbf{k}) \phi(n, \mathbf{k}). \tag{1}$$

The wave function of a GaAs/AlAs superlattice ψ can be written as an expansion in terms of the complete set of Bloch states

$$\psi = \sum_{n, \mathbf{k}} A(n, \mathbf{k}) \phi(n, \mathbf{k}). \tag{2}$$

This function must satisfy the Schrödinger equation for the superlattice Hamiltonian $H = H_0 + V$,

$$H\psi = E\psi, \tag{3}$$

where the superlattice potential V is the difference between the potentials of AlAs and GaAs in the space where there is the barrier (AlAs) material, i.e.,

$$V = \sum_{j} [V_j(\text{AlAs}) - V_j(\text{GaAs})], \tag{4}$$

where j runs through all atomic sites in the barrier layer. Since V is a periodic potential, the values of $\mathbf{k}$ in the expansion for ψ in Eq. (2) are uniquely determined by symmetry. For example, if the superlattice period d in the $\langle 001 \rangle$ direction (the width of the GaAs well plus the width of the AlAs barrier) is 140 Å, there are 51 bulk Bloch states for each band. The number (and position) of these $\mathbf{k}$ points in the bulk Brillouin zone is obtained by dividing the zone width $2\pi/a$ (a is the bulk lattice constant) into smaller units of the length of the superlattice reciprocal lattice vector $2\pi/d$. We say that the bulk band structure is folded into the small (superlattice) Brillouin zone.

We can solve Eq. (3) by substituting for ψ from Eq. (2), by multiplying from the left by the complex conjugate of $\phi(n', \mathbf{k}')$, and by integrating over all real space coordinates. For each $n', \mathbf{k}'$ we obtain a linear algebraic equation of the form

$$A(n', \mathbf{k}')[E(n', \mathbf{k}') - E] + \sum_{n,\mathbf{k}} A(n, \mathbf{k}) \langle n', \mathbf{k}'|V|n, \mathbf{k}\rangle = 0. \quad (5)$$

The corresponding eigenvalue problem can be solved by direct diagonalisation. The computer returns eigenvalues E and a matrix of coefficients $A(n, \mathbf{k})$ for each state ψ. Since the energies E are closely spaced, the matrix formulation must be preferred to a determinantal (e.g., Green function) approach.

The number of bands required in Eq. (2) depends on the nature of V and must be increased until a fully convergent result is achieved. For example, from our discussion above, in the case of GaAs/AlAs, we must expect the mixing of states between different bands to be weak except in those energy regions where states of different momenta cross. However, in other systems, where states of different momenta cross. However, in other systems, where the breakdown of the EMA is of different origin (e.g., Si/Ge and CdTe/HgTe), the number of bands required (Eq. (2) is expected to be large (10–30). Also, in the latter case, there must be a large number of finite coefficients $A(n, \mathbf{k})$ from bulk $\mathbf{k}$ points lying further from the band edge. If, however, the EMA is valid for the structure in question (e.g., for the ground state of a 70 Å GaAs–$Ga_{0.7}Al_{0.3}As$ well), only a couple of coefficients $A(n, \mathbf{k})$ are significant (e.g., two plane waves are needed to represent a standing wave in the particle-in-a-box picture).

A brief comparison of the computational scheme based on Eqs. (1–4) with other methods will be given in Section III.A. However, any full treatment can be rewritten in the form given above, and we shall use this formulation throughout this text. It will provide us with a simple means of identifying the physical process in question in terms of the origin and strength of momentum mixing, which is well represented by the magnitude and distribution of the coupling coefficients $A(n, \mathbf{k})$. The calculation of transition rates reduces in this representation to a calculation of matrix elements of the relevant operators between the bulk Bloch states. The properties (e.g., selection rules) of these bulk wave functions are familiar from numerous texts on solid-state physics (e.g., Bassani and Pastori-Parravicini, 1975; Callaway, 1974).

III. Microscopic Theory of Ordered Superlattices

A. COMPUTATIONAL TECHNIQUES

Following the discussion in Section II, we can divide problems concerning ordered superlattices into two conceptually quite distinct categories. In the first, we have problems in which the electronic structure of the superlattice is such that the simple particle-in-a-box model holds good. In such problems, the bulk linear momentum is still a good quantum number, and the changes in the bulk band structure due to the confinement effect can be accounted for by first-order perturbation theory, in the spirit of the effective-mass approximation discussed in Section II.C. The vast majority of research literature on physics and applications of superlattices have, in fact, been concerned with problems belonging to this category, and there are a number of excellent reviews of the research (pure and applied) literature (e.g., Bauer et al., 1984, 1986; Kelly and Weisbuch, 1986; Capasso and Margaritondo, 1987; Dingle, 1988). A simple tutorial account of the relevant physics and applications is also available (Jaros, 1989).

On the other hand, there is a class of structures in which the difference between the constituent materials is large enough or their configuration is chosen so as to give rise to mixing of bulk states of different linear momentum. We identified quite specific conditions for the existence of this effect in Section II.C. In general, this occurs whenever constituent bulk levels associated with different parts of the bulk Brillouin zone cross, and when the atomic potentials or the crystal structure of the constituents are significantly different. It is this class of systems that is of interest here. The observable properties of such systems depend on the strength and the spectral form of the Fourier components of the microscopic superlattice potential V defined in Eq. (4). This determines the degree of coupling between the bulk Bloch functions used as a basis set in the quantitative model of the superlattice wave function. These considerations will assist us in establishing the criteria determining the suitability of a given computational procedure for modelling such systems.

First, we want to model structures whose characteristic dimension is 0–10 nanometers, since this is the actual layer width on which electronic-structure engineering can be done. Second, we must be able to establish a safe reference point for the assessment of our electronic-structure calculations; for instance, we want to make sure that the band offset seen by our calculation really corresponds to a given empirical value. Third, we must also ensure that the calculation can reproduce quite accurately the band structure of the host semiconductors, at least in the region near the forbidden gap that contributes strongly to the observed transitions.

The most obvious technique for full-scale calculations is the supercell method pioneered by Cohen and collaborators and applied widely first to surface and then also to interface studies (e.g., Cohen, 1980). In this method, the superlattice is viewed simply as a new crystal with a large unit cell. The only reference point in such a calculation is the electronic structure of a free atom. This means that the potential must be accurate over at least the full energy range covering all occupied states spanning many rydbergs! However, potentials that ensure this degree of accuracy for the ground state of most crystalline systems have been obtained by allowing the wave function in the core region to imitate the real solution of the Schrödinger equation as much as possible. This leads to deep momentum-dependent (nonlocal) potentials, which in turn demand the best computational facilities (e.g., see Bachelet *et al.*, 1982; Bylander and Kleinman, 1986; Froyen *et al.*, 1988, and references therein).

The main approximation in this first-principles approach is the use of a statistical average for the exchange and correlation terms in the many-body-crystal Hamiltonian and turn it into a simple function of electron density (hence the label *local density approximation*, hereafter referred to as LDA). The magnitude of this term is difficult to establish, and in practice, this terms is treated as a parameter. For example, it is often used to correct the magnitude of the forbidden gap (a 40% error is typical). The main strength of the LDA is to reproduce correctly the properties of the valence band. The excited states (conduction band) are not so well accounted for. This makes the method particularly attractive for calculations of structural properties (e.g., to determine minimum-energy nuclear coordinates) that depend only on the ground state. The method also provides information about elastic constants and band offsets that are needed in simpler methods used to model superlattice electron and phonon properties. The method yields bulk lattice constants with better than 0.1% accuracy. The results for the bulk modulus and the total valence-band energy are not so impressive, and a typical error is of order 10%, though still better than results obtained by any other approach.

The application to superlattice calculations is limited by the failure of the LDA to represent the conduction bands and the forbidden gap, and by the limited (<25) number of atoms in the unit cell, which is reasonable to use even with the largest computers. The former limitation has been largely removed by including many-body corrections (Hybertsen and Louie, 1985, 1986; Godby, Schlüter and Sham, 1986; Hybertsen and Schlüter, 1987, and references therein). This state-of-the-art quasiparticle approach naturally increases computational requirements and reduces further the possibility of studying larger blocks of atoms. Superlattice calculations on zinc-blende semiconductor–semiconductor systems based on the LDA and quasiparticle approaches have generally not exceeded periods of two to three lattice constants (12 atomic planes).

One way to achieve a better representation of bulk-band-structure parameters and to increase the number of atoms in the unit cell is to use the formulation in Eq. (5). It has been demonstrated that this formulation represents a realistic self-consistent method for electronic structure calculations in a wide range of low-dimensional problems (for point and line defects, see Jaros, 1982, Kirton *et al.*, 1984; for superlattices, Andreoni and Carr, 1980; Gell *et al.*, 1987). A computer diagonalises Eq. (5) and returns differences between the electronic band structure energies of the superlattice and those of the bulk constituents. In this way, a meV accuracy in positioning superlattice states with respect to the bulk band edges can be achieved (Ninno *et al.*, 1986).

The input Bloch functions in the expansion for ψ can be constructed either in the pseudopotential (plane-wave) representation as in many supercell calculations (Pickett *et al.*, 1978) or in any other way. So far, calculations based on Eq. (5) employed local and nonlocal empirical pseudopotentials. In the empirical pseudopotential calculation, the atomic potentials in V of Eq. (4) are obtained by fitting the first few form factors (Fourier components of the potential taken at the shortest bulk reciprocal lattice vectors) in order to reproduce the observed forbidden gaps and energy surfaces near the band edges. However, the formulation in Eq. (5) implies that the computer evaluates the potential at the superlattice reciprocal lattice vectors, i.e., the potential is needed at a number of wave vectors lying between the values determined from the observed gaps. This raises the important question about the error due to the uncertainty with which the potential is set up at such points.

First of all, we must recall that the actual value of the matrix elements in Eq. (5) depends on the product of the Fourier component associated with atomic potentials and the structure factor that is a sum of complex exponentials over all atomic positions in the unit cell (i.e., over the full length of the superlattice period). Hence, when the layer in question is thick, this sum reduces all contributions other than those peculiar to the bulk crystal. Only a few contributions lying very close to the bulk reciprocal lattice vectors are large. It is easy to verify that for layers whose thickness is larger than one lattice constant, this reduction is very considerable. In fact, once the layer thickness exceeds two lattice constants, particular care must be taken to represent atomic coordinates and other input constants as accurately as possible, since they may lead to a more significant error. This is because the structure factor sum is the only way the matrix obtains information about nuclear coordinates, i.e., about the layer thickness and symmetry relative to the chosen origin. It is well known from dimensional analysis (i.e., particle-a-box picture) that quantum well and superlattice calculations are very sensitive to these coordinates.

In layers consisting of less than four atomic planes, this reduction is not so significant, and the details of the potential away from bulk terms are important. This has led to discrepancies in band-structure properties of monolayer superlattices generated from empirical pseudopotentials as pointed out, for example, by Bylander and Kleinman (1986). However, it is important to realise that the first-principles pseudopotentials and the empirical pseudopotentials are very similar for the first few shortest reciprocal lattice vectors. This can be exploited to remove uncertainties from the form of the empirical fit in the region where interpolation between bulk values is necessary. The most important interactions in the short-period superlattices are between the bulk bands folded into the superlattice (smaller) Brillouin zone. These interactions are dominated by the new potential components associated with the first few superlattice reciprocal vectors i.e., in the long-wavelength region where the above improvements can easily be made.

Finally, there is the question about the degree of self-consistency required in superlattice calculations. As demonstrated by Pickett et al. (1978), Kirton et al. (1984), and others, self-consistency in the usual sense of quantum theory (charge density and potential) can be achieved without any difficulties whether one uses local or nonlocal (momentum and wave-vector-dependent) empirical pseudopotentials. However, any attempt to implement the calculation in structures consisting of highly dissimilar materials thoroughly would lead us towards the local-density or quasiparticle formulation of the Hamiltonian. It is therefore necessary to check the degree to which the bulk character of the charge-density distribution is preserved in a first-principles calculation. Fortunately, local-density and quasiparticle calculations have been performed on a number of most important superlattices accessible to present-day experimentations (Van de Walle and Martin, 1985, 1986a, b, 1987; Hybertsen and Schlüter, 1987; and others). These calculations have demonstrated that in semiconductor–semiconductor systems of similar crystal structure (this includes all structures considered in this chapter), the *charge transfer at interfaces is small and confined to the interface layers, and that it can be accounted for in terms of the superposition of bulklike neutral pseudoatoms.*

The neutral pseudoatom is precisely the concept on which we depend in semiempirical pseudopotential calculations. It is assumed that the bonds formed by atoms in the diamond and zinc-blende lattice retain their form in the superlattice, with only minor corrections that can be treated by second-order perturbation theory. The "charge transfer" at the interface occurs only to the degree to which it is necessary in order to generate what must always be a "new" bond across the interface. One might therefore expect semiempirical methods to perform well even for some ultrathin structures. For instance, in Si/Ge superlattices, this bond closely resembles an average of the bulk Si and Ge bonds. If this were not the case, the corrections to the empirical potential would become large. Indeed, results reported from empirical pseudopotential calculations for four-monolayer Si_4Ge_4 superlattices are in an excellent agreement with the state-of-the-art quasiparticle calculations for the same system (Section IV.C.2). A similar agreement has been acknowledged for $(GaAs)_1/(AlAs)_1$ (Ciraci and Batra, 1987).

There are notable examples of interfaces where charge transfer is important and truly self-consistent treatment is essential. For example, it is worth reminding the reader that the supercell approach has been used first of all to model surfaces. Here the interface is between (e.g., six-monolayer) slabs of, say, Si and vacuo. This must be the most frequently studied "superlattice." The disruption at the interface is probably as strong as it ever can be, and the charge density representing the dangling bonds does differ from that of the bulk bond. However, even here, the bulk behaviour is re-established within two atomic planes from the interface (Cohen, 1980, and references therein). The same conclusion is obtained in the case of lattice vacancies and deep impurities (see review by Jaros, 1982, and references therein). This, alas, does not mean that individual quantum states are necessarily well confined within a couple of monolayers. Whereas the total valence-charge density is well accounted for in a small cell normally used in first-principles calculations, the individual quantum states may not be. An example of such states can be found at polar interfaces such as GaAs/Ge (001) (see a seminal paper by Kunc and Martin, 1981). This explains why polar interfaces are so poorly understood both from theoretical and from experimental standpoints.

By far the most popular technique for modelling superlattices is the tight-binding method (TB) pioneered by Schulman and McGill (1977). Chang and Schulman (1985, and references therein) have presented state-of-the-art calculations using this technique for a wide range of systems and provided the most complete tests of the EMA. They were also very successful in identifying and interpreting a number of novel observable properties of superlattices. In fact, they dealt with the first and simplest example of momentum mixing, which occurs when light- and heavy-hole bands cross. However, they concentrated on medium-wide and thick layers. Attempts have also been made to introduce some degree of self-consistency into the TB superlattice calculation for more dissimilar materials (e.g., Munoz *et al.*, 1987; see also Van de Walle and Martin, 1988).

It is worth pointing out that the TB scheme is normally implemented so that the Hamiltonian parameters are fitted to represent the bulk band structure with a minimum basis set. This procedure excludes the contributions to the wave functions of atomic states with higher angular momentum (e.g., d-states) irrespective of whether next nearest-neighbour parameters are included or not. This means, for example, that the wave functions derived from states associated with the bulk X minima are not properly represented. Similarly, when the contribution of higher bands to the superlattice states is important, the TB representation in its usual form is suspect. In this case, the truncated basis set may also misrepresent the meaning of the band offset, and consequently the degree of momentum mixing. It is easy to understand the essence of this problem in the language of the "impurity" model introduced in Section II.B. When the potential difference between the constituents forming the superlattice is large, an electron approaching the interface is scattered into states of different bulk momenta. If the basis set is truncated, this scattering process cannot be correctly accounted for, and the effective strength of the barrier at the interface is altered. Naturally, the importance of this discrepancy is diminished if the role of the microscopic potential is weak and the corrections to the EMA small. In this regard, the tight-binding model stands somewhere between the EMA and the empirical pseudopotential formulation.

B. Heterojunction Band Lineups

The heterojunction band offset is an indispensable parameter in studies of semiconductor superlattices. Since it is sufficient to know only one of the (two) offsets, in the following we shall refer to the valence-band offset ΔE_v. In general, the evaluation of ΔE_v is a formidable problem, since it requires that the valence-band edges of the constituent materials be positioned on an absolute scale and in a self-consistent manner. Only the first-principle methods are likely to be suitable for carrying out such a research program. Many authors have made accomplished studies of ΔE_v, e.g., Van de Walle and Martin (1987), Cardona and Christensen (1987), Christensen (1988), Massida et al. (1987), Lambrecht and Segall (1988), to mention just a small fraction of these attempts (see Capasso and Margaritondo (1987) for a comprehensive review of experimental data and theory). All these theories of heterojunction discontinuities involve computational or conceptual manipulations of great complexity, and their accuracy and limits of applicability are still being debated. This is because the band offset, which is of order 0.1 eV, must be obtained by subtracting the energy of the valence band of the two constituents, each of which is of the order of several rydbergs. Furthermore, as pointed out by Van de Walle and Martin (1987), there is no simple way of extracting the magnitude of the band offset from the LDA supercell calculation, and a model is needed for that purpose. An accurate experimental determination of the valence-band discontinuity is no less demanding

(Wolford et al., 1986). However, it turns out that a good account of the magnitude of ΔE_v *for many heterojunctions can be obtained from bulk properties of the constituents*, without recourse to a numerical study (Jaros, 1988, 1989). The fact that the simplest means give results at least as satisfactory as the largest calculations is not surprising, since those very calculations predict only a small deviation from bulk charge-density distribution for the heterojunctions in question.

The optical dielectric constant ε of a semiconductor can be expressed as

$$\varepsilon(\mathbf{q}) = 1 + (e^2/\varepsilon_0 q^2) \sum_{\mathbf{k},\mathbf{g}} |\langle \mathbf{k} | \exp(i\mathbf{q}\cdot\mathbf{r}) | \mathbf{k}+\mathbf{q}+\mathbf{g} \rangle|^2$$
$$\times [f(\mathbf{k}) - f(\mathbf{k}+\mathbf{q}+\mathbf{g})]/[E(\mathbf{k}+\mathbf{q}+\mathbf{g}) - E(\mathbf{k})]. \qquad (6)$$

Here $\mathbf{k}$ is the reduced wave vector and $\mathbf{g}$ is the reciprocal vector; f is the occupation number of the state in question. When $\mathbf{q} \to 0$, the only empty states available are the states in the conduction band in the vertical direction above the occupied valence states. If it is then assumed that $E(\mathbf{k}+\mathbf{q}+\mathbf{g}) - E(\mathbf{k})$ is the same for all $\mathbf{k}$ (the rationale of this approximation has been fully discussed elsewhere, see Phillips, 1973) and equal to the forbidden gap E_g, the expression in Eq. (6) at $q=0$ reduces to the well-known formula

$$\varepsilon = 1 + (\hbar\omega/E_g)^2; \qquad (7)$$

$\omega = (ne^2/\varepsilon_0 m)^{1/2}$ is the plasma frequency; n is the electron density, which in the diamond lattice is $32/a^3$; a is the lattice constant. The experimental values of ε and a are readily available. E_g is an average gap or simply the gap obtained in the most elementary version of the nearly free electron model of an isotropic semiconductor (Fig. 4). In this model, the forbidden gap opens at $k=k_F$ and remains the same everywhere. The magnitude of E_g can be obtained from Eq. (7) by using the empirical values of the dielectric constant. Since the position

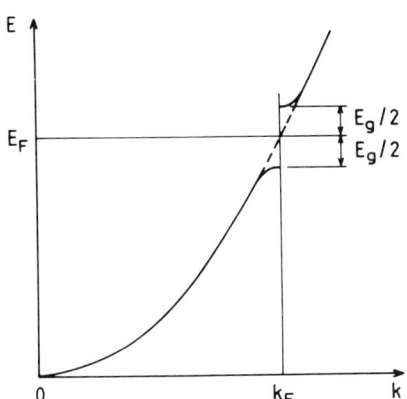

FIG. 4. The bandgap E_g at the free-electron Fermi energy E_F obtained in the nearly-free electron model calculation. k is the wave vector [Reprinted with permission from the American Physical Society, Jaros, M. (1988). *Phys. Rev.* **B37**, 7112.]

of the band edges is given relative to the free-electron Fermi energy (E_F), which is the same for materials with identical lattice constants, the band discontinuity at a heterojunction formed by two lattice-matched model semiconductors A and B is

$$\delta E = (E_g(A) - E_g(B))/2. \tag{8}$$

It is well known from full-scale band-structure calculations that the lowest conduction band is a rapidly varying function of **k**. This is particularly so in the region near the center of the Brillouin zone. Although this variation does not affect significantly the density of states and, consequently, the average bandgap, it does make it unlikely that any realistic predictions concerning the conduction-band discontinuity can be made in terms of the model in Fig. 4. On the other hand, the uppermost valence band is rather flat. Although its width is finite and decreases with increasing ionicity of the crystal, this variation in width does affect the density of states, and it is reflected in the value of the dielectric constant and consequently in E_g. Therefore, δE of Eq. (8) might provide useful information concerning the actual heterojunction valence-band offset, i.e., $\delta E \simeq \Delta E_v$. The magnitude of ΔE_v obtained from Eq. (7) and Eq. (8) is given in Table I for a number of heterojunctions, in good

TABLE I

VALENCE-BAND OFFSETS[a]

	$\delta a(\text{Å})$	ΔE_v	Exp	Tersoff	SCIC	LMTO	TB
GaAs–Ge	0.00	0.47	0.56	0.32	0.63	0.51	0.61
GaP–Si	0.01	0.41	0.8	0.45	0.61	0.53	0.64
AlAs–GaAs	0.01	0.43	0.42–55	0.55	0.37	0.43	0.32
AlAs–Ge	0.01	0.89	0.95	0.87	1.05	0.87	0.94
ZnSe–Ge	0.00	1.51	1.52	1.52	2.17	1.46	1.70
ZnSe–GaAs	0.00	1.04	0.96	1.20	1.59	0.99	1.01
CdTe–α–Sn	0.01	1.22	1.0	–	–	0.99	–
CdTe–InSb	0.01	0.89	0.87	0.84	–	0.66	–
CdTe–HgTe	0.00	0.35	0.35	0.51	0.23	0.64	0.38
InAs–GaSb	0.05	0.20	0.46	0.43	0.38	0.55	–
Si–Ge	0.22	0.48	0.2	0.18	0.58	–	–
Si–GaAs	0.22	0.01	–	–0.14	–0.13	–	–
GaAs–InAs	0.40	0.37	0.17	0.20	–	–	–

[a]The valence band offset ΔE_v calculated from Eqs. (7) and (8) in eV (Jaros, 1988), a is the difference in the constituent bulk lattice constants. The experimental results (Exp), the theoretical values under Tersoff, SCIC (self-consistent interface calculation), LMTO (linear muffin-tin orbital method), and TB (self-consistent tight-binding calculation) are taken from Cardona and Christensen (1987), Tersoff (1985), and Munoz et al. (1987). The sign of ΔE_v is such that the compound with the deeper valence–band edge is listed first. In strained structures, where the top valence-band levels are split, the average value obtained in SCIC is quoted, except InAs–GaSb where the effect of strain is ignored, and the difference between top valence levels is referred to in both Exp and SCIC.

agreement with experiment and with sophisticated calculations. It must be stressed that the experimental accuracy is not known for many of these values, so that agreement with experiment is not always very meaningful, although there are a few notable exceptions (e.g., GaAs/AlAs, HgTe/CdTe). However, it is gratifying that the differences between full-scale calculations are larger or of the same order of magnitude as the difference between the simple theory outlined here and the most sophisticated models.

Naturally, this simple model is inappropriate for strained structures. The ΔE_v calculated from Eq. (8) for strained systems is only an approximate value of the "average" valence-band offset, since the model does not account for the splitting of the cubic states by the strain. For instance, in Si–Ge (001) superlattices grown on Si substrates, the axial field splits the sixfold degenerate Δ-valleys into twofold and fourfold degenerate states. Although this splitting can be quite large, it can be estimated from the simple bulk deformation potentials (People, 1985; this procedure is described in some detail elsewhere in this book), in good agreement with the LDA calculations of Van de Walle and Martin. This is again to be expected if the deviation of the superlattice charge density from the bulk form is small. Finally, it should be noticed that there are nontrivial differences between the splittings reported in first-principles calculations (e.g., Van de Walle and Martin, 1985, 1986a, b; Froyen *et al.*, 1987; Hybertsen and Schlüter, 1987).

It is interesting to note that there is an overall similarity between ΔE_v obtained here and Tersoff's (1985) predictions for well-lattice-matched structures. Both models invoke only bulk properties of the constituent materials. However, in Tersoff's theory (the idea originates in Tejedor *et al.*, 1977; Flores and Tejedor, 1979) the band lineup is achieved via a "neutral-point energy" that is obtained from a detailed calculation of the complex band structure. The existence of this neutral point is explained by assuming that charge flows across the interface and leads to the formation of interface dipoles that equilibrate the charge-neutrality levels peculiar to the constituent semiconductors.

The notion that the atoms at the interface layers do not preserve their bulk-charge-density form and that the resulting dipole makes a significant contribution to the band offset lies at the heart of the subject and has been frequently voiced in the past (e.g., Frensley and Kroemer, 1977). In the model presented here, the absolute energy scale is obtained directly from the nearly free electron theory and without invoking interface dipoles. This is well in keeping with the observation made earlier in the text concerning the absence of any significant charge transfer in the reports on *a priori* self-consistent calculations. Strictly speaking, the success of the model of Eqs. (6)–(8) shows that, whatever the size of the dipole at the interface of lattice-matched heterostructures, its variation across a wide range of systems of different ionicities remains small enough to be neglected. Hence, it would appear that the basic physics underlying the formation of semiconductor–semiconductor heterojunction band lineups has now been understood. However, it must also be stressed that there is still a great deal of uncertainty about the precise magnitude of ΔE_v for most systems. More experimental and computational work will have to be done before a quantitative appreciation of ΔE_v for a wide range of heterojunctions is achieved. In heterojunctions formed by dissimilar materials, the role of interface dipoles and other phenomena may prevail (e.g., Perfetti et al., 1986, on Si–SiO$_2$).

C. STRUCTURAL STABILITY AND ORDERING EFFECTS

Questions concerning growth kinetics, and chemical and thermodynamic stability of superlattices are addressed elsewhere in this book in some detail. We shall confine our discussion here to a few key points of particular relevance to the main theme of this chapter, i.e., to "band-structure engineering" in ordered superlattices. This is because in some cases, structural and ordering effects lead to significant changes in the electronic-structure and transition rates.

The theoretical effort concerning structural properties of superlattices has been concentrated, so far with only a few notable exceptions, on ultrathin (GaAs)$_n$(AlAs)$_m$ infinite free-standing superlattices with $n, m \simeq 1\text{-}2$ (note that (GaAs)$_1$ signifies two atomic planes, but Si$_1$ only one (i.e., a layer of Si$_4$ is one bulk-lattice constant thick). In order to establish the stable configuration, one must compute terms such as the enthalpy of mixing. These quantities are of order 10 meV, i.e., very small (Bylander and Kleinman, 1986; Wood et al., 1987, 1988; and others) compared to the total energy of the system, and necessarily do depend on the technical details of what are very extensive calculations indeed. It is, therefore, hardly surprising that a general consensus concerning quantitative assessment is slow to develop. However, this is a field where quantitative theory and first-principles calculations in particular have a great deal to contribute in the future.

Although the LDA predicts bulk lattice constants very accurately (0.1%), it gives cohesive energies that are 10–20% too large. This is because (Bhattacharya et al., 1985) the underestimate of the exchange and correlation term by the LDA for crystals is smaller than for free atoms. However, the stability of a superlattice is measured by the heat of formation of the superlattice from the bulk constituent crystals of equal chemical composition whose valence-charge distribution is very similar (certainly much more so than the charge distribution in a free atom and covalent crystal). Consequently, the error is expected to be smaller. It is therefore quite possible that numerical errors due to convergence and other technical factors dominate those incurred due to the LDA itself. The figure quoted, for example, by Bylander and Kleinman is

−9.2 meV/unit cell for $(GaAs)_1(AlAs)_1$. Others quote negative or positive figures of the same order of magnitude. In any case, it is generally agreed among theorists that the heat of formation is very small (0–10 meV) and that the random alloy phase is the room-temperature stable state. However, experimental data show that $(GaAs)_1(AlAs)_1$, $(GaAs)_1(InAs)_1$, $(GaSb)_1(InAs)_1$, $(InAs)_1(InSb)_1$, $(GaP)_n(InP)_n$, etc. exhibit a high degree of crystalline order and a relatively high critical temperature (e.g., 850°K in $(GaAs)_1(AlAs)_1$) for order–disorder transition. These results are highlighted by the report of Kuan et al. (1985) showing that $(GaAs)_1(AlAs)_1$ represents an equilibrium and thermodynamically stable room-temperature configuration. In fact, the monolayer superlattice is established during continuous deposition of Al, As, and Ga, without any attempt to impose a growth sequence. This spontaneous ordering has been reported for other monolayer superlattices (e.g., $(InAs)_1(GaAs)_1$, $(GaAs)_1(GaSb)_1$, Fukui and Saito, 1984, Jen et al., 1986, respectively). It is possible that this result is due to stabilisation effects not included in the theory (Bylander and Kleinman, 1986; Wood et al., 1987, 1988; Wei and Zunger, 1988), i.e., kinetics rather than energetics might be relevant here. If we exclude kinetic effects, we can speculate that the superlattice is established in a metastable state separated from the ground state by a barrier much higher than the actual energy difference between the metastable state and ground state. Such a built-in barrier may owe its existence to imperfections and other deviations from ideal interfaces. In that case, the barrier should vary depending on sample preparation. As the growth control and characterisation techniques advance, such propositions could be tested in a systematic way. It remains to be seen to what extent the predictions of structural studies based on a priori theory survive such tests. For example, it has recently been shown (Kondow et al., 1989) that the observed ordering in $Ga_{0.7}In_{0.3}P$ alloys actually contradicts the theory.

Stability has also been studied as a function of layer thickness (see Wood et al., 1988, and references therein). All researchers agree that bulk properties are established rapidly and that only the interface layers show small deviations. Wood et al. (1988) also predict that unstable thin superlattices become more stable with increasing thickness, whereas the stable ones become less stable. None of these predictions have been tested experimentally.

The stability of Si–Ge superlattices is of special interest. The chief motive (Bevk et al., 1986) for studying Si/Ge is the possibility of using strain and ordering effects to alter the magnitude of the optical transitions across the gap. We shall see in Section IV.C.2 in some detail that in these superlattices, small structural changes may lead to qualitative changes in the electronic structure. This is not so in heterostructures whose bulk constituents are direct-gap materials.

The structural stability of Si/Ge has been studied by Van de Walle and Martin (1985, 1986a, b, 1987), Froyen et al. (1987), Hybertsen and Schlüter (1987), and Froyen et al. (1988). All calculations predict nuclear coordinates in these strained layers to lie within 0.1% of those predicted from the classical elasticity theory. The interlayer Si/Ge distance is predicted to be within 0.005 Å of the exact average of bulk Si and Ge bonds. The bond is slightly longer on the Ge side of the interface compared to Ge/Ge bonds inside the Ge layer, which are bulk-like. The presence of tetragonal displacements and their approximate magnitude has been confirmed experimentally by Feldman et al. (1987). However, recent experimental evidence indicates that the lattice separation in the direction perpendicular to the interface tends to be more bulk-like (Chambers and Loebs, 1989), and that the role of defects may be crucial in determining the nature of ordering (Muller et al., 1989). The formation enthalpy obtained by Froyen et al. (1988) for the epitaxial configuration (the lattice constant in the interface plane is fixed by the Si substrate in all these models) as the difference between Si_nGe_n and the equivalent amount of Si and Ge with the same tetragonal coordinates is of order 0–10 meV (the uncertainty of 30 meV is quoted, which is related to uncertainty in Ge equilibrium energy). We must assume that the formation energy differences between ordered Si/Ge superlattice and either strained or random alloy of corresponding composition are of the same order of magnitude. The formation enthalpy of Si/Ge thus appears to be of the same order as that in III–V superlattices as expected from a dimensional analysis.

Structural properties of semiconductor alloys used to build superlattices have been investigated with a view to establishing mechanisms leading to breakdown of the virtual crystal approximation (see review by Jaros, 1985). We noted in Section II that this breakdown leads to broadening of optical spectra and an increase in carrier scattering. The effect has also been studied theoretically in Si–Ge superlattices in terms of the coherent potential approximation, with very much the same result (Ting and Chang, 1986). The effect of short-range order in SiGe alloys can be found in the earliest literature on the subject (e.g., Braunstein et al., 1958) as well as in more recent reports (Agrawal, 1980). In the bulk form, SiGe might be regarded as an ideal alloy, since it is capable of withstanding prolonged annealing at temperatures as high as 900°C without exhibiting long-range order. Neither short- nor long-range order effects have been noticed in the optical spectra. The indirect edge absorption obeys nicely the Macfarlane–Roberts (Macfarlane and Roberts, 1955) expression for electron–phonon interactions. There is evidence of short-range order from the analysis of phonon spectra and from calculations and photoemission measurements of the density of states. As we also noted in Section II, such changes have been seen in II–VI- and III–V-compound semiconductor alloys. The question of interest here is whether a new ordering occurs in superlattices and if so, in what way it differs from bulk phenomena.

In contrast to the bulk properties of SiGe alloys, Ourmazd and Bean (1985) reported order–disorder transition in $Ge_{0.4}Si_{0.6}(75\,\text{Å})$–$Si(225\,\text{Å})$ superlattices grown on Si (001), which led to doubling of periodicity of the bulk lattice in a $\langle 111 \rangle$ direction. This creates a structure in which pairs of widely spaced $\{111\}$ crystal planes are occupied with the same atomic species. The ordering seems to favour the possibility that each Si atom has three Ge and one Si neighbour, and vice versa. The Si layers in this configuration can remain unstrained relative to the substrate so that all strain is taken up preferentially by Ge atoms. Ciraci and Batra (1987) reported LDA calculations that lend further support to this picture. They show that in their calculation, the phase reported by Ourmazd and Bean in fact lowers the total energy of the superlattice. The minimum is, however, only a secondary one, and the lowest energy configuration corresponds to the segregation into bulk-Si and strained-Ge constituents (see also discussion by Flynn, 1986, and Martins and Zunger, 1986). The absence of any ordering in the bulk form of the alloy suggests that the effect is related to the presence of strain in the superlattice. Note that Ciraci and Batra (1987) in fact conclude that the ordered phases of both $Ga_{1-x}Al_xAs$ and $Si_{1-x}Ge_x$ alloys are metastable. The energetics clearly favour segregation into constituents.

Finally, it is worth pointing out that the analysis of structural stability and ordering effects has been based on the assumption that the interface between layers exhibits perfect two-dimensional translational symmetry, that it is abrupt and free of any defects and imperfections, and that the superlattice is infinite (periodic boundary conditions are used along the growth axis). Indeed, it is only under these conditions that the first-principles methods can be applied, since any deviation from these idealisations would require a prohibitive increase in the size of the unit cell. However, some important strained-layer systems (such as the Si–Ge superlattice on Si (001)) consist by necessity (e.g., Pearsall et al., 1987) of only a few periods and should be treated as finite structures. In models of growth of such structures, it is necessary to invoke the existence of surface/interface defects (e.g., Kobayashi and Das Sharma, 1988). The apparent stability and spontaneous formation (ordering) of ultrathin (e.g., monolayer) superlattices remains unexplained within the above-mentioned constraints. The spectroscopic evidence for GaAs–AlAs superlattices discussed in Section IV indicates that cross-interface recombination might be affected by subtle deviations from ideal interfaces. It remains a major challenge to develop a framework in which real geometry structures could be modelled.

D. Nonlinearity in Semiconductor Superlattices

The conclusion one can draw from the preceding paragraphs is that, for most structures of interest, the key ingredients for microscopic theory, namely the components of the superlattice potential V of Eq. (5)—including the band lineup, the nuclear coordinates, and the basis set $\{\phi(n, \mathbf{k})\}$—can be derived from the bulk properties of the constituent semiconductors. However, it is also clear that there are significant corrections whose precise magnitude remains to be established. In particular, one must expect these corrections to grow in importance when more dissimilar materials are joined to form the interface.

It is important to reiterate that the statement about the bulklike character of the electron-charge distribution at semiconductor–semiconductor interfaces refers to the total valence-charge density, which determines the total energy of the system and, consequently, the stable lattice configuration. We argued in preceding paragraphs that in metallic and covalent crystals, the total charge-density changes brought about by even the strongest perturbations of the lattice, e.g., in the vicinity of a vacancy or surface, disappear at about the nearest-neighbour distance from the defect (see Jaros, 1982). It is therefore not very surprising that a much smaller perturbation, such as the presence of an interface with a not too dissimilar material, causes no significant changes in the total charge density. However, this does not mean that *individual quantum states* possess the same qualities. For example, there are states extending several lattice separations away from a defect. Thus, in structures where the particle-in-a-box picture breaks down, interactions across the interface may greatly alter the form of certain superlattice states even though the total valence-charge density and the lattice structure retain their bulk character. The strength of excitations into such complex wave packets is not a linear function of the applied field inducing the excitation. The response of such systems is therefore highly nonlinear.

More significantly, since the (vertical in k-space) separations between the energy levels obtained by folding the bulk bands into the small Brillouin zone of the superlattice are much smaller than their counterparts in the bulk Brillouin zone, even the changes in the potential V of Eq. (5) due to minor deviations from, say, bulk lattice coordinates, may lead to significant changes in the electronic structure of the superlattice. This relationship lies at the heart of the momentum mixing, which makes the semiconductor superlattice a novel nonlinear system and which is responsible for most of the phenomena of interest here.

Let us consider this relationship in some detail. Covalent crystals are a highly linear system in that they are characterised by parabolic bands, in which the electron velocity is a linear function of momentum, and which are separated by large gaps. Their second- and third-order response functions (e.g., susceptibility) are many orders of magnitude smaller than the first-order one. Whereas the first-order response of a quantum system (e.g., the first-order susceptibility in Eq. (6)) depends on virtual excitations corresponding to second-order (two-level) processes, the higher-order (nonlinear) effects depend on virtual excitations of electrons involving third- and fourth-order ones (i.e., three or more levels are involved). Such processes will be discussed in detail in Section IV.B.7 and IV.C.3. For this effect to be efficient, the energy separation between adjacent states must be small, and the probability of the jump (i.e., the corresponding matrix element that depends on the form of the wave functions) must be large. In bulk crystals, the (vertical) separation between levels strongly coupled by optical fields is large, and consequently, the optical nonlinearity is small.

In superlattices, both the level separation and the matrix element can be tuned. The coupling between the states can be built in so that the level position and the suitable momentum composition of the wave functions are engineered into the stationary states of the superlattice, or it can be switched on by an external field (electric, magnetic, pressure, etc.). Hence, in such a system, the electronic response, which is a quadratic or cubic function of the applied field, may be significantly enhanced (Morrison *et al.*, 1989). An even greater opportunity for creating nonlinear response arises in connection with quantum wires and dots, since reduction of dimensionality increases the oscillator strength. This nonlinearity can be observed in luminescence spectra, in wave mixing, as well as in transport experiments of different kind. The same virtual excitations influence fast transients and collective phenomena, although it must be stressed that in this context, their contribution has not been explored in any detail. Indeed, much of what has been said in the literature about nonlinear response in semiconductor superlattices rests on the manipulation of occupation numbers (band filling) and its effect upon the exciton gas (see Schmitt-Rink *et al.*, 1985, and references therein). The nonlinear response due to band filling can be understood without reference to the effect of microscopic forces upon the superlattice band structure and consequently will not be discussed here.

IV. Qualitative Assessment of Microscopic Phenomena

A. INTRODUCTION

In Section III, we outlined the microscopic formulation that is needed when the simple particle-in-a-box picture breaks down and that can be used to provide an accurate description of physical properties of any semiconductor superlattice. We discussed briefly the qualitative character of some key observables associated with the effects of momentum mixing. In Section IV, we shall concentrate on a few cases where at least a limited quantitative appreciation of some of these effects has been achieved. Three very different materials were chosen for this purpose. They will provide us with a vehicle for demonstrating the archetypal effects of momentum mixing in semiconductor superlattices currently accessible to experimentalisation.

These examples are in many ways quite pedestrian and, as emphasised in the very first paragraphs of this chapter, share some principal concepts with earlier efforts in achieving tunable electronic structure. The new feature is the ordered character and dimensionality of superlattice systems. It is also important to see that the principles governing momentum mixing in these structures are quite general and apply in more complex situations and in more boldly designed structures. As the growth and processing techniques improve, such structures—perhaps composites involving semiconductors, metals, as well as dielectrics—will become available and will be the subject of fresh investigations. A great variety of interfaces other than the semiconductor–semiconductor layers of nanometer thickness of very dissimilar materials have yet to overcome the formidable difficulties associated with the mismatch between electronic and mechanical properties of the constituents discussed in Section II.B.

What is sorely missed is an example of a full-scale quantum-mechanical study of the dynamics of propagation of electron and photon beams in such systems. It is really these dynamic processes (e.g., Chen and Mills, 1987) that will highlight the novelty of nanostructure systems, their nonlinearity, and its tunable character. As pointed out in Section III.D, the physics underlying dynamic effects is related to the same features in the electronic structure as those discussed here in connection with transitions between stationary states.

B. GaAs–Ga$_{1-x}$Al$_x$As

1. Level Crossing in Wide Well Structures

The most apparent examples of the role of momentum-mixing-related zone folding on band structure and optical properties of quantum well structures can be obtained when the alloy composition is altered so that the confined levels derived from the lowest and the secondary conduction-band minima cross. In that situation, one can observe a continuous change of the momentum wave function and its manifestations in both the nodal and envelope character of the charge density as a function of the alloy composition.

To demonstrate this effect, we have calculated (Ninno et al., 1985) the electronic structure of (001) GaAs–Ga$_{1-x}$Al$_x$As superlattice of period 140 Å, at the centre of the Brillouin zone, with the aluminium fraction x ranging from 0.2 to 0.5. We limited our attention to the conduction-band states in the range of energy 1.55–2.01 eV (the zero of energy is the top of GaAs valence band). The results of these calculations are shown in Fig. 5, where the curves labelled 1 and 2 correspond to the Γ-like confined states lying below the top

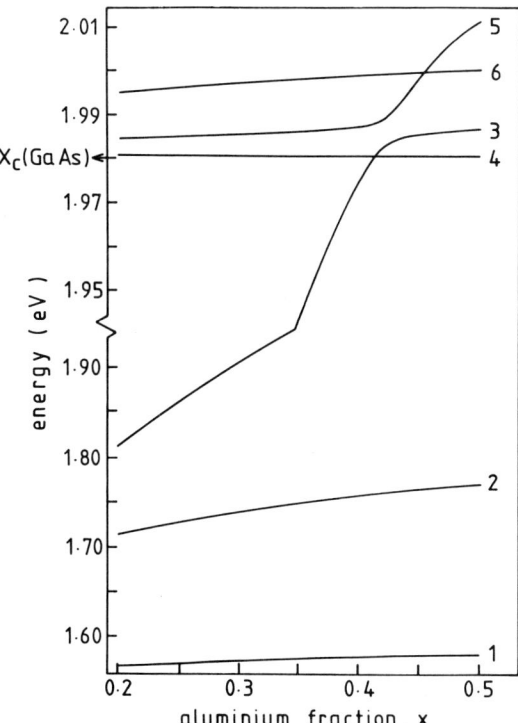

FIG. 5. The energy of the lowest minibands, at the center of the Brillouin zone, obtained in a pseudopotential calculation for a GaAs–Ga$_{0.8}$Al$_{0.2}$As superlattice of 140 Å period. 1 is the lowest conduction state of Γ character, while 4 is the lowest state of X character (Ninno et al., 1985). [Reprinted with permission from the American Physical Society, Ninno, D., Wong, K.B., Gell, M.A., and Jaros, M. (1985). Phys. Rev. **B32**, 2700.]

of the conventional confining barriers, curve 3 is a Γ-like resonance lying just above the barrier, and 4, 5, and 6 are X-related resonances, i.e., superlattice states lying above the X-minimum of the GaAs band structure and made up of bulk X-like Bloch states. Following Dingle (1975), the band lineup chosen here is such that X (GaAs) lies below X (GaAlAs). A summary of the charge densities along the superlattice axis (z) illustrating the evolution of the Bloch components of the superlattice wave function corresponding to state 5 is presented in Fig. 6. As the aluminium fraction is increased, state 3 is pushed towards the GaAs X valley, crossing the first X-related state 4 at $x=0.42$. This behaviour can be easily understood on intuitive grounds. With the increase of the aluminium concentration, the conduction-band minimum of Ga$_{1-x}$Al$_x$As rises in energy. The wave function of state 3 can be thought of as the ground state of an inverted well, since it peaks in the alloy layer.

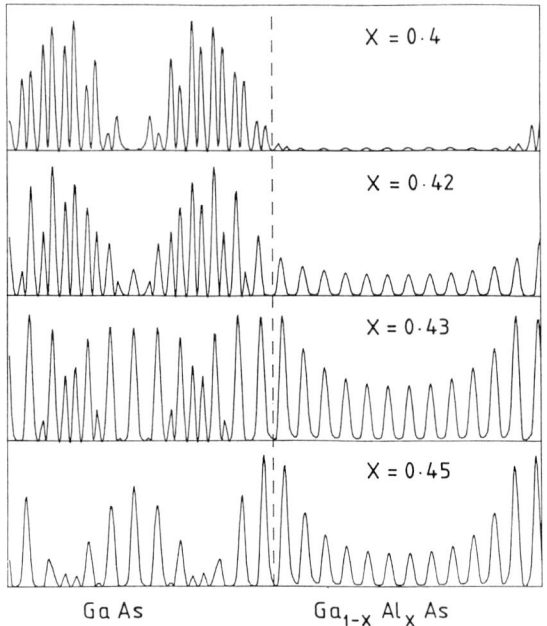

FIG. 6. The charge density in real space along the superlattice axis for state 5 shown in Fig. 5, as a function of alloy composition x [Reprinted with permission from the American Physical Society, Ninno, D., Wong, K.B., Gell, M.A., and Jaros, M. (1985). *Phys. Rev.* **B32**, 2700.]

Therefore, when the edge of the alloy rises in energy, it carries state 3 with it. It is of interest to note that state 3, after crossing state 4, does not cross state 5, the first excited X-related state. Indeed, manipulations exploiting the symmetry properties of Bloch functions show that state 3 and 5 have the same type of point-group symmetry. This means that in the interval $x=0.4$–0.45, the wave functions corresponding to states 3 and 5 are mixed. This effect can be seen directly by inspecting the evolution of the nodes of the charge densities presented in Fig. 6. In order to provide more insight into the degree of this mixing, we show in Fig. 7 the modulus squared of the $A(n,k)$ coefficients corresponding to state 5 of Fig. 5, as a function of k along the $\langle 001 \rangle$ line of the Brillouin zone of bulk GaAs. For simplicity, only contributions from the first GaAs conduction band are indicated. Figure 7 shows that state 5 is dominated by the X-valley before the crossing ($x=0.4$), and by the Γ-valley after the crossing ($x=0.45$), whereas at points of maximum mixing with state 3, it takes contributions from both Γ and X.

We calculated the optical matrix elements $T_{i,j}=|\langle\psi_i|\hat{p}|\Psi_j\rangle|^2$, where Ψ_j and ψ_i are valence and conduction wave functions generated at the centre of the superlattice Brillouin zone; $\hat{p}$ is the momentum operator whose direction

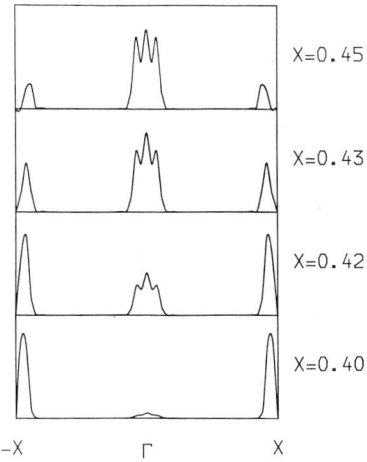

FIG. 7. The modulus squared of the wave function coefficients $A(n, k)$ associated with state 5 of Fig. 5, as a function of the alloy composition x [Reprinted with permission from the American Physical Society, Ninno, D., Wong, K.B., Gell, M.A., and Jaros, M. (1985). *Phys. Rev.* **B32**, 2700.]

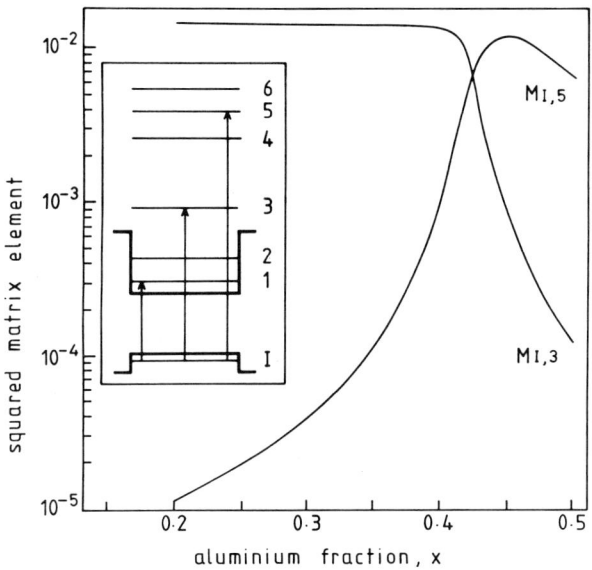

FIG. 8. The squared optical matrix elements for transitions from state 5 and 3 to the valence-band state I, normalised to the principal transition across the gap from 1 to I, as a function of the alloy composition x [Reprinted with permission from the American Physical Society, Ninno, D., Wong, K.B., Gell, M.A., and Jaros, M. (1985). *Phys. Rev.* **B32**, 2700.]

has been chosen to lie in the x, y (interface) plane. In Fig. 8, we show results concerning transitions from the top heavy-hole state to state 3 (curve $M_{1,3}$) and to state 5 (curve $M_{1,5}$) as a function of composition x. We have chosen, as unit of the squared matrix elements, the transition $T_{1,1}$ from the top heavy-hole state to the first conduction state (see inset in Fig. 8). Since the top heavy-hole (state I) is of p-like character, the magnitude of the matrix elements is determined by the amount of s-like character contained in the conduction states. As Fig. 8 shows, transitions involving state 3 are quite strong in the range 0.2–0.4 ($T_{1,3}$ is only two orders of magnitude weaker than $T_{1,1}$). This is because state 3, in this range of composition, is dominated by contributions from the Γ-valley and, valley and, therefore, has predominantly s-like character. A transition involving an X-related state like 5 has small matrix elements in the same range of composition ($M_{1,5}$ is less than 10^{-3} for $x \cong 0.4$). Finally, at high aluminium fraction ($x > 0.4$), the mixing of states 3 and 5 causes a change of s-like character resulting in an increase of $M_{1,5}$ at the expense of $M_{1,3}$.

2. Pressure-Induced Level Crossing

An analogous effect occurs when the lowest level localised in GaAs wells is driven up in energy by the application of hydrostatic pressure, towards the

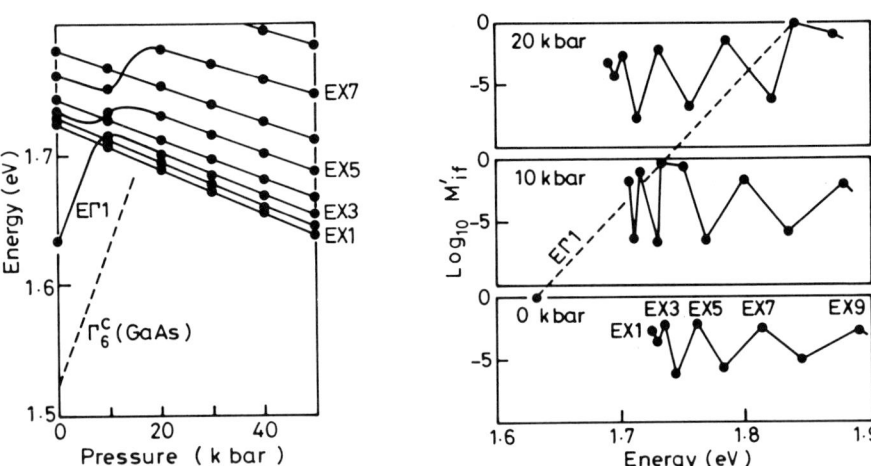

FIG. 9(a). The pressure dependence of the lowest confined levels of Γ and X character described in the text (Gell et al., 1987). (b) $\log_{10}$ of the optical matrix element for transitions to the top of the valence band as a function of pressure (GaAs–AlAs quantum wells 51 Å thick separated by barriers 102 Å thick). EX are energy levels of X character [Reprinted with permission from the American Physical Society, Gell, M.A., Ninno, D., Jaros, M., Wolford, D.J., Kuech, T.F., and Bradley, J.A. (1987). Phys. Rev. **B35**, 1196.]

states derived from the secondary X minima of the $Ga_{1-x}Al_xAs$ alloy (Gell et al., 1987).

Figure 9(a) shows the variation with hydrostatic pressure of the energies of the lowest conduction-band states generated at the centre of the superlattice Brillouin zone, in a GaAs/AlAs structure with wells 51 Å thick and barriers 102 Å thick. The solid dots correspond to the results of full calculations. The interrupted line indicates the edge of the conduction band of bulk GaAs.

Figure 9(b) shows plots of $\log_{10}$ of the squared optical matrix element M_{if} as a function of pressure for the system in Fig. 9(a). The initial state i is the ground heavy-hole state. The position of the zone-centre state is marked by the interrupted line. All other states are X-like states. The polarisation vector is chosen to lie in the $\langle 110 \rangle$ direction. In this calculation, the band lineup is such that X (GaAs) lies above X (GaAlAs). The coupling across the bulk Brillouin zone is strong near the crossing point in spite of the fact that this state derived from the centre of the bulk Brillouin zone is spatially confined in GaAs, whereas the state of X character is confined in the alloy. In fact, when the calculation is repeated with the "Dingle" band offset, the strength of the mixing is of the same order of magnitude, although in this case, the spatial overlap of the interacting levels is nearly 100%. Clearly, the difference in the phase of the Bloch function across the interface in the former case (staggered band offset) compensates for the lack of spatial overlap and enhances the coupling as well as the optical matrix element.

3. Direct Optical Determination of Band Offsets

The early work by Dingle and coworkers established a consensus of opinion that prevailed for nearly a decade and identified the fraction of the energy-gap difference between GaAs and $Ga_{1-x}Al_xAs$ appearing at the conduction-band edge as 0.85. This was achieved by fitting the barrier height (band offsets) to account for exciton-recombination energies. However, closer examination showed that other fractions may fit the data better and, more significantly, that the fit is quite insensitive to the choice of band offset. Fractions ranging from 0.85 to 0.50 have been quoted by authors using the optical fitting approach. More recently, charge-transfer studies, internal-photoemission studies, thermal transport over barriers, C–V profiling techniques, and Hall measurement in modulation-doped superlattices have all been used to deduce the conduction- or valence-band offsets (see Capasso and Margaritondo, 1987). The errors quoted by most authors are generally $> \pm 50$ meV (e.g., when $x = 1$, $\Delta E_v \sim 0.52 \pm 0.05$ eV). This is a composite error involving assessment of both experimental and model-dependent variables, and its value is difficult to establish accurately. In general, results involving electrical measurement lead to estimated offsets of $\sim 0.6:0.4$ distributed between the conduction and valence bands for the GaAs/AlAs system (e.g.,

Batey and Wright, 1986). However, a direct and more accurate measurement of the barrier height is possible if high hydrostatic pressure is used, at low temperatures, to drive the lowest confined states associated with the Γ-minimum of the conduction-band edge of GaAs up in energy, to the crossing point with the indirect (X) edge of the $Ga_{1-x}Al_xAs$ barriers (Wolford et al., 1986). This crossing occurs in a narrow range of energies in which the confined states, pinned to the secondary X-minima of the $Ga_{1-x}Al_xAs$, hydridize with the zone-centre (Γ) states of the GaAs, and as a result, yield observable optical spectra. This enables us to deduce the barrier heights, or the so-called band offsets, *directly* without introducing model-dependent parameters and with a meV accuracy. It also provides experimental evidence concerning physical properties of confined states having an X character, e.g., their energies, wave functions, and transition probabilities.

Figure 10 shows pressure results for a 70 Å well GaAs–$Ga_{0.71}Al_{0.28}As$ superlattice for pressures from 21.7 kbar to well beyond the bulk GaAs $\Gamma = X$ crossover. The narrow (7.4 meV) line at 21.7 kbar results from the Γ_{1e}–Γ_{1hh} transition across the gap, whereas the weak accompanying line results from substrate emission. For this sample, intensity remains strong and virtually independent of pressure from room pressure to ~30 kbar. Above 30 kbar, however, intensity decreases sharply with pressure (note scale factor of × 125 at 31.3 kbar), and a set of new lines appears, together with relative strengthening of the substrate emission. Occurring first as a shoulder on the main line at 31.3 kbar, this new emission becomes more prominent as excitation power density is reduced. Note that at this pressure, these new processes lie above the bulk GaAs band edge, but below the band edge of the Γ-confined quantum states. As pressure is increased still further, these lines shift towards lower energies, eventually passing through the GaAs band edge, while the Γ_{1e}–Γ_{1hh} line continues to follow the direct gap to higher energies.

There are many interesting features in the X-luminescence: (1) The superlattice potential splits the bulk four-fold degenerate X-levels in the interface plane (x, y) as suggested by Finkman et al. (1986), Ihm (1987), and others. However, this interaction is *not* strong enough to lower these states below the zone-folded X-levels in the z direction (Ting and Chang, 1987; Moore et al., 1988; Brown et al., 1989). (2) The Γ–X coupling along the growth direction (superlattice axis z) is *always* finite whether the number of monolayers is odd or even in either constituent. This is because the interface atomic site has a different microscopic potential from those inside the barrier and well (Gell et al., 1987)—even if they all belong to the same chemical species, e.g., As in GaAs/AlAs structures! (3) The Γ–X coupling increases with decreasing superlattice period. However, even in short period structures, the Γ–X gap is of order 10 meV (Gell et al., 1987; Pulsford et al., 1989). This implies that the transfer time from Γ to X levels is in excess of several

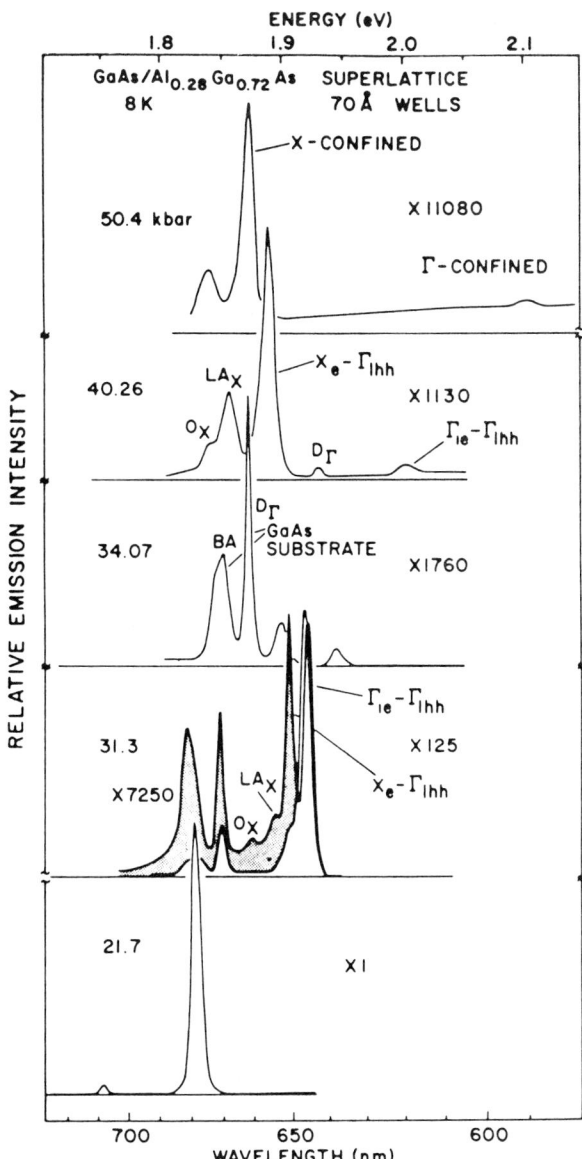

FIG. 10. Photoluminescence spectra at 8°K of a GaAs–Ga$_{0.72}$Al$_{28}$As superlattice showing the variation with pressure of the direct ($\Gamma_{1e} \to \Gamma_{1hh}$) and indirect transitions across the gap. The substrate emission is also included [Reprinted with permission from the American Institute of Physics, Wolford, D.J., Kuech, T.F., Bradley, J.A., Gell, M.A., Ninno, D., and Jaros, M. (1986). *J. Vac. Sci. Technol.* **B4**, 1043.]

picoseconds, i.e., much *longer* than the time an electron spends in the barrier. Hence, the effect of Γ–X coupling on electron *transport* across *perfect* interfaces is negligible. (4) The crossing of Γ and X levels can be brought about by an external electric field (Meynadier et al., 1988). However, the field merely *shifts* the confined levels of Γ and X character so that they become degenerate in energy. The level proximity increases the coupling. Even a very strong (10^5 V/cm) field is not capable of altering directly the wave function, i.e., the magnitude of the coupling responsible for the strength of the mixing (Hagon et al., 1989).

4. Extrinsic Nature of Cross-Interface Recombination

The reported characteristics of the recombination luminescence from type-II (indirect-gap) GaAs/Ga$_{1-x}$Al$_x$ superlattices vary widely. Danan et al. (1987) observed 40–100 nsec lifetime luminescence decays from short-period GaAs/AlAs superlattices, whereas Finkman et al. (1986) reported long-lived (μ sec) *nonexponential* cross-interface recombination from similar short-period GaAs–AlAs superlattices. Dawson et al. (1986) observed 50 nsec decays and low external quantum efficiencies in their 120-Å-period, compositional type-II superlattices with alloy wells. The theory (Gell et al., 1987) predicts that the radiative lifetime of electrons and holes in a type-II superlattice is expected to be very long compared to the lifetime in a direct-gap (type I) superlattice, since the oscillator strength is diminished for two reasons. The first factor responsible for the diminished oscillator strength is the greatly reduced spatial overlap of the wave functions. The second is the reduced k-space overlap of the wave functions that arises because the conduction-band minimum in the barrier has X-character, and the oscillator strength is proportional due to band mixing. In fact, the time-resolved luminescence experiments reveal that extrinsic interface-related phenomena strongly influence the detailed nature of the observed spectra. The oscillator strength of the indirect, cross-interface transition must therefore be extremely sensitive to the tails of the electron and hole functions in both real and k-space.

A model accounting for the observed luminescence processes (Steiner et al., 1988) is outlined schematically in Fig. 11. The laser-pumping excitation excites electrons to unconfined states high in the conduction band (schematically indicated as the resonance levels in Fig. 11). From there, the electrons rapidly thermalize to the Γ- and X-levels. These levels represent the lowest energy superlattice-confined states in the GaAs and GaAlAs, respectively. Some of the Γ-electrons recombine radiatively with confined holes from the lowest valence-band state resulting in the P_Γ luminescence before the remainder transfer across the interface to the conduction-band-minimum X-levels in the GaAlAs with characteristic transfer time τ. The lowest energy

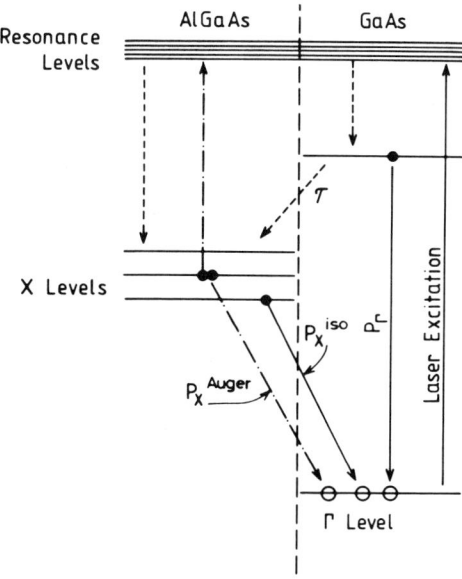

FIG. 11. A level diagram showing radiative recombination (solid lines) across the gap, and nonradiative (Auger) recombination. The heterointerface is indicated by the vertical line in the middle. Dashed lines indicate the path of electrons descending from higher levels towards the band edges [Reprinted with permission from Academic Press Inc. (London), Steiner, T.W., Wolford, D.J., Kuech, T.F., and Jaros, M. (1988). *Superlattices and Microstructures* **4**, 227.]

state P_x^{ISO} behaves as an isoelectronic trap, i.e., it has a long lifetime and high quantum efficiency, while the higher energy X-line has a predominantly nonradiative Auger decay mechanism. In this case, the recombination energy of one of the pair of electrons recombining with hole at Γ of the valence band is imparted to the remaining electron that re-excites it to the resonance levels from which it can return to X or Γ. The later possibility accounts for the observed P^{Long}-component.

The observation of the long-lived component in the P_Γ-luminescence that tracks the P^{Auger}-lifetime as the sample temperature and pressure are changed indicates an Auger-mediated decay of the P_x-state. The ejected hot electrons thermalize to the direct-gap local conduction-band minimum where some recombine. This results in the slow P^{Long}-luminescence before electrons transfer to the conduction-band minimum in the barrier material.

The Auger lifetime is remarkably short. In fact, if one compares with the lifetime of indirect excitons bound to shallow donors in bulk GaP—which is also an indirect-gap material—one finds the cross-interface process *faster* by at least one order of magnitude. This is so in spite of the *smaller* binding energy of defects in question. A similar conclusion can be obtained by comparing with theoretical estimates based on calculations of single-particle wave functions described above. It follows that even if one accepts the impurity model of Fig. 11, one has to assume an additional interface-related source of momentum (i.e., Γ–X) mixing in order to account at least qualitatively for the observed rates. This result can also be deduced from transport measurements in heterojunction structures (Section IV.B.8).

5. Quantum Wires

Recently, it has become possible to fabricate quantum-wire structures (QWS) of submicron dimensions. The electronic properties of QWS have been modelled in the framework of approximate schemes in which the coordinates perpendicular to the wire axis are assumed to be independent. The confined levels are then found by solving the effective-mass Schrödinger equation for a particle-in-a-box with infinite barriers. The energy levels can be characterised by quantum numbers n_x and n_y derived from the uncoupled one-dimensional square well problem. However, in a realistic model of the QWS, the potential representing the individual layers is finite, and motion in the two quantisation directions is not separable. In pseudopotential calculations (Wong et al., 1987), GaAs/GaAlAs (001), QWS are realised by embedding wires of GaAs in a regular fashion in the barrier material. The method of calculation is a natural extension of the approach used to model conventional superlattices. Let us begin with QWS where the alloy contains 30% of Al, and where the GaAs wells are 102×102 Å wide. The energy of the ground-conduction state agrees with the assumption that there is no coupling between the two directions perpendicular to the quantum-wire axis ("infinite" well). In a calculation for a conventional superlattice of exactly the same alloy composition and well widths, we find that the ground state is 28 meV above the bottom of the conduction-band edge of GaAs. This predicts a QWS ground state at energy 2×28 to 56 meV, which is also the energy obtained in our QWS calculation. However, this agreement is not obtained in thin wires. The differences are 3 and 25 meV in 68×68 and 34×34 Å quantum wires, respectively. There are two reasons for these differences. First, the well depth is not infinite. Second, there are interactions between the two quantisation directions that give rise to additional reflections and contribute to the energy of the confined states. The evidence for the latter can be found if we look at the wave function coefficient $A(n, \mathbf{k})$. In the case of thick wires, the ground state is derived from bulk states lying along the quantisation directions, i.e., $\langle 100 \rangle$ and $\langle 010 \rangle$. There are no contributions from off-axial terms. These off-axial contributions, which lie outside the conceptual framework of the particle-in-a-box model, become important for thin wires. However, the effect of the additional reflections upon the zone folding is small relative to that due to increase in localisation because of change in dimensionality. This increase reduces the effective strength of the superlattice potential so that the coupling between the states derived from the zone-centre and zone-edge (X) states is significantly reduced. Even levels only a couple of meV from each other do not interact! In conventional superlattices, such an interaction would be very strong indeed. This result has important consequences for transport processes in QWS. We have also calculated effective masses along the principal directions in the QWS band structure. We find that the masses in $\langle 100 \rangle$, $\langle 010 \rangle$, and $\langle 001 \rangle$ directions at the bottom of the conduction band are 0.134, 0.134, and 0.094 m, respectively, in 34×34 Å wires. The corresponding result for a conventional superlattice of identical parameters are 0.087, 0.087, and

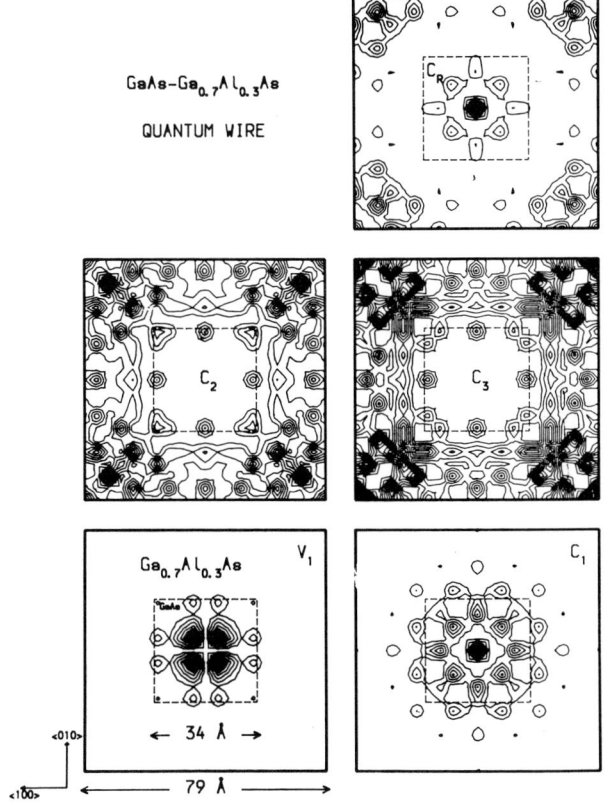

FIG. 12. The electron charge densities of the uppermost valence state V_1 and the lowest conduction states ($C_1, C_2, \ldots$) in a GaAs quantum wire. The cross section of the unit cell is shown for each state [Reprinted with permission from the American Physical Society, Wong, K.B., Jaros, M., and Hagon, J.P. (1987). *Phys. Rev.* **B35**, 2463.]

0.182 m. (The barriers are 45 Å wide in this calculation; note that $\langle 001 \rangle$ is the superlattice as well as the QWS axis.)

Figure 12 shows the electron-charge densities of the lowest confined levels in a QWS. The wire axis is perpendicular to the plane of this picture. Since the conduction-band states C_1-C_3 are entirely confined to the wire (GaAs), only the wire cross-sectional area is shown. State C_R lies above the semiclassical confining barrier and is localised mainly in the barrier material. This state is also solely derived from the zone-centre valley of GaAs. The Al concentration is 30%. State V_1 is the lowest level derived from the valence band. The material available for experimentation has not been of the quality and dimensions permitting a direct comparison with theory.

6. *Ultrathin Layers*

Ischibashi *et al.* (1985, 1986) (see also Nagle *et al.*, 1987, and others) have recently shown that it is possible to grow high-quality $(GaAs)_N(AlAs)_N$ (001) superlattices with $1 < N < 24$ using metalorganic chemical vapour deposition. The results provide evidence for the crossing of zone-centre- and zone-edge-related conduction states with decreasing layer width. Significantly, the energetics of this crossing can be linked to the relative alignment of the GaAs and AlAs band structures. The experimental data are also useful in that they are derived from superlattices ranging from those in which the width of the layers is a monolayer ($N = 1$), defined here as one layer of anions plus one

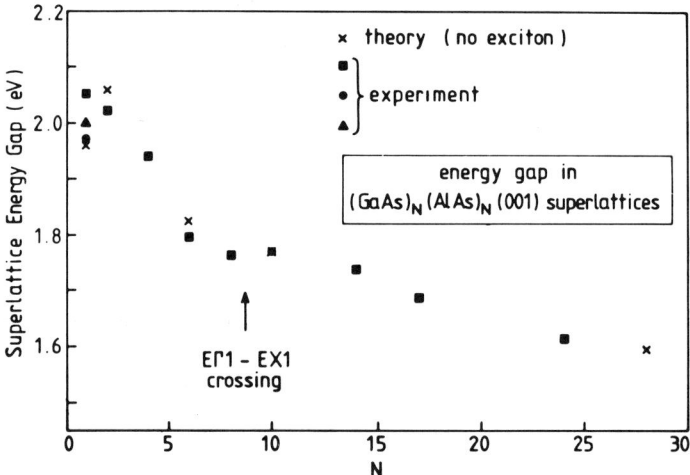

Fig. 13. This figure shows the variation of the superlattice energy gap as a function of the number (N) of monolayers in the superlattice period. The experimental results are also indicated [Reprinted with permission from the American Physical Society, Gell, M.A., Ninno, D., Jaros, M., and Herbert, D.C. (1986). *Phys. Rev.* **B34**, 2416.]

layer of cations, to those in which the superperiodicity is much larger than the underlying natural periodicity of the lattice.

In particular, one of the significant features of the experimental data presented by Ischibashi et al. (1985, 1986) is the plateau at and near $N = 8$ in the plot of the superlattice energy gap against N (Fig. 13). Such a plateau would not be expected if the lowest conduction state was always derived from the centre (Γ) of the bulk Brillouin zone. It has been shown that the plateau arises because the lowest conduction state at the centre of the superlattice Brillouin zone in the structures with $N < 8$ is derived primarily from the zone edges of AlAs. The pseudodirect transition across the fundamental gap to Γ from the ground heavy-hole-like state HH1 to the lowest zone-edge-related state (EX1) is allowed owing to a strong mixing of bulk zone-centre components into the conduction state. Owing to the very strong coupling between the zone-centre and zone-edge bulk states, the oscillator strength for the transition from HH1 to EX1 (for $1 < N < 8$) is comparable with or only about one to three orders of magnitude less than that for the same transition to the lowest zone-centre-related state. The theoretical result is very sensitive to the band offset and to the relative position of the X and Γ minima. For example, in order to account for the plateau in Fig. 13, the relative alignment of the band structures of GaAs and AlAs, at least near $N = 8$ in the ultrathin regime, must be such that X_6 (AlAs) lies about 0.3 eV below X_6 (GaAs) (Gell et al., 1986, and references therein).

From these results, it would seem that the effective band offset in the $(GaAs)_2(AlAs)_2$ superlattice is quite meaningful and remains close to the offset present in systems comprised of thicker layers. It follows that the mechanism that controls band offsets and that is active in the thick-layered systems must also be a dominant mechanism in the $(GaAs)_2(AlAs)_2$ superlattice, even though the layers are only one lattice constant in width. There is a noticeable but surprisingly small discrepancy for the monolayer structure.

7. Band Nonparabolicity

Mobile electrons in a nonparabolic band have a nonlinear velocity-momentum relation. If an external electromagnetic field is applied, the electron momentum follows the frequency of the applied field. The nonlinear velocity component causes the induced current to contain mixed frequency components. The magnitude of this nonlinearity is measured by the third-order nonlinear susceptibility that is proportional to the fourth derivative of the band energy versus wave vector. The bulk conduction-band nonparabolicity in, say, GaAs is quite small (e.g., Patel et al., 1966, see also a summary paper by Chang, 1985). However, we have argued that momentum mixing associated with the breakdown of the particle-in-a-box model of confinement affects the position of the confined levels and the composition of the momentum wave function. Consequently, it alters the optical matrix elements between confined states. In particular, it also alters the matrix elements involving higher-lying states associated with the primary and secondary conduction-band minima above the semiclassical confining barrier. The third-order susceptibility depends on the strength of the virtual excitations involving these higher states, and is proportional to d^4E/dk^4. It follows from the account given above that the nonlinear optical constants of semicon-

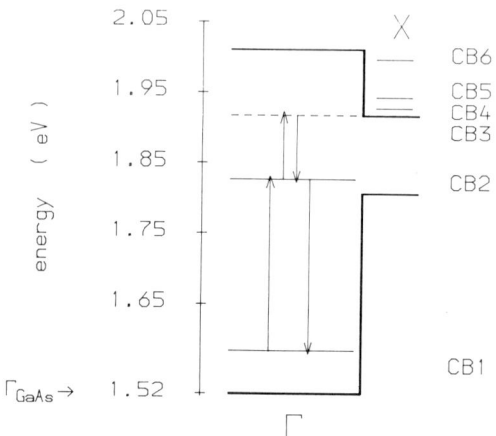

FIG. 14. The alignment of the principal and secondary (X) minima for a GaAs (56 Å)–$Ga_{0.7}Al_{0.3}As$ (22 Å) superlattice. The lowest confined levels in the conduction band are shown. A four-stage virtual excitation process, which determines the fourth derivative of the lowest miniband, is also indicated (arrows)[Reprinted with permission from the American Physical Society, Brown, L.D.L., Jaros, M., and Ninno, D. (1987). *Phys. Rev.* **B36**, 2935.]

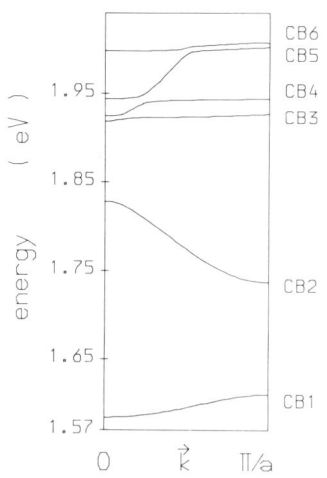

FIG. 15. The band structure of the superlattice described in Fig. 14. CB1, CB2,... are the lowest conduction minibands in the small Brillouin zone [Reprinted with permission from the American Physical Society, Brown, L.D.L., Jaros, M., and Ninno, D. (1987). *Phys. Rev.* **B36**, 2935.]

ductor microstructures can be "tuned" by manipulating the degrees of momentum mixing in the system involving the relevant states. We shall demonstrate this tuning on a simple example (Brown et al., 1987).

To help visualize the virtual process associated with the nonparabolicity, the Γ–X well alignment for GaAs/Ga$_{0.7}$Al$_{0.3}$ is shown in Fig. 14, together with the confined levels and a typical four-state virtual excitation. The levels linked to the well minimum lie in the GaAs layer, whereas those linked to the X well minima lie in the alloy. In Fig. 15, we can see the miniband dispersion across the superlattice Brillouin zone for the conduction states shown in Fig. 14. In this example, we considered GaAs (56 Å)–Ga$_{1-x}$Al$_x$As (22 Å) superlattices. These relationships between localisation, dispersion, and momentum components are familiar from early considerations in which the momentum mixing was assessed with a view to evaluating the optical matrix elements across the gap. We saw there that as the aluminium fraction in the barrier material is changed, with it the well depth and, consequently, the position of the minibands change, too. This change in turn alters the magnitude of the momentum mixing, and the optical matrix elements between higher minibands that enter the expression for the fourth derivative acquire new values. This is pictured in Fig. 16. Around $x = 0.35$, where the bulk and X conduction-band edges cross, the fourth derivative exhibits a minimum. Reduction or increase in the X energy increases the Γ–X separation, and the magnitude of d^4E/dk^4 falls. At the maximum, the conduction-band nonparabolicity is six times larger than that in bulk GaAs.

The behaviour outlined above is contrasted with that predicted by the semiclassical model. The lower curve in Fig. 16 was obtained by the Kronig–Penney method. It shows the decreasing fourth derivative with decreasing x. This is not surprising, since for a fixed well and barrier width, and fixed well material (GaAs), the only compositionally dependent parameter is the barrier

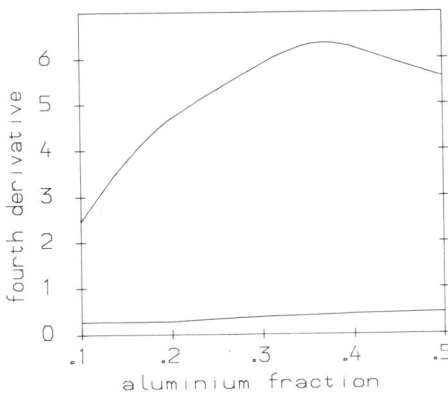

FIG. 16. The fourth derivative of the lowest conduction miniband for the superlattice shown in Fig. 15, as a function of Al fraction. This derivative is normalised to that of bulk GaAs. The upper curve corresponds to pseudopotential results, and the lower curve to a Kronig–Penney calculation [Reprinted with permission from the American Physical Society, Brown, L.D.L., Jaros, M., and Ninno, D. (1987). Phys. Rev. **B36**, 2935.]

height. This simple model cannot reproduce the structure obtained by the full-scale calculation.

It follows that even in structures where the position of the energy levels in the confining wells appear to be well represented by the particle-in-a-box model, the nonlinear response functions is not accurately accounted for, and the short-wavelength rapidly varying component of the microscopic potential must be fully represented. The nonlinear response functions of these superlattice structures depend on the virtual excitations involving higher-lying states in the conduction and/or valence band, which can only be accounted for in a full-scale calculation.

8. Vertical Transport

There have been numerous reports of anomalous electron transport in $GaAs/Ga_{1-x}Al_xAs$-GaAs barrier structures in the direction perpendicular to the interface plane. For example, Solomon et al. (1986) and Moezawa et al. (1986) deduced from the temperature dependence of the thermally ionised current in this structure that for $x > 0.5$, the barrier height decreases with increasing x, as expected from the indirect character of the fundamental gap of $Ga_{1-x}Al_xAs$. The prefactor in the Richardson formula reported in these papers is remarkably large. This can only be explained if it is assumed that the coupling between the Γ and X states is stronger than that predicted for an idealised $GaAs/Ga_{1-x}Al_xAs$ interface. Since the strength of the observed Γ–X coupling varies continuously from $x = 0.5$ to 1.0, effects such as alloy disorder can be excluded (such effects have, in fact, often been invoked to account for unusual features of cross-interface recombination and vertical transport). An analogous conclusion emerged from the luminescence data of Steiner et al. (1988) discussed briefly in Section IV.B.4.

In analysing the thermoionisation (or tunnelling) current data, one takes as a first-order approximation the barrier to be given by the separation between the Γ minima of GaAs and $Ga_{1-x}Al_xAs$, with an approximate correction for carrier concentration, space charge potential, temperature, and external field effects. For $x > 0.5$, this approach obviously fails, since the data shows that the barrier height follows the $\Gamma(GaAs)/X(Ga_{1-x}As_xAs)$ separation, and the prefactor in the Richardson formula gradually drops. This is often accounted for by retaining the semiclassical model and by simply regarding the effective mass in the barrier as an adjustable parameter. Unfortunately, neither the WKB (Wigner-Kramers-Brillouin) nor the effective mass approximation remain valid in this regime because of the change of bulk momentum at the interface (from Γ to X), and the parametrisation is misleading. The reduction in the prefactor with increasing x follows the trend predicted by Gell et al. (1987)—and expected on intuitive grounds—for the Γ–X coupling in this structure. This coupling decreases with increasing separation of X and Γ in $Ga_{1-x}Al_xAs$ and gives a microscopic meaning to the prefactor in the Richardson formula. It is this coupling, which is a sensitive function of the crystal potential at the interface, that is required to be anomalously large to fit the data (Zohta, 1988). The data point to a significant deviation from ideal interface structure, but its precise nature remains to be identified.

C. $Si-Si_{1-x}Ge_x$

1. *Superlattices Grown on an Alloy Buffer (Symmetric Strain)*

Let us consider $Si-Si_{1-x}Ge_x$ structures in which the substrate is an alloy of Si and Ge whose composition is chosen so that it has the same lattice constant as the average of the lattice constants of the materials constituting the superlattice. In such circumstances, the strain is taken up by both constituents (i.e., compressive strain in the alloy and tensile strain in Si). Such symmetrically strained structures are particularly suitable for applications, since the net strain per period is zero, and at least in principle, a large number of periods can be grown without creating dislocations. The growth of such structures was pioneered by Kasper, and it has recently been brought to considerable perfection (e.g., Kasper *et al.*, 1988, and references therein).

In this section, we shall approach the problem from the simplest point of view (particle-in-a-box picture). Then it is possible to visualise electrons confined by a barrier whose height is given by the separation of the $X(\Delta)$ minima of bulk Si and Ge crystals corrected for strain-induced splitting. This is because the strain lowers symmetry from tetrahedral to tetragonal. Similarly, the states at the top of the valence band are split by this axial field. The magnitude of these splittings was estimated from bulk-deformation potentials (People, 1985). The full-scale calculations of Van de Valle and Martin (1986a, b) provided the magnitude of the valence-band offset and confirmed the magnitude of the splitting at the top of the valence band obtained from bulk-deformation potentials.

Two types of calculations were performed to model the electronic-band properties of these structures at the conduction- and valence-band edges (Morrison *et al.*, 1987; Morrison and Jaros, 1988), which will be referred to as the strained and unstrained calculations. In the unstrained calculations, the positions of the Si and the alloy atoms are the same as in the bulk of the substrate material. This represents a superlattice made from bulk silicon with a slightly larger lattice constant and bulk $Si_{0.5}Ge_{0.5}$ with a slightly reduced lattice constant. This enlargement and compression of the lattice constant is the same in all directions. In the strained calculation described here, the positions of the silicon and alloy atoms were shifted from the bulk lattice sites of the substrate material (which in this case is $Si_{0.75}Ge_{0.25}$) in the $\langle 001 \rangle$

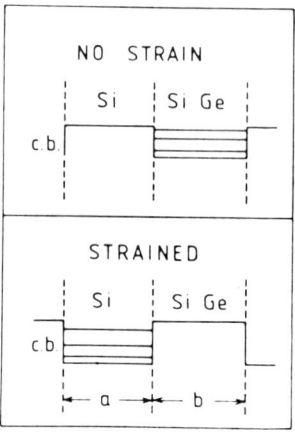

FIG. 17. The schematic diagram of the effect of strain upon the electron confinement in Si–SiGe superlattices matched to the substrate lattice constant. [Reprinted with permission from the American Physical Society, Morrison, I., K.B. (1987). *Phys. Rev.* **B35**, 9693.]

direction, and their overall position was adjusted so that it corresponds to the free-standing minimum energy configuration normally considered in the literature. In the unstrained case, we obtained well-confined electron states lying in the alloy layers. The effect of strain is to shift the energies of the confined states so that the separation between levels is substantially increased. The most important feature is the shift in the position of the X-valleys, which makes the Si valley lie below that of Ge. Consequently, the electron states lying near the conduction-band edge are localised in the silicon layers (Fig. 17). This localisation has been demonstrated in Raman spectra (Abstreiter *et al.*, 1985).

The oscillator strength F_{ij} of the transition across the superlattice gap increases significantly with decreasing superlattice period (Fig. 18). This is

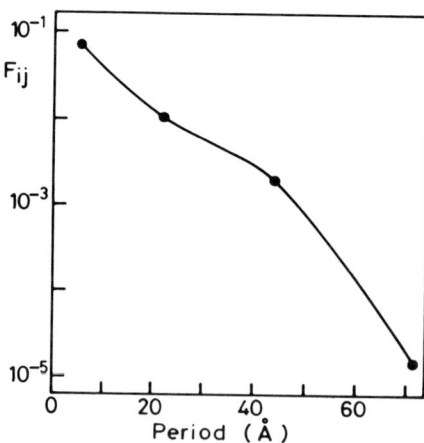

FIG. 18. The oscillator strength of the transition across the gap as a function of superlattice period for Si–SiGe structures lattice matched to the substrate [Reprinted with permission from the American Physical Society, Morrison, I., Jaros, M., and Wong, K.B. (1987). *Phys. Rev.* **B35**, 9693.]

due to the breakdown of the particle-in-a-box model associated with a combined effect of strain and ordered-superlattice potential referred to in Section III. The short-wavelength components of this microscopic potential mix bulk states of Eq. (5) across the bulk Brillouin zone and introduce a substantial Γ component into the conduction-band wave function. For ultrathin layers, the optical matrix element actually approaches values that lie within one order of magnitude of that for bulk GaAs! As pointed out in Section III, in thick layers, the structure factor multiplying the atomic potential contributions reduces the magnitude of the Fourier components of the microscopic potential, and only terms with wave vectors in the proximity of bulk reciprocal lattice vectors are large. The enhancement of the optical transition probability is negligibly small, and the conduction-band wave function retains its Δ-character peculiar to the Bloch states derived from the bottom of the conduction band of bulk silicon. However, the increase in the separation of the ground and excited confined levels, and consequently, the effective confining power of the barrier material is significant even in large-period structures. This change in separation reflects the broadening of the superlattice wave function in the momentum space. However, in this case, the broadening involves only higher-lying bulk states that are mainly derived from the silicon X-valley. It does not reach across the Brillouin zone, and the corresponding optical matrix element is therefore weak.

2. Zone Folding and Dispersion Relations in Symmetric Strain Superlattices

The understanding of transport and optical properties of superlattices depends on the dispersion relations (band structures) of the low-lying conduction states and the effective masses associated with the various minima. A natural starting point for such considerations is to take bulk bands of the constituents and fold them into the small (superlattice) Brillouin zone. We can then switch on interactions between bands that cross or lie close to each other. The zone-folding exercise is perfectly straightforward and quite instructive to study qualitative changes in the band structure as a function of the superlattice period. The period determines the size of the small Brillouin zone and, consequently, the way bulk bands are broken up into minibands (see People and Jackson, 1987). In fact, it turns out that even in small-period superlattices, the bulk origin of these minibands is easily detected!

Let us have a look at the dispersion relations for two superlattices of period 22.0 Å and 27.5 Å, corresponding to unit cells containing 8 Si and 8 $Si_{0.5}Ge_{0.5}$ atoms, and 10 Si and 10 $Si_{0.5}Ge_{0.5}$ atoms, respectively.

In Fig. 19, we can see the dispersion relations of the lowest conduction bands in the direction perpendicular to the interfaces, for both periods, generated in a full-scale pseudopotential calculation. The solid lines refer to the smaller-period system. In Fig. 20, there is the dispersion in the direction

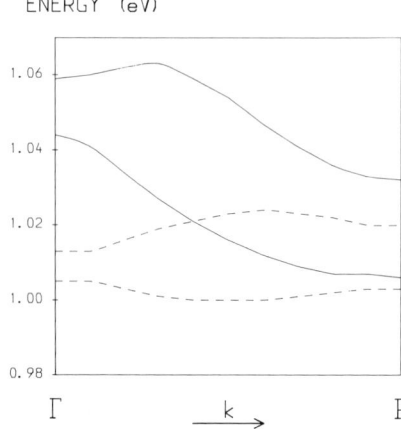

FIG. 19. The dispersion of the lowest two-conduction minibands in Si–SiGe superlattices of 22 Å (solid line) and 27.5-Å- (interrupted line) period superlattices, in the direction perpendicular to the interface. Energies are measured with respect to the top of the valence band. [Reprinted with permission from the American Physical Society, Morrison, I., and Jaros, M. (1988). *Phys. Rev.* **B37**, 916.]

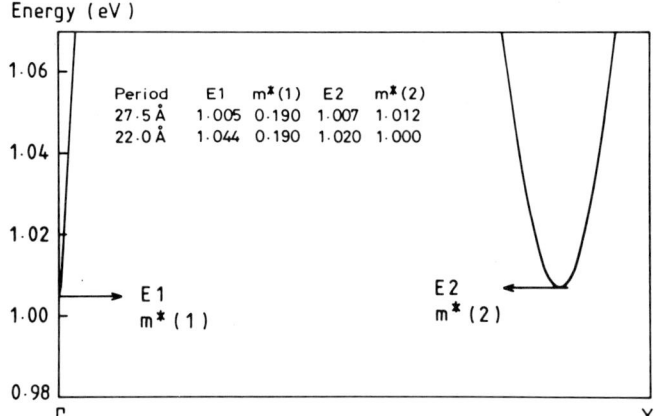

FIG. 20. The bottom of the conduction band of a Si–SiGe superlattice in the direction parallel to the interface plane. The energies measured with respect to the top of the valence band, and the corresponding masses in units of free-electron mass are also shown [Reprinted with permission from the American Physical Society, Morrison, I., and Jaros, M. (1988). *Phys. Rev.* **B37**, 916.]

parallel to the interfaces, in the 27.5-Å-period case. The picture for the 22.0-Å-period case, in this direction, is very similar and is not shown here. The energies of these minima and the effective masses m^* associated with them are also shown. All energies presented here are measured from the highest-lying valence band in each case. m^* is in units of free-electron mass.

To gain deeper insight into the origin of this dispersion, we must recall the method of calculation. In this method, the superlattice eigenfunctions, ψ, are

5. MICROSCOPIC PHENOMENA IN ORDERED SUPERLATTICES   225

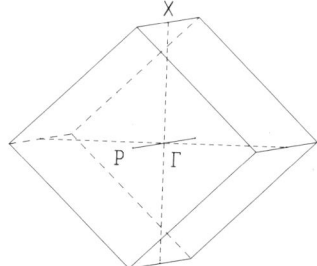

FIG. 21. The superlattice Brillouin zone with the symmetry points referred to in the text.

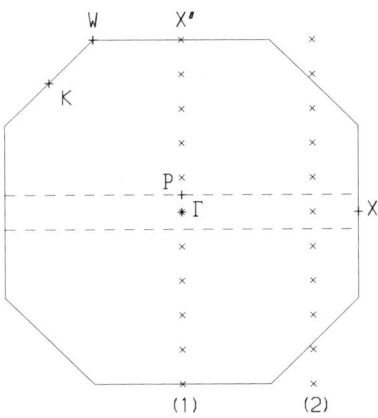

FIG. 22. A cross-section of the bulk Brillouin zone, which shows the way in which the superlattice Brillouin-zone (dashed lines) maps inside the bulk zone for a 27.5-Å-period superlattice [Reprinted with permission from the American Physical Society, Morrison, I., and Jaros, M. (1988). *Phys. Rev.* **B37**, 916.]

obtained as an expansion of the eigenfunctions of the buffer material, $\phi(n, \mathbf{k})$, in terms of coefficients $A(n, \mathbf{k})$. The translational symmetry associated with the superlattice defines the $\mathbf{k}$ points in the expansion as those coupled by a superlattice reciprocal lattice vector, i.e., $\mathbf{k} = \mathbf{k}_s + \mathbf{g}$, where $\mathbf{k}_s$ is the reduced wave vector in the superlattice Brillouin zone (Fig. 21); $\mathbf{g}$ is a superlattice reciprocal lattice vector. In the picture where the bulk band structures are "folded" inside the superlattice Brillouin zone, the bulk bands at the k points defined above are all folded back to the same point, $\mathbf{k}_s$. This is shown in Fig. 22. We see there a cross section of the bulk Brillouin zone, of the $Si_{0.75}Ge_{0.25}$ buffer layer, which is lattice matched to the superlattice period, with points of symmetry marked. Also shown is the way in which the superlattice Brillouin zone maps inside the bulk zone, in the 27.5-Å-period case (the interrupted lines). Two sets of sampling points needed in the expansion of the superlattice eigenfunctions are indicated (x); (1) for the calculation performed at Γ, and (2) for the calculation performed at a secondary minimum away from Γ along the line $\Gamma-X$. These points are inside the superlattice Brillouin zone. For the calculation performed away from Γ along the line Γ–P, the points indicated for the calculation at Γ are displaced vertically along the line Γ–X'. The sampling points seen to lie outside the bulk Brillouin zone can be translated through a bulk reciprocal lattice vector to lie inside.

Consider the dispersion relations in the direction perpendicular to the interface planes. The position of the minima in this direction depends upon the period of the superlattice. In the 22-Å-period case, the minimum occurs at the point P, whereas in the 27.5-Å-period calculation, the minimum is midway along the line Γ–P. The corresponding states are highly confined in the silicon layers. As we traverse the superlattice Brillouin zone in the direction Γ–P, the sampling points coupled to the reduced wave vector in the superlattice Brillouin zone are displaced vertically along the line Γ–X from those needed at Γ. The minimum will occur at the point in the superlattice Brillouin zone that is coupled to the reciprocal lattice vector. As the lengths of the superlattice reciprocal lattice vectors, in this direction, are dependent on the period, the position of this minima is also dependent on the period. The quasidouble degeneracy of these lowest two bands originates from the camelsback-type structure of the bulk conduction (the bulk minima lying away from the bulk X-point along the line Γ–X). For states lying below the energy of the bulk X point, there will be two bulk states, on either side of the minima, with similar energies, which are coupled to the same point inside the superlattice Brillouin zone. The superlattice eigenfunctions along these bands are constructed mainly from states around the bulk minima along the line Γ–X.

Now consider the dispersion relations in the direction parallel to the interface planes. Here we see two minima with quite different effective masses. As explained above, the minimum at Γ originates from the folding in of the bulk minimum from the Γ–X' direction to inside the superlattice Brillouin zone, with the corresponding superlattice states highly localised in the silicon layers. The second minimum near the superlattice X point originates from the equivalent bulk minima along Γ–X (the bulk X-point maps directly onto the superlattice X point). The superlattice states here are constructed mainly from states around this bulk minimum and do not exhibit any confinement in either constituent.

3. *Superlattice Grown on a Si Substrate (Asymmetric Strain)*

When a Si/Ge superlattice is grown on a Si (001) substrate, all strain is taken up by the Ge layers, and the Si layers retain their bulk coordinates. This means that the effect of strain is maximised for a given chemical composition. The axial field and the corresponding splittings of the Ge bulk levels are much larger. Here we again assume that the layers are thin enough for a homogeneous strain model to apply. Since in this arrangement the strain is building up with the addition of new layers, we cannot increase the number of layers or periods indefinitely (Bevk et al., 1986). For example, if the thickness of the Si and Ge layers is one bulk lattice constant (four atomic planes per layer), we can afford only five periods. Before such a five-period structure is repeated, a thick (e.g., > 200 Å) layer of Si must be used to separate it from the other five-period structures. Otherwise the formation of dislocations would destroy any useful properties of our sample. However, in most models of these superlattices, their finite character has been ignored and they have been treated as infinite superlattices, i.e., in terms of the usual periodic boundary conditions along the growth axis invoked in the preceding sections. This makes it possible to use the concept of band structure in the analysis. We shall adopt this attitude here too and return to consider the finite character of these superlattices in Section IV.C.5. Under such circumstances, no new concepts are needed, and we can give an overall description of the structures grown on Si in terms of band offsets, axial field splittings of bulk states, zone folding, and momentum mixing in the same manner as in Sections IV.C.1 and IV.C.2 (People and Jackson, 1987). The band offsets are shown in Fig. 23, which also gives the magnitude of the splitting of the bulk Ge conduction-band minima derived from the Δ-point in the bulk Brillouin zone.

The band structure of a Si_4Ge_4 superlattice near the band edges, obtained by Wong et al. along the $\langle 001 \rangle$ direction in the superlattice Brillouin zone, is shown in Fig. 24. In this calculation, the nuclear coordinates are those predicted for the ideal infinite superlattice from classical elasticity theory and in first-principles calculations of Van de Walle and Martin (1986a, b) discussed in Section III. States V1 and V2 are the uppermost valence-band states. States C1 and C2 are the lowest conduction-band states. These states originate from the degenerate bulk conduction-band minima at the bulk X-character. The states C3 and C4 are also derived mainly from the X-valley.

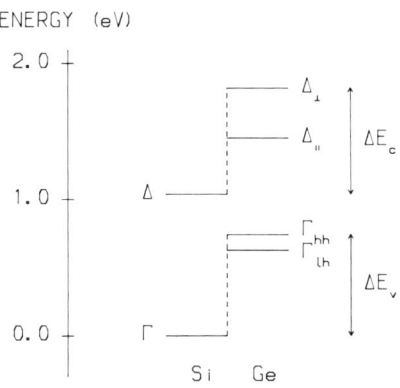

FIG. 23. The valence band offset ΔE_v and the conduction band offset ΔE_c for a Si/Ge (001) heterojunction structure grown on Si. Γ and Δ are the bulk symmetry points. The values corresponding to directions parallel and perpendicular to the interface are shown. hh and lh indicate the heavy and light hole levels, respectively [Reprinted with permission from the American Physical Society, Wong, K.B., Jaros, M., Morrison, I., and Hagon, J.P. (1988). *Phys. Rev. Lett.* **60**, 2221.]

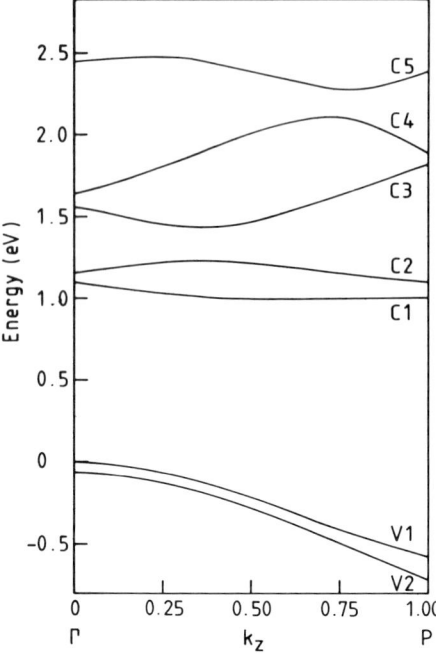

FIG. 24. The band structure of the four monolayer ideal (infinite and defect free) Si/Ge superlattice in the small Brillouin zone, along the direction of the superlattice axis, described in the text.

C5 is the lowest state of bulk Γ-origin. If we take into account the relevant corrections (e.g., the band gap is underestimated by about 0.5–0.6 eV in the LDA, and overestimated in the quasiparticle calculation by about 0.1–0.2 eV, whereas in the empirical calculations where bulk gaps are fitted, an uncertainty of order 0.1 eV might nevertheless be expected in the representation of the axial field splittings; see Gell, 1988, for a detailed assessment of such corrections), this band structure is in fact the same within 0.1 eV as the band structure obtained with the ideal coordinates in the quasiparticle (Hybertsen and Schlüter, 1987), LDA (Froyen et al., 1787), and tight-binding calculation (Brey and Tejedor, 1987).

4. *Role of Ordering and Substrate Composition upon the Electronic Structure of Ultrathin Superlattices*

The chief motive for electronic structure manipulations concerning Si–Ge is the possibility of enhancing the optical transition probability across the gap and of optimising the energy at which the transition occurs. It is clear from Fig. 18 that, to achieve this objective, we must work with layers whose

5. MICROSCOPIC PHENOMENA IN ORDERED SUPERLATTICES 229

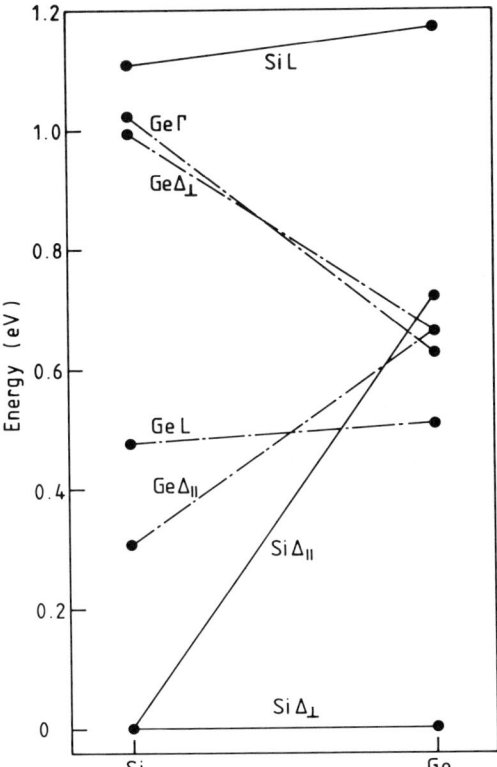

FIG. 25. The energy levels of bulk Γ, Δ, and L symmetry of the Si and Ge superlattice layers, measured from the bottom of the conduction band of Si, as a function of the choice of substrate (Si to Ge). (The data are given by Gell, 1988.)

thickness is of order one bulk lattice constant. We must also aim at maximising the role of strain to increase the momentum (Γ–X) mixing.

The strain has two effects. By compressing Ge, the states of Γ (and L) character are raised in energy relative to X and Δ. The tetragonal field splits the sixfold degeneracy at the X point and lowers the states oriented in the plane of the interface. The X- and Δ-states lying along the superlattice axis are pushed up. This is illustrated in Fig. 25. Thus if we grow on Si, the sign of the tetragonal field is such that the Ge Δ-minimum along the superlattice axis is raised relative to its counterpart in Si; this increases the conduction-band offset and the degree of electron confinement in Si layers. The top of the valence band is always in Ge (i.e., holes are localised mainly in the Ge layers), so that by increasing the conduction-band offset, we are reducing the electron–hole overlap and the transition probability. Since the net effect of strain on Ge states of Γ-character is to increase the energy separation between the band edge and the Γ states, this has an unfavourable effect on the Γ–$X(\Delta)$ mixing (and the optical transition probability). Note that this is the relationship familiar from our account of Γ–X mixing in GaAs-related structures: In the spirit of second-order perturbation theory, the mixing between these two states is inversely proportional to their separation in energy.

These considerations provide simple guidelines for optimising the choice of substrate. For example, it follows that if we grow on Ge, the axial splitting would be reversed, and the Δ-minimum along the superlattice axis would lie at the band edge. This also reduces the separation between the Ge states of bulk Γ, and the $X(\Delta)$ origin.

The effect of ordering can be seen in Fig. 18. This figure shows that the optical transition probability increases with decreasing superlattice period. This is always the case, provided that we reduce the period by scaling a given structure as in Fig. 18, i.e., for a fixed strain, chemical composition, and ratio of layer widths. By choosing the period, we choose (1) the way bulk bands are folded into the small Brillouin zone, and (2) the shift of the confined levels from the bulk band edge, since, according to the particle-in-a-box picture, that increases as the inverse of the square of the well width. Thus, apart from changing the degree of momentum mixing in the conduction-band wave function, we also change the magnitude of band gaps. This is illustrated in Fig. 26. Both the fundamental gap and the separation between Γ and $X(\Delta)$ levels are affected. For example, it is easy to see that while in an alloy of equal chemical composition the first (direct) optical transition occurs (at Γ) at an energy larger than 2 eV, the 2-, 4-, and 6-monolayer Si/Ge superlattices on Si exhibit transitions at Γ at much lower energies.

The alloy bandgap shown in Fig. 26 is that of a bulk alloy with 50% of Si. The electronic structure of this alloy is also affected by the 4% difference in the bulk lattice constants of Si and Ge. There are corrections to the virtual crystal approximations that lead to broadening of the density of states near the band edges and, consequently, to a shift of about 50 meV of the absorption edge (i.e., bandgap reduction; see coherent potential studies by Hass et al., 1984; Krishnamurthy et al., 1986; and for the effect on superlattice band structure, Ting and Chang, 1986, and references therein). However, the effective masses are not significantly affected. An additional reduction of bandgap results from straining the alloy by growing it on a Si substrate (Froyen et al., 1988). The tetragonal field splits the degenerate bulk levels as indicated in the above paragraphs. However, the changes are of order 0.1 eV, i.e., small compared to the large difference (about 1 eV) between the alloy and superlattice direct gaps.

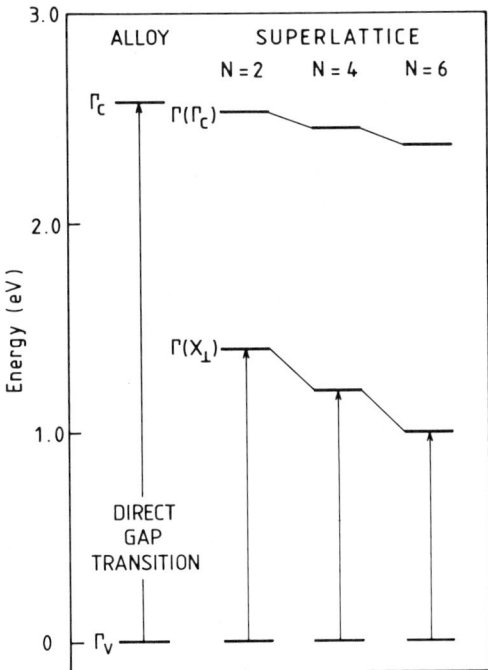

FIG. 26. A sketch of the direct band gaps at the centre of the Brillouin zone (Γ) of a SiGe bulk alloy and $Si_N Ge_N$ superlattices with N = 2, 4, and 6. c, v indicate the conduction and valence band, respectively. Note that states of bulk X symmetry in the direction perpendicular to the interface map on the Γ point. The top of the valence band of all structures is placed at zero energy to facilitate comparison.

It is interesting to note that this qualitative difference in bandgaps of an ordered heterostructure and the corresponding alloy, and the possibility of achieving a quasidirect transition was considered theoretically many years ago by Gnutzmann and Clausecker (1974). However, the effect of strain needed for a realistic estimate of the oscillator strength was not incorporated into their calculation.

The levels representing the superlattice conduction-band edge are obtained by zone-folding bulk states of Δ (X) character. For transitions from such states to be observable without phonon assistance, there must be a considerable momentum mixing, and a proper appreciation of the strength of such transitions requires a detailed calculation. Although a good estimate of these energy levels can be made by simple means (e.g., Kronig–Penney model, see People and Jackson, 1987), detailed calculations are needed to establish the precise positions of the superlattice states near the band edges. Morrison et al. (1987) and Morrison and Jaros (1988) addressed the question of the origin of the enhancement of the momentum mixing and succeeded to some extent in identifying the individual contributions due to ordering and strain. They adjusted the band offsets in order to account fully for the effect of strain there and used the usual periodic boundary conditions to model the superlattice. However, they did not include the contribution to the Hamiltonian of the terms in the potential (V of Eqs. (4) and (5)) due to the fact that atoms in a strained layer are displaced from their bulk positions. They computed the superlattice wave function and found that the optical transition probability was reduced by nearly two orders of magnitude compared to the full calculation. Although such a separation cannot be unique (since both are manifestations of the difference in periodic crystal potentials), their results show quite clearly that an explicit account of strain in terms of nuclear coordinates of the atoms in the unit cell is essential for successful modelling.

A detailed study of quantitative trends on which this discussion is based has been compiled by Morrison et al. (1987), Morrison and Jaros (1988), Froyen et al. (1988), and by Gell (1988).

Most theoretical and experimental studies of ultrathin-period superlattices have concentrated on structures where the widths of the constituent layers are the same. The effect of changes in the layer width upon the energy-level position follows the intuitive trend: Reducing the well width pushes the confined levels up towards the top of the confining barrier, and vice versa. This permits further tuning of energy gaps. However, it will transpire in the following paragraphs that, particularly in structures with layers whose widths is less than one lattice constant, practical and conceptual uncertainties are such that the link between the strength of the optical transitions and the structural properties is difficult to appreciate.

5. Optical Properties of $\{(Si_4Ge_4)_5Si\}$ Structures

The four-monolayer superlattice is of particular interest. This is because in this structure, strong optical transitions have been observed by Pearsall et al. (1987) at 0.76 and 1.25 eV. These transitions are stronger than those normally expected in bulk indirect alloys of SiGe and occur at energies well below the direct gap of the corresponding alloy. They must therefore arise as a result of superlattice order in this structure.

The electronic structure obtained in theoretical studies (Fig. 24) can be understood in terms of a simple band-offset diagram presented in Fig. 23, along the lines of the argument developed in Sections IV.C.1 and IV.C.2. The magnitude of the conduction-band offset ΔE_c depends on the difference in the bulk bandgaps of Si and Ge in the $\langle 001 \rangle$ direction, and on the magnitude of the splitting of the conduction-band minima of strained Ge. In the direction of the interface plane (x, y), there is no confinement. The conduction-band minimum lies away from the zone centre, and its position is given by the Δ-minimum point of bulk Si. In the $\langle 001 \rangle$ direction, the confinement effect pushes the electron states away from the bottom of the corresponding "well," and the bandgap at the zone centre (Γ) increases. This increase is substantial, since both the conduction- and valence-band offsets are large. As a result, the principal gap of the superlattice is indirect and lies between the lowest Si Δ-point and the top of the valence band at Γ derived from the Ge layer. The magnitude of this gap is predicted to be about 0.8–0.9 eV. However, since this transition is forbidden, it can only be observed with phonon assistance. Such transitions are familiar from studies of bulk Si (Braunstein et al., 1958). The first direct transition (V1 to C2 in Fig. 24) at the centre of the small Brillouin zone is predicted to lie much higher at 1.1–1.3 eV.

In all models referred to above, the five-well system studied by Pearsall et al. (1987) in electroreflectance was represented as an infinite superlattice, and the external electric field was ignored. It is intuitively obvious that when the periodic boundary conditions are removed, the bands are reduced to a set of discrete levels. The overlap between the adjacent quantum states associated with the individual wells and barriers may then be modified, and with it the coupling to external fields. This is indeed what was found in pseudopotential calculations (Wong and Jaros, 1988) in which the actual five-period superlattice studied by Pearsall et al. (1987) was modelled for the first time. However, it turns out that the energy and oscillator strength of the lowest observable transition remain practically the same as those for the ideal infinite system. The effect of an external field of up to about 10^4 V/cm, characteristic of electroreflectance experiments, on the optical spectra is negligible.

The wave functions obtained in this calculation are shown in Fig. 27. As indicated at the bottom of this figure, we modelled the five-period structure—embedded in Si buffer about 25 Å thick on each side—as a large *isolated* block of length 100 Å. (Note that in a calculation invoking periodic boundary conditions the superlattice states localized in the Si buffers would dominate the solution outside the five-period structure.) The key novel features of the solution *peculiar to the finite character* of this structure are: (1) The lowest conduction level (top plot in Fig. 27) responsible for the optical transition observed at 0.76 eV is localized in the *middle* of the five-period structure. The state strongly reflects the system's symmetry (of the four Si wells the middle two must "share" the ground state). (2) The uppermost valence states "see" a large Si/Ge well confined by the barriers of the Si substrate. Hence, these states also exhibit an envelope function (lower part of Fig. 27) characteristic of the *larger* well. (3) Because of the momentum mixing at the interface, the states *not* confined inside the five-period structure exhibit a certain degree of localization at the interfaces with the Si buffer. In a system free of defects,

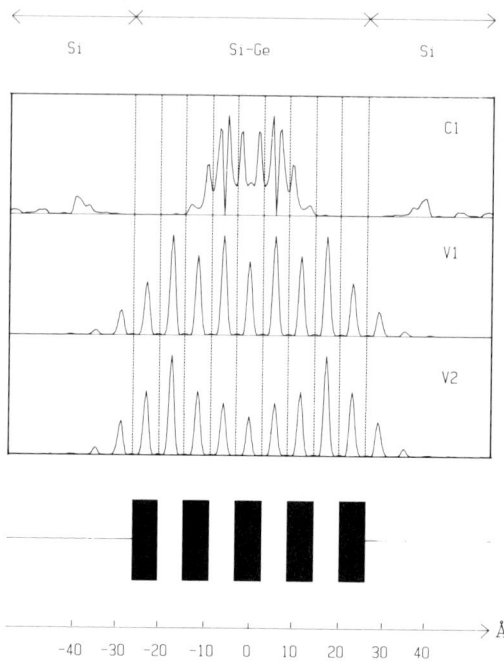

FIG. 27. This figure shows the charge densities of the lowest conduction state (C1) and the uppermost valence states (V1, V2) confined in the five-period structure studied by Pearsall et al. (1987). This structure is illustrated at the bottom (dark blocks indicate the Ge layers each containing four atomic monolayers) [Reprinted with permission from the American Institute of Physics, Wong, K.B. and Jaros, M. (1988). *Appl. Phys. Lett.* **53**, 657.]

these states form a *continuum*, although in a calculation such as Wong and Jaros, i.e., with a discrete basis set, only a few such levels can of course be recovered. The wave packet nature of these continuum states has so far been ignored in modelling Si/Ge heterojunction devices.

6. *Origin of the Strength of the Optical Spectra in Ultrathin Superlattices*

There are two questions that must be addressed: (1) Why it is that the $Si_N Ge_N$ structure with $N=4$ (four monolayers) yields strong electroreflectance spectra (Pearsall *et al.*, 1989) symptomatic of ordering whereas no such spectra are seen in other ultrathin structures? (2) Since the transitions predicted in all calculations to lie at about 0.8 eV are indirect and therefore forbidden, how can we account for the apparent strength of the electroreflectance signal? It may take some time before we know enough about ultrathin superlattices to be able to formulate a definitive answer to these questions.

However, it is possible to identify the relevant issues and point to the competing views concerning the possible ways of addressing them.

The experimental results for $N=1$ and $N=6$ structures strongly resemble those taken for an alloy of equal composition. When $N=1$, the Si–Ge looks like a zinc-blende bulk crystal, so that the alloylike electronic structure is only to be expected, with splittings due to the ordering effect that are smaller than the linewidth. The samples with $N=6$ are unlikely to possess high-quality superlattice features, since even with only three periods, the critical thickness limit is exceeded (Bevk et al., 1986). In the case of $N=2$, ten periods can be grown, and a good ordering effect is expected. However, the signal in the 1 eV range is weak and resembles that obtained for an alloy as it does in the case of $N=1$ and $N=6$ structures. The signal for $N=4$ at 0.76 eV is at least two orders of magnitude larger. There is no simple explanation that could be deduced from band-structure calculations for this large difference in observed spectra. However, as we noted in Section III, the $N=4$ superlattice has layers whose width is equal to one cubic lattice constant. Such a layer is more likely to retain its bulk form. When $N=2$, each atom has next-nearest neighbours of the other species. It is well known (e.g., Bassani and Pastori-Parravicini, 1975) that the bulk band structure at the $X(\Delta)$ point is sensitive to the terms in the Hamiltonian representing next-nearest-neighbour parameters. This means that properties of the conduction-band states that are derived from the bulk X- or Δ-points are particularly sensitive to structural details. We might speculate that in the presence of defects, new degrees of freedom not accounted for in idealised models are involved, and the effect of ordering upon the conduction-band states is significantly reduced.

As for the $N=4$ result, it may be argued that the electron–phonon coupling leads to an increase in the transition probability for the indirect transition predicted to lie near 0.8 eV by band-structure calculations. However, as pointed out by Hybertsen and Schlüter (1987), this interpretation is unsatisfactory. The amplitude of the reflectivity signal reported by Pearsall et al. (1987) is only an order of magnitude smaller than the first $\Gamma \to \Gamma$ bulklike transition at 2.3 eV. Furthermore, as pointed out earlier, the signal is much stronger than the indirect-edge transition in an alloy of comparable composition.

The transition probability for the first direct transition at Γ, which lies at about 1.2 eV predicted in all existing calculations, is quite large, only about one order of magnitude smaller than the value needed to account for the observed spectra. It follows that it might be possible to solve the problem by identifying a mechanism that could lead to lowering of the superlattice bandgap at Γ by about 0.4 eV. We can hypothesise that this might happen as a result of the imperfect nature of the superlattice. Of course, we have seen that in the idealized (infinite and defect-free) Si–Ge superlattice, the lattice constant of Ge along the $\langle 001 \rangle$ axis reported in first-principles calculations agrees very well with the value obtained by minimizing the macroscopic elastic energy and used in all existing models. However, a quantitative assessment of the equilibrium nuclear configuration of an imperfect five-period structure such as that studied by Pearsall et al. (1989) lies outside the scope of first-principles techniques. Indeed, recent experiments indicate that neither the lattice separation nor the ordering has been predicted correctly in certain systems (Chambers and Loebs, 1989; Müller et al., 1989). It would appear that in real (i.e., most likely imperfect) systems the lattice separation in the strained layer in the growth direction remains more bulk-like and that the structure may strongly reflect the role of defects.

Accordingly, Wong et al. (1988) considered two models, both of which led to the desired reduction of the bandgap. They fitted the Si and Ge empirical pseudopotentials so as to reproduce the bulk band structure along the $\langle 001 \rangle$ direction of perfect bulk Si and tetragonally strained Ge. When they chose the nuclear coordinates in exactly the same manner as did the previous authors, they obtained a superlattice electronic structure (Fig. 24) in good agreement with the state-of-the-art quasiparticle calculations. However, when the pseudopotential form factors at finite wavelengths representing Ge atoms in the strained layers are chosen to be those required to compute the band structure along the $\langle 100 \rangle$ direction of a bulk cubic Ge, the direct bandgap of the superlattice at Γ is reduced from 1.2 to about 0.9 eV. A similar result is obtained when it is assumed that the Ge/Ge bonds relax to acquire their bulk length. Since the lattice constant in the interface plane is fixed by the substrate (Si), this relaxation must be achieved by increasing the Ge/Ge separation in the direction perpendicular to the interface.

In Table II, we can see the oscillator strengths and the transition energies of the lowest three transitions at the Γ point predicted by Wong et al. (1988), and compare with them those observed by Pearsall et al. (1987). It is worth pointing out that the ratio of the reflectivity amplitudes for the transitions at 2.3 eV and 0.76 eV is only about ten times larger than that predicted

TABLE II

Optical Transitions in Si_4Ge_4[a]

Transition	E(theor)	E(exp)	F
V_1-C_1, C_2	0.9	0.76 ± 0.14	2×10^{-3}
V_1-C_3	1.4	1.25 ± 0.13	3×10^{-2}
V_1-C_5	2.3	2.31 ± 0.12	3×10^{-1}

[a]The experimental and theoretical transition energies E in eV, from Pearsall et al. (1987) and Wong et al. (1988), respectively. F is the calculated oscillator strength. The quantum states in question are shown in Fig. 24.

theoretically. These oscillator strengths are typical of all theoretical estimates for these transitions available in the literature within the numerical uncertainty normally expected of large calculations; the only difference is the energy at which they are predicted to occur.

The view that the 0.76 eV transition seen by Pearsall et al. (1987) is due to an indirect-gap process was given a strong support by the interpretation of photocurrent experiments. Hybertsen et al. (1988) presented photoconductivity data fitted to $(\hbar\omega - E)^l$, where $\hbar\omega$ is the photon energy and $E = E_g + \hbar\omega_p$. E_g is the (indirect) bandgap, and ω_p is the phonon frequency; l turns out to be 1.5–2, as expected for indirect-band-edge absorption in, say, bulk Si. However, the results in Fig. 27 imply that a similar form is expected for a quasidirect transition proposed by Pearsall et al. (1987) and accounted for by Wong et al. (1988). Although for a direct transition in a bulk material one expects $l \sim 0.5$, Wong and Jaros (1988) predict that photoexcited electrons end up in the middle Si wells of the five-period structure. Transitions into electron states localised nearer to the thick Si layer are ineffective. In order to be collected, photoexcited electrons must tunnel through the adjacent Si/Ge barrier. The transmission probability increases with increasing $\hbar\omega$, thereby giving the observed $l \sim 1.5$ dependence. Hence, if we take account of the finite character of the structure in question, the argument put forward by Hybertsen et al. (1988) in fact appears to support the direct process. However, the analysis of the photocurrent data based on the formulae for bulk indirect edge is most likely to be too crude to give a convincing interpretation of the spectroscopic evidence one way or the other. This is because the assumptions leading to these formulae (e.g., constant transition matrix element, bulk band structure) are not really valid in this case. Also the photoconductivity lineshape is broad, and like all transport measurements, it is a sum of several processes. This makes it difficult to find an unabiguous interpretation.

Finally, it is worth drawing attention to other uncertainties that might have some bearing upon the origin of the optical spectra epitaxial strained-layer superlattices. (1) We know that in the case of GaAs/AlAs heterojunction, the predictions of first-principles calculations greatly underestimated the magnitude of the valence-band offset, and it was not until after the offset had been established experimentally that the theoretical value was corrected. This suggests that the valence-band offset for Si/Ge must be carefully examined experimentally before its magnitude can be regarded as a reliable parameter for interpretation of experimental data. All of the above discussion is based on theoretical results. (2) The optical properties of Si grown by molecular beam epitaxy (MBE) are notorious for the defect-related luminescence bands (e.g., see Robbins et al., 1985, and references therein). Similar features have been seen in Si–Ge heterojunctions and superlattices (e.g., Eberl et al., 1987), and strained-SiGe alloys (Northrop et al., 1988).

Unlike other optoelectronic materials, band-to-band luminescence spectra are not readily available for characterisation of Si/Ge superlattices. (3) Kobayashi and Das Sharma (1988) found in their growth-simulation calculations that defects play an important part in the process of formation of Si/Ge interfaces. (4) Wong and Jaros (1989) considered the possibility that some Si sites in the interface monolayer are occupied by Ge atoms, and vice versa. This breaks the translational symmetry in the interface plane and enhances the no-phonon indirect-transition probability. However, they found that even when every eighth atom is exchanged at random in one monolayer, the enhancement is insufficient to account for the strength of the observed spectra. It would appear that a detailed understanding of optical properties can only be achieved after a program of careful characterisation of epitaxial Si and Si-related materials has been completed.

Let us summarise our discussion in this section. There are two ways to explain the strength of the electroreflectance signal at 0.76 eV. One is to assume that the transition is a phonon-assisted indirect process, because it occurs at the energy corresponding to the indirect gap of an infinite ideal (defect-free) superlattice predicted in all existing calculations. The other is to assume that it is a quasidirect transition whose energy is affected by anomalous properties of the Ge layers. The main weakness of the latter is that it attempts to infer structural properties from an approximate model of the electronic structure. The main weakness of the former proposition is that the coupling to the lattice would have to be anomalously strong. In fact, both arguments lead to the same conclusion: One must either invoke some form of deviation from perfect superlattice structure or look for a more subtle interpretation of electroreflectance experiment peculiar to this structure in order to account for the observed spectra. It is a challenge to future research to develop experimental and theoretical techniques capable of identifying the signature of the condition of strained-layer interfaces.

7. Band Nonparabolicity in Strained-Layer Superlattices

The band curvature in $Si/Si_{1-x}Ge_x$ strained-layer superlattices with periods of 10–30 Å has been described qualitatively in terms of dispersion curves and the relevant virtual excitations, and quantitatively as effective masses and nonparabolicities (i.e., the second- and fourth-order derivatives of energy versus wave vector, respectively) in the manner outlined in Section IV.B.7 (Turton et al., 1988).

In the valence band, the nonparabolicity and effective mass are approximately the same magnitude in the superlattice as in bulk $Si_{0.5}Ge_{0.5}$. In the conduction band, the differences between the superlattice and the bulk are far more noticeable. Along the superlattice axis, three main differences are observed. (1) The position of the minimum within the superlattice Brillouin zone varies as a function of the superlattice period. (ii) The bulk minima are virtually parabolic, whilst in the superlattice the nonparabolicity is marked and is approximately equivalent to that in bulk GaAs. (iii) The longitudinal effective mass in the superlattice is noticeably smaller than in the bulk.

The differences in (ii) and (iii) indicate that the curvature of the superlattice bands differ significantly from that predicted for the bulk material. This is due to the virtual transitions between the lower conduction states, which are allowed in the superlattice but not in the bulk. The importance of these terms in affecting the band curvature is enhanced by the small energy separations involved. The effective mass and the nonparabolicity are proportional to the inverse of the energy separation, and to the inverse of the energy separation cubed, respectively. In the systems considered in Fig. 19, the lowest two conduction states are separated by energies between 10 meV and 200 meV, whereas transitions across the fundamental gap involve an energy change of about 1 eV. Thus the virtual excitations between the conduction-band states dominate the band curvature. In fact, the nonparabolicity is so strongly dependent on the separation of the lower conduction states that the value can vary by two orders of magnitude for comparatively small changes in the superlattice parameters. This sensitivity suggests that 'tuning' of the superlattice would be possible to obtain a desired value of nonparabolicity.

It is interesting to compare the results for Si/SiGe with the conduction-band nonparabolicities obtained for $GaAs/Ga_{1-x}Al_xAs$ and $GaAs/GaAs_{1-x}P_x$ (Brown *et al.*, 1987; Brown and Jaros, 1988). The principal differences can be summarised as follows.

(i) In the $Si/Si_{1-x}Ge_x$ system, the minima are away from the major symmetry points, and therefore the position of the minimum within the superlattice Brillouin zone varies with the period, whereas the minimum in the GaAs systems is at Γ and so remains in this position.

(ii) The bulk GaAs conduction band is significantly nonparabolic, and the enhancement in the superlattice is less than one order of magnitude. The bulk silicon conduction band is virtually parabolic near the extrema, but the effects of strain and zone folding enhance the nonparabolicity by many orders of magnitude, making it roughly equivalent to that of bulk GaAs.

(iii) The mechanisms responsible for the conduction-band curvatures are considerably different in the three superlattice systems. In the silicon systems, the nonparabolicity is entirely due to virtual transitions within the conduction band. These transitions are between states that originate from the same region of the bulk Brillouin zone, namely the minimum, therefore the overlap between these states is almost complete. The nonparabolicity of the $GaAs/Ga_{1-x}Al_xAs$ superlattices is also due solely to virtual transitions within the conduction band; however, the principal transitions are between the Γ minimum and the zone-folded states originating from the bulk X-minimum. The overlap between these states, and therefore the matrix element between them, is made finite by the momentum mixing of the superlattice potential. The $GaAs/GaAs_{1-x}P_x$ superlattices are strained, as in the $Si/Si_{1-x}Ge_x$ systems, to accommodate the lattice mismatch. The band curvature is affected by virtual transitions between the conduction states and across the superlattice bandgap. For small concentrations of P, the principal transition is again that between the Γ-minimum and the folded X-minimum. As the strain is increased, i.e., for larger concentrations of P, the momentum mixing is further enhanced, leading to an increase in the nonparabolicity. For x greater than about 0.53, the X minimum moves from the GaAs layer into the barrier, and the transition across the superlattice gap becomes dominant.

(iv) In directions parallel to the interface planes, the band curvature in the GaAs superlattices is unchanged from that of the bulk material. In the silicon superlattices considered here, the strain splits the degenerate minima such that the minima along the superlattice axis are about 150 meV below those in the other directions. Consequently, the dispersion, and therefore the band curvature, parallel to the interface planes is significantly different to that of the bulk material.

D. $Hg_{1-x}Cd_xTe/Hg_{1-y}Cd_yTe$ SUPERLATTICES

1. Introduction

There are two novel features in the electronic structure peculiar to HgTe/CdTe: (1) interface states, and (2) enhanced localisation due to mixing of bulk states of different momenta via large differences in atomic potentials of the constituents. The purpose of this section is to identify the signature of these effects in the context of observable properties of such structures and their applications.

The main practical interest concerning the small-gap HgTe-related materials stems from their application in infrared imaging. The alloy of $Hg_{1-x}Cd_xTe$ is the material most often used to fabricate infrared detectors. However, the small effective-electron mass is responsible for large tunnelling currents at junctions. Since the magnitude of the bandgap is a sensitive function of the alloy composition, the difficulty in controlling the composition changes the cutoff wavelength of the device. It was suggested by Schulman and McGill (1979) that both the magnitude of the effective mass and the bandgap control can be improved if the alloy is replaced by a HgTe/CdTe superlattice. Although HgTe is a semimetal, the large-gap CdTe provides barriers for electrons and holes to be confined in HgTe layers. The valence-band offset can be predicted as outlined in Section III. The confinement effect lifts the states localised in HgTe away from the bulk band edges in a manner familiar from the particle-in-a-box picture. The width of the corresponding layers can then be adjusted to achieve the desired magnitude of the superlattice gap. The energy levels have been studied in some detail within the particle-in-a-box picture adjusted to account for the details of the relativistic bulk band structure at the centre of the Brillouin zone (Γ), and by other methods (e.g., Bastard, 1982; a detailed review of the theoretical and experimental literature on HgTe-related heterostructures was compiled by Herman and Pessa, 1985). In order to achieve bandgaps of 100–150 eV, it turns out that layer widths of order 30–40 Å are needed.

Clearly, there are many similarities between modelling HgTe-related structures and other heterostructures considered earlier in this chapter. For example, the momentum-mixing effects in the electronic structure that occur as a result of a reduction of the superlattice period or layer widths can be modelled in a similar fashion in HgTe/CdTe structures. These effects are familiar from preceding sections and will not be dealt with here. In any case, the separation of the primary (Γ) and secondary (X, L) conduction-band valleys is much larger in this structure. This reduces the magnitude of observable quantities (e.g., transition rates) symptomatic of Γ–X mixing in zone-folded conduction-band states and in the related optical spectra. This effect is therefore of no interest here.

Then there are problems peculiar to the material in question. For example, the field has been plagued by large uncertainties in the valence-band offset. Initially, a small (40 meV) offset was used (e.g., Guldner et al., 1983), only to be replaced later by a larger value (>350 meV; see Tersoff, 1984, 1989; Kowalczyk et al., 1986; and others). Unlike GaAs/AlAs, interdiffusion is a much greater problem, and the poor definition of layer width is also an obstacle to accurate comparison between theory and experiment. In general, the characterisation of this class of materials is far from complete, and we shall make no attempt to review the literature on this subject. Instead, we shall focus on new microscopic concepts peculiar to these structures.

We have already pointed out that as far as the practical objectives are concerned, it is not necessary to make ultrathin layers in which the effect of ordering leads to enhanced momentum mixing. In view of the strong diffusion of atoms across the interface, such layers would be very difficult to make, too. The lattice constants of HgTe and CdTe are very similar (see Table I), and effects of strain are limited to a few meV splittings of the confined levels (Schulman and Chang, 1986). Hence strain-induced momentum mixing is also unimportant. Indeed, in Section II.C, we singled out HgTe/CdTe as an example of a system in which there is a significant difference between the atomic potentials of Hg and Cd. We argued there that this difference must be responsible for mixing of bulk states of different momenta in the superlattice and that it must affect the localisation of confined states. In fact the nonrelativistic potentials and energy levels of free atoms of Hg and Cd are almost identical, and much of the effect we are referring to is due to the large relativistic correction at Hg and the difference in the strength of the covalent bond in these two materials.

The effect of the difference in the short-range potentials of Hg and Cd was recognised in the work on the electronic and structural properties of HgCdTe alloys. It leads to a breakdown of the virtual crystal approximation (for references see review by Jaros, 1985; see also Hass et al., 1984). An isolated Hg atom (impurity) in CdTe behaves as a "deep" impurity, since as a result of momentum mixing at Hg, the wave function of the localised states is a wave packet with contributions spanning a large volume of momentum space (Jaros and Guimaraes, 1985). In a superlattice, each atomic site of Hg and Cd contributes to this momentum-mixing process. Hence, unlike all preceding examples of the breakdown of the EMA, there is no critical length to associate with the breakdown of the EMA.

The case of Cd and Hg (i.e., HgTe and CdTe) is by no means the only example of "deep" superlattice potential in the sense of Eq. (5). For instance, we can think of ZnS/ZnSe/ZnTe structures that are also being used to make optoelectronic materials (e.g., Capasso and Margaritondo, 1987). Here again the difference in atomic potentials of, say, S and Te is large enough to give rise to significant momentum mixing. Similar pairs of isoelectronic substituents can be found in III–V compounds (e.g., P and Bi, P, and N) although the corresponding ordered structures have not been studied. The divide between "shallow" and "deep" isoelectronic impurities is not sharp, but the impurity model does serve as an approximate guide. The classification of impurities in this regard is available (Jaros, 1982, and reference therein; see also Vogl, 1981).

Having identified a new effect *peculiar* to HgTe/CdTe, we must now specify its meaning and likely magnitude more precisely as an observable quantity in superlattices. First of all, the potential difference between Hg and Cd is not large enough to give rise to levels localised in the gap of CdTe: The impurity states of CdTe:Hg are really resonances lying far from the band edges in the conduction and valence bands, and their precise position has not been determined. It is well known from detailed deep-level calculations that such states are not strongly linked to the conduction band Γ valley where the mass is small and density of states low. On the other hand, all relevant bulk states contributing to the formation of the superlattice states near the gap are derived from the region in momentum space near Γ. Furthermore, the bandgap in structures of interest is small, and its magnitude must strongly reflect uncertainties in layer width, effective masses, and band offset, etc. Similarly, the link between the band offset and level depth is affected by momentum mixing, since the scattering at interfaces involves states of different momenta and not just a reflection, but it will be difficult to distinguish its contribution from other competing corrections of comparable magnitude. It follows that any influence upon the position of confined levels due to the difference in atomic potentials of Hg and Cd will be difficult to establish within the limitations imposed by the circumstances in question.

The localisation of confined levels does reflect the degree of momentum mixing, and this should, in turn, manifest itself in the width of the minibands. However, the miniband width is difficult to measure in any microstructure, not to speak of HgTe/CdTe, and the separation of the bulk states of Γ- and X-character is of order 1 eV, i.e., so large that the effect of any mixing of X and Γ upon the levels near the band edges must be weak. We must therefore look for the signature of the breakdown of the EMA due to the difference in atomic potentials in those localisation features of the electronic structure that are either decoupled from the periodic order parameter (superlattice period, layer widths) or depend on higher-order properties (e.g., effective masses, third-order susceptibility; while the former depends on second-order virtual excitations, the latter depends on fourth-order ones).

So far, we have dealt with structures that have no states localised at the interfaces. We expect such states only when the abruptness of the change of the potential at the interface is stronger in its influence upon the boundary condition than the order parameter due to periodicity. The question has been extensively covered in the literature, e.g., Zhu and Kroemer (1983), Gell et al. (1986). The HgTe/CdTe interface is very special in this regard. Since HgTe is a semimetal, the uppermost valence band has in fact an opposite curvature to that of its counterpart in CdTe. This means that the matching condition at the interface requires the coupling of the state with the inverted positive mass (Γ_8) at the conduction edge of HgTe and the "normal" negative mass of the light-hole (Γ_8) band of CdTe. Then the wave function must have a large amplitude at the interface (Chang et al., 1985; Lin and Sham, 1985). Note that the existence of this interface state does not stem from some breakdown of the EMA. On the contrary, the EMA provides a clear existence condition for such a state. However, the degree of localisation at the interface should reflect the strength of the momentum-mixing process. If the mixing is strong, we obtain a well-localised level. Such a level then behaves at least to some extent as a "deep" impurity state in that it reflects strongly its local environment. This makes the interface state sensitive to the condition of the interface. Hence, for example, a local charge transfer due to deviations from perfect interface condition (e.g., presence of defects breaking the periodic boundary condition in the interface plane) would lead to bound states or resonances and consequently to changes in transport properties. This is by far the strongest and most important consequence of the "deep" character of the potential for experimental characterisation of these heterostructures.

In the following sections, we shall present an outline of some key features in the electronic structure, effective masses, and localisation in HgTe–CdTe and related structures, with a view to pointing out the novel aspects of the link to the microscopic potential peculiar to this class of structures. As far as observable properties are concerned, the relation between theory and experiment is too rough to present us with an opportunity to demonstrate clearly the degree to which the deep character of the superlattice potential affects the band structure. Given the existing error bars, the results for energy levels are very similar to those reported in the literature, even though the approximations underlying the individual theoretical models are quite different. In any case, as in the case of other superlattice structures considered above, the particle-in-a-box model is expected to predict the overall level structure correctly, and the main difference lies in the form of the wave function. Alas! the spectroscopic features sensitive to the details of the momentum wave function (e.g., miniband widths) are still buried in the uncertainties of the material parameters. We shall also discuss the conceptual link between the band offset and the way approximate band-structure calculations are carried out under the condition of the breakdown of the EMA peculiar to HgTe/CdTe. This may shed some light upon predictions obtained by different computational methods.

2. Band Structure, Interface States, and Effective Masses

Whereas CdTe is a wide-gap semiconductor, HgTe is a semimetal with a negative gap (Fig. 28). All existing calculations of the optical spectra of this structure have been based on the assumption that the valence-band offset is small ($\Delta E_v = 40$ meV), in accordance with the so-called common anion rule

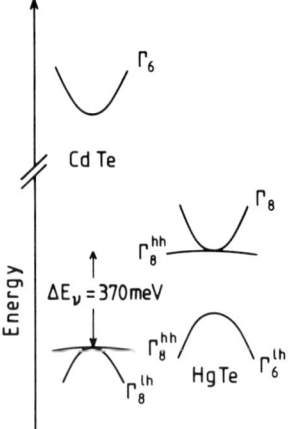

FIG. 28. The band edge states of bulk HgTe and CdTe indicating the band offset ΔE_v.

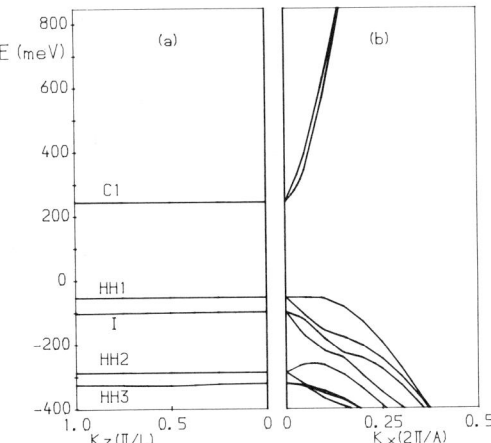

FIG. 29. Sub-bands of a superlattice consisting of 19 Å HgTe and 39 Å CdTe. (a) Perpendicular to the interface, (b) parallel to the interface. A is the bulk lattice constant. State HH1 is the uppermost valence state. C1 is the lowest conduction band state, and I is the interface state. All energies are measured relative to the bulk HgTe valence band maximum. L is the superlattice period (Zoryk and Jaros, 1988).

(see Herman and Pessa, 1985; Faurie, 1986, and references therein). In pseudopotential calculations (Jaros et al., 1987), a large valence-band offset is found of 0.3–0.4 eV, in good agreement with recent experimental studies (Kowalczyk et al., 1986) and theoretical (Tersoff, 1986; Van de Walle and Martin, 1987, 1988).

In Fig. 29(a) and (b), we can see the band structure for a superlattice comprising 19.4 Å HgTe and 38.8 Å CdTe layers (Jaros et al., 1987; Zoryk and Jaros, 1988). In the plane parallel to the interface (b), the spin–orbit coupling splits the doubly degenerate subbands. The spin–orbit splitting is large, and the valence subbands are highly nonparabolic, with significant hybridisation away from the zone centre. The real-space charge densities, calculated at the centre of the Brillouin zone, for states I, HH1, and C1 are shown in Fig. 30 in the xz plane drawn through the anion sites, covering an area of three lattice constants in the x-direction, and nine lattice constants in the z-direction. The positions of the anions are shown by the solid circles. The state C1 is the lowest empty (conduction-band) state, state HH1 is dominated by the heavy-hole bulk contribution, and I is the interface state.

The figure shows the confinement of states HH1 and C1 in the HgTe layer and the strong localisation of state I at the interface. This localisation in real space is reflected by the delocalisation of the momentum wave function away from the Brillouin zone centre. The main difference between the pseudo-

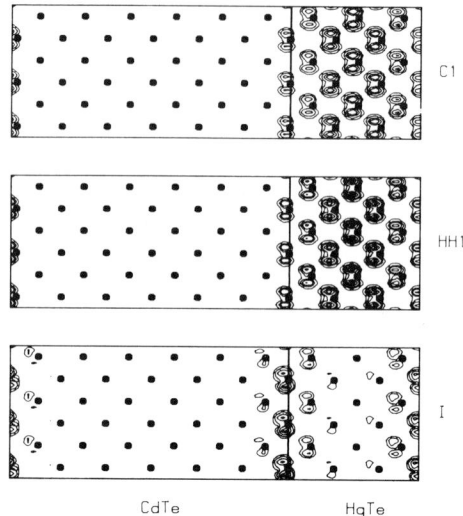

FIG. 30. A contour plot of the electron charge density of states I, HH1 and C1 introduced in Fig. 29, in an area spanning three by nine lattice constants in an xz plane passing through anions which are marked by solid circles. z is perpendicular to the interface (Zoryk and Jaros, 1988).

potential band structure shown here and those obtained by the EMA and tight-binding models (see Schulman and Chang, 1986) is the width of the minibands near the band edges. The pseudopotential calculation gives smaller widths, and the localisation of the interface state in real space is stronger. In addition to the light-hole state contribution familiar from the EMA, there is a significant participation of heavy-hole bulk waves in the formation of this state. The "deep" character of the interface state predicted by Jaros *et al.* (1987) has been confirmed in first principles calculations (Christensen, 1988).

It is worth pointing out that some care is needed in interpreting the meaning of momentum wave function coefficients, particularly when a relationship is sought between different computational techniques. Although the charge density provided in this manner is unique, the relative value of the momentum coefficients across the bulk Brillouin zone is not, since it depends on the choice of the expansion set. For example, a different set of coefficients $A(n, \mathbf{k})$ is obtained depending on whether one uses the bulk band structure of HgTe, CdTe, or of an alloy of these materials as a starting point in the calculation.

The tight-binding calculations in fact also yield a highly localised interface state of this character, but its energy level lies deeper below the valence-band edge. It is important to realise that the uncertainty in the predictions of energy levels of such "deep" states is significant, i.e., greater than the bandgap in this case. This is so, irrespective of the choice of Hamiltonian. The problem is well known from calculations of deep levels at point defects. Indeed, pseudopotential calculations show that although the overall character of the wave function remains unchanged, the energy level position with respect to the band edge is sensitive to the details of the potential. The discrepancy between the tight-binding and empirical pseudopotential calculations is therefore not very surprising.

The "deep" character of localization at the interface becomes particularly apparent when the perfect translational symmetry in the interface plane is removed. Beavis and Jaros (1989) modelled an HgTe/CdTe superlattice with interfaces where on average every eighth atom of Cd in the interface layer is replaced with Hg. In such an Hg-rich interface the interface state I becomes a level in the gap (Fig. 31). The effect on the superlattice band gap between the "normal" confined states at the band edges (C1, V1) is very small. The localization of this interface state is shown in Fig. 32. An analogous effect is obtained for superlattices with Cd-rich interfaces. There is every reason to think that the Hall data reported in the literature (Hoffman et al., 1988, and refs. therein) can in fact be interpreted in terms of such levels.

McGill and collaborators have made detailed comparisons of the ordered superlattice with alloys (e.g., Smith et al., 1983). They used a simple two-band $k \cdot p$ model to demonstrate theoretically the increase in effective masses perpendicular to the interface and the corresponding reduction in tunnelling length. They also studied the relations between the bandgap and order parameters (layers 30–140 Å thick). These results as well as results of more sophisticated tight-binding and $k \cdot p$ calculations by other authors are very

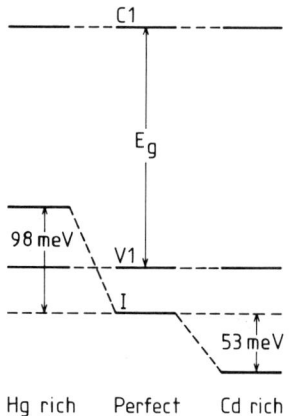

FIG. 31. Middle section: The position at the center of the Brillouin zone of the lowest conduction state (C1), the uppermost valence state (V1), and the state localized at the interface (I) in an ideal symmetric HgTe-CdTe superlattice of period 25.8 Å. The magnitude of the superlattice band gap $E_g = 228$ meV is indicated. Left and right sections (the dashed lines are added to guide the eye only) show that the interface state is raised in energy in structures with Hg-rich interfaces and lowered in those with Cd-rich interfaces (Beavis and Jaros, 1989).

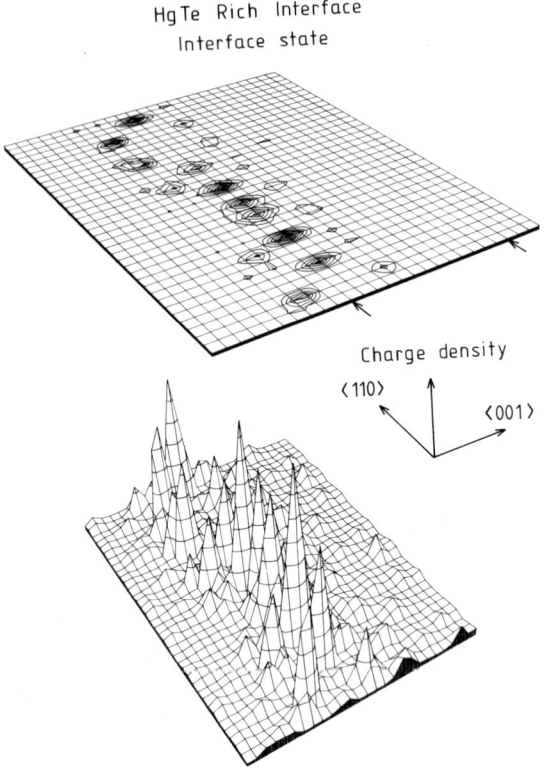

FIG. 32. The plot of the electron charge density associated with the interface state in the Hg-rich structure whose energy is given in Fig. 31. The arrow indicates the position of the interface. The area of the plot has the dimensions of the superlattice period (25.8 Å) in the growth direction (z) and 32.3 Å in the $\langle 110 \rangle$ direction parallel to the interface plane. It is seen that the state is well localized inside the 16 atom supercell in the interface plane.

similar to those expected from the particle-in-a-box model, and their interpretation can be made along the lines familiar from preceding sections. Calculations of the optical transition probability show that the absorption coefficient near the band edge is, for any practical purposes, the same in both the superlattice and the alloy of the same gap. There is no sufficiently detailed spectroscopic evidence with which to compare these theoretical results.

There are large uncertainties concerning layer widths and composition profiles, which make a detailed comparison with experiment difficult to achieve. For example, recent experimental analysis has shown that the HgTe/CdTe superlattices do not necessarily consist of pure HgTe and CdTe layers, but rather consist of HgTe and $Hg_{1-y}Cd_yTe$ layers, with y varying from 0.65 to 0.90 (Perkowitz et al., 1987). Accordingly, to compare with these

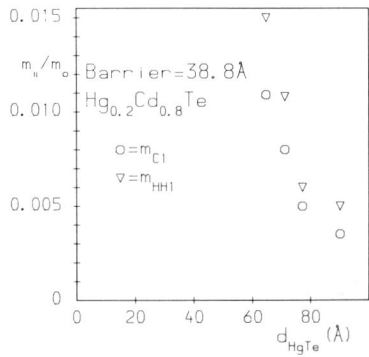

FIG. 33. The variation of the effective masses of states HH1 and C1 in the plane parallel to the interface, as a function of HgTe layer width, for HgTe/Hg$_{0.2}$Cd$_{0.8}$Te narrow gap superlattices (Zoryk and Jaros, 1988).

experiments, let us consider HgTe/Hg$_{1-y}$Cd$_y$Te superlattices with $y=0.80$. Figure 33 shows the general trend in the change of the effective masses of states C1 and HH1 in the plane parallel to the interface as a function of HgTe layer thickness. The Hg$_{0.2}$Cd$_{0.8}$Te layer thickness is 38.8 Å. These values are calculated at the centre of the Brillouin zone and represent an average of the doubly degenerate states there. In particular, for a 25 Å HgTe/52 Å Hg$_{0.2}$Cd$_{0.8}$Te superlattice, empirical pseudopotentials yield for the electron effective mass $m^*/m = 0.053$. The experimental value is 0.048. In view of the uncertainties in both theory and experiment, this is as good an agreement as can be expected.

Measurements in p-type and n-type superlattices have revealed very high Hall mobilities for the lowest conduction and higher valence subband in narrow-gap superlattices. It has also been demonstrated that if the transition from a type-III HgTe/CdTe (semimetal–semiconductor heterostructure) superlattice to a type-I (direct-gap semiconductor–semiconductor heterostructure) Hg$_{1-x}$Cd$_x$Te/CdTe ($x > 0.14$) superlattice is made, then there is an observed drop by a factor of about 50 in the measured hole mobility (Faurie, 1986). Pseudopotential calculations show that the changes in the effective mass are well accounted for by microscopic theory and that they can be explained simply by changes in the electronic band structure versus composition (i.e., gaps). When we calculate the effective masses for a superlattice consisting of 71.2 Å HgTe and 38.8 Å CdTe, a structure similar to that used in the experimental measurements, we find that in making the transition from a HgTe/CdTe to a Hg$_{0.72}$Cd$_{0.28}$Te–CdTe superlattice, the effective mass in the plane of state HH1 changes by almost a factor of 10, from 0.012 to 0.117 m (see also Fig. 33).

The effective mass can be obtained from the $k \cdot p$ theory in terms of a familiar expression (Callaway, 1974),

$$\{m/m_L^*\}_j = 1 + \sum_{L'}' f_{L',L}^j, \tag{9}$$

where $f_{L',L}^j = 2|M_{L',L}^j|^2/(m(E_L - E_{L'}))$ is the oscillator strength and $M_{L',L}^j$ is the matrix element of the jth component of the linear momentum. In our formulation,

$$M_{L',L}^j = \langle L|\hat{p}_j|L'\rangle = \sum_{n,n'}\sum_{\mathbf{k}} A_{L'}(n',\mathbf{k})A_L^*(n,\mathbf{k})\langle \phi(n,\mathbf{k})|\hat{p}_j|\phi(n',\mathbf{k})\rangle. \qquad (10)$$

In the EMA $|L\rangle \sim F_L\phi(n,0)$, where F is a smooth (envelope) function and $\phi(n,\mathbf{k})$ is the bulk Bloch state associated with band n and wave vector $\mathbf{k}$. Hence, in full theory for the case of interest, we expect a large number of bulk momentum component contributions to $|L\rangle$ (Zoryk and Jaros, 1988). This affects both, the normalisation of the wave function and the integral in $M_{L',L}^j$. Equations (9) and (10) illustrate the origin of quantitative corrections that may be needed as a result of the breakdown of the EMA. Such corrections could be incorporated into the EMA solution for example by $\mathbf{k}\cdot\mathbf{p}$ perturbation theory, since the coefficients $A(n,\mathbf{k})$ associated with bulk states lying further from the band edges are small. For example, we obtain from a pseudopotential calculation for a superlattice comprising 100.3 Å HgTe and 38.8 Å CdTe, a heavy-hole effective mass in the plane parallel to the interface $m^*/m = 0.013$, whereas the EMA (Johnson et al., 1988a) predicts $m^*/m = 0.014$–$0.046\,\Delta E_v(\text{eV})$. Using the large band offset $\Delta E_v = 0.315\,\text{eV}$, the EMA predicts $m^*/m < 0.001$. The cyclotron resonance result is 0.017 (Berroir et al., 1986). The result by Johnson et al. (1988a) gives a satisfactory result only if a smaller value (40 meV) is used for the band offset.

This small valence-band offset was inherited from the early EMA models of the electronic structure of HgTe/CdTe. It was argued that in the framework of the EMA, the offsets $> 100\,\text{meV}$ were inconsistent with the existence of the forbidden gap in these superlattices and were in good accord with magneto-optical measurement (see Faurie, 1986, and references therein). However, more recently, Johnson et al. (1988b) reconsidered the effective-mass solution and reported that a desired gap can be recovered for offsets in the region of 350 meV. This is the value deduced from photoemission spectra (e.g., Kowalczyk et al., 1986) and in theoretical models discussed above. Whatever the remaining quantitative uncertainties, it would appear that the main obstacle for reconciling competing views on the magnitude of the valence-band offset in this system has now been removed.

3. Consequences of the Breakdown of the Semiclassical Barrier Concept

In Section III, it was pointed out that the concept of band offset in the form of a steplike potential difference retains its simple meaning only if the EMA is valid in that energy range. It is the case, for example, for electron states near the band edges of $GaAs/Ga_{0.7}Al_{0.3}As$ microstructures. This is because, as argued in Section II.B, the states of $Ga_{0.7}Al_{0.3}As$ can be obtained, to a good approximation, from those of GaAs by first-order perturbation theory. In that approximation, the potential difference representing $V(\text{Al})-V(\text{Ga})$ merely shifts energies and leaves bulk-wave functions intact. The situation changes only when levels of different bulk momenta, e.g., states of Γ- and X-character, cross as for examples discussed in Sections IV.B.1, 2 and IV.B.4–8. Consider now, for the sake of simplicity, a heterojunction formed by CdTe and $\text{Hg}_{0.7}\text{Cd}_{0.3}\text{Te}$. For any k, n, we have

$$E_{n,k}(\text{Hg}_{0.7}\text{Cd}_{0.3}\text{Te}) = \langle \psi | V | \psi \rangle + E_{n,k}(\text{CdTe}),$$

where

$$\langle \psi | V | \psi \rangle = \sum_{n',\mathbf{k}',n'',\mathbf{k}''} A^*(n',\mathbf{k}')A(n'',\mathbf{k}'')\langle \phi(n',\mathbf{k}') | V | \phi(n'',\mathbf{k}'') \rangle;$$

$\phi(n, \mathbf{k})$ are bulk Bloch functions of CdTe. This spread of the superlattice wave function (ψ) in k-space reduces the overlap $\langle \phi(n, \mathbf{k}) | \psi \rangle$ between the states at the bottom of the conduction band of the two constituents forming the heterojunction compared with the case of $\text{GaAs}-\text{Ga}_{0.7}\text{Al}_{0.3}\text{As}$ by $|A(n, \mathbf{k})|^2$, where $n, \mathbf{k}$ label the bulk band edge (note that the wave-function normalisation requires $\Sigma_{n,\mathbf{k}} |A(n, \mathbf{k})|^2 = 1$).

It follows that higher-order virtual excitations may successfully compete with the usual transition probability obtained from the conventional lowest-order perturbation treatment. This applies to optically as well as thermally excited transitions and to tunnelling. For example, electrons can be excited via a second-order virtual process involving higher-lying resonances. In such a process, the transition is proportional to $\langle \psi | \delta h | \psi' \rangle \langle \psi' | \delta h' | \phi(n, \mathbf{k}) \rangle / \Delta E_v$. Here $\delta h, \delta h'$ are the relevant perturbation Hamiltonians. This is analogous to the well-known second-order process in which phonons give rise to the indirect absorption edge in Si or Ge. However, here the overlap between ψ, ψ' and $\phi(n, k)$ takes on a more complicated form depending on the atomic microscopic potential difference V. For example, such processes may be particularly relevant in studies of weak tunnelling currents, which have so far been modelled in terms of the effective-mass approximation underlying the WKB method. Of course, the spectra related to the $\Gamma-X$ coupling in $\text{GaAs}/\text{Ga}_{1-x}\text{Al}_x\text{As}$ (or the band-edge spectra of Si/Ge structures) discussed in preceding sections are just special cases of the more general situation considered here.

These considerations have yet to be taken into account in quantitative studies of HgTe/CdTe and the related structures. However, the consequences for modelling are intuitively easy to appreciate. For example, if one uses the barrier height (i.e., the band offset) as a steplike potential and represents the wave function with a truncated basis set, this is equivalent to setting some of the $A(n, \mathbf{k})$ to zero and modifying those that are retained. Although it is often quite easy to obtain from a simple theory a good fit for a few energy levels in a limited range of energies, the transition rates that depend on the details of the momentum wave function (i.e., $A(n, \mathbf{k})$ coefficients) may not be correctly represented. The situation is analogous to the differences between approximate methods used for localised defects in semiconductors and the microscopic models.

This analysis is relevant to the discussion of differences in predictions obtained by various semiempirical methods for superlattice band-structure calculations (e.g., tight binding, pseudopotentials). In particular, the procedure often used in semiempirical methods for obtaining the correct band lineup in GaAs/Ga$_{1-x}$Al$_x$As structures—namely, fitting the bulk gaps and then shifting rigidly the bulk band structures of constituent materials relative to each other—is no longer appropriate here. This is, for example, the procedure employed in all existing tight-binding and $k \cdot p$ band-structure calculations of HgTe/CdTe superlattices. No such adjustment was made in empirical pseudopotential calculations. Indeed, in this case, our tests indicate that had we made such a rigid shift, we would have been unable to recover the accurate bulk band structure of HgTe using $\phi(n, \mathbf{k})$ of CdTe as a basis set (or vice versa)! Even in systems where perturbation theory is a good approximation for energy levels, rigid shifts of bands may lead to spurious localisation effects (e.g., incorrect interface states) and incorrect shape of wave functions. A detailed account of this problem can be found in Jaros et al. (1985).

The energies and wave functions of states exhibiting "deep" localisation such as the interface state reported above are particularly difficult to describe accurately. The position of such a level is no longer stabilised by the combined effect of dimensionality and magnitude of the steplike band-offset potential (e.g., Zhu and Kroemer, 1983; Gell et al., 1986a. Small changes in the short-wavelength components of the potential affecting mainly bulk bands lying further from the band edges may lead to changes in the position of the "deeply" localised levels that are significant on the scale of energies in question. This is particularly important in HgTe/CdTe structures with a small bandgap (< 150 meV). In the EMA, the character of the interface state is almost solely due to the inverted curvature of the relevant bands in HgTe and CdTe, and the largest coefficients in the expansion of the wave function are therefore of light-hole character, i.e., reflecting simply the directional properties of the (001) interface. This state is pinned to lie just below the valence-band edge. In full treatment (e.g., semiempirical pseudopotentials, tight binding, local density, quasiparticles), the uncertainty in the interface state position is larger than the bandgap.

E. Concluding Remarks

With the exception of one- and two-monolayer structures, the microscopic potential of ideal superlattices (i.e., structures with abrupt or graded interfaces with two-dimensional translational symmetry, periodic boundary conditions characteristic of an infinite structure, and no defects) can be viewed as a superposition of bulk atomic potentials with bulk coordinates. A good estimate of the energy levels near the superlattice band edges can be obtained from the steplike potential representing the relative position of bulk bands. In lattice-mismatched (strained) systems, the change in nuclear coordinates and the corresponding level splittings obtained from classical elasticity theory and bulk deformation potentials, respectively, are in good agreement with first-principles calculations. However, to obtain a quantitative description of the electronic structure, the particle-in-a-box picture is often inadequate, and a mixing of bulk states of different momenta must be fully accounted for in terms of microscopic theory. The degree of this mixing strongly depends on the order parameters (superlattice period and layer widths) and on the change in the strength and symmetry of crystal potential across the interface. Although the corrections to the particle-in-a-box model concerning energy levels are not so significant, the wave functions and transition rates are strongly affected. This mixing gives a quantitative meaning to the concept of electronic-structure engineering.

The critical layer width below which the bulk parametrisation breaks down is one bulk lattice constant. In particular, first-principle calculations show that one-monolayer ordered superlattices are better viewed as new compounds. However, a detailed comparison between theory and experiment concerning these superlattices has not been achieved.

The archetypal momentum-mixing processes—induced by a combined effect of reduction of superlattice period, crossing in energy of states of different bulk momenta, by the effect of strain, and by significant differences in the depth of atomic potentials in adjacent layers—have been demonstrated in this chapter on selected examples of GaAs/AlAs, Si/Ge, and HgTe/CdTe (001) superlattices. At the heart of all these momentum-mixing processes is the coupling of bulk states from the zone centre (Γ) and the zone edge (e.g., X, Δ points). The above structures are chosen in order to illustrate this process in three distinctly different forms. The strength of this process in each case reflects the role of the characteristic order length in renormalising certain short-wavelength Fourier components of the microscopic potential. Similar relations occur in other superlattice systems.

Although the idealised, infinite, and defect-free ordered superlattice is a useful tool for modelling the key features in the electronic structure, there is growing experimental evidence that there are small but notable deviations from this idealisation. These deviations are of crucial importance for material evaluation and applications. The Γ–X mixing in GaAs-based as well as in Si/Ge indirect-gap structures is a sensitive probe of the interface quality, since it can turn indirect transitions into quasidirect ones, i.e., from forbidden to allowed transitions. In HgTe/CdTe, the momentum mixing increases the localisation of the interface states and makes them more susceptible to changes in local environment. It transpires that in all three systems under consideration here, a quantitative contact with experiment will be made only when the finite and imperfect character of these structures is fully accounted for.

Although there is agreement in general about the need to go beyond the idealisations inherited from the earliest stage of development in this field, there is no consensus about the precise nature of deviations from ideal structures. What is needed are experimental and theoretical efforts highly specific to individual processes affecting the condition of the interfacial layers and to the real geometry of the system. Careful characterisation of defects in materials prepared by novel epitaxial growth techniques is a necessary precondition for successful accomplishment of this task.

Acknowledgments

I would like to thank the S.E.R.C. (United Kingdom), R.S.R.E., Malvern (M.O.D., United Kingdom) and Office of Naval Research, United States (contract No. N00014-85-C-0868) for financial support. It is a particular pleasure to thank Dr. D. J. Wolford and his colleagues at the IBM T.J. Watson Research Center, Yorktown Heights, New York, for stimulating conversations and for their hospitality.

References

Abstreiter, G., Brugger, H., Wolf, T., Jorke, H., and Herzog, H. J. (1985). *Phys. Rev. Lett.* **54**, 2441.
Agrawal, B. K. (1980). *Phys. Rev.* **B22**, 6294.
Altarelli, M. (1983). *Phys. Rev.* **B28**, 842.
Ando, T., Fowler, A. B., and Stern, F. (1982). *Rev. Mod. Phys.* **54**, 437.
Andreoni, W., and Carr, R. (1980). *Phys. Rev.* **B21**, 3334.
Bachelet, G. B., Hamann, D. R., and Schluter, M. (1982). *Phys. Rev.* **B26**, 4199.
Bassani, F., and Pastori-Parravicini, G. (1975). "Electronic States and Optical Transitions in Solids." Pergamon, New York.
Bastard, G. (1981). *Phys. Rev.* **B24**, 5693.
Batey, J., and Wright, S. L. (1986). *J. Appl. Phys.* **59**, 200.
Bauer, G., Kuchar, F., and Heinrich, H. (Eds.) (1984). "Two Dimensional Systems, Heterostructures and Suplerlattices." Bauer, G., Kuchar, F., and Heinrich, H. (Eds.) (1986). "Two Dimensional Systems, Physics and New Devices." Springer, New York.

Beavis, A. W., and Jaros, M. (1989). Unpublished.
Bergh, A., and Dean, P. J. (1976). In "Light Emitting Diodes," Clarendon Press, Oxford.
Berroir, J. M., Guldner, Y., and Voos, M. (1986). *IEEE Journal of Quantum Electronics* **QE-22**, 1793.
Bevk, J., Mannearts, L. C., Feldman, L. C., Davidson, B. A., and Ourmazd, A. (1986). *Appl. Phys. Lett.* **49**, 286.
Bhattacharya, B. K., Bylander, D. M., and Kleinman, L. (1986). *Phys. Rev.* **B33**, 3947.
Braunstein, R., Moore, A. R., and Herman, F. (1958). *Phys. Rev.* **109**, 695.
Brey, L., and Tejedor, C. (1987). *Phys. Rev. Lett.* **59**, 1022.
Brown, L. D. L., and Jaros, M. (1988). *Phys. Rev.* **B37**, 4306.
Brown, L. D. L., Jaros, M., and Ninno, D. (1987). *Phys. Rev.* **B36**, 2935.
Brown, L. D. L., Jaros, M., and Wolford, D. J. (1989). *Phys. Rev.* **B40**, 6413.
Bylander, D. M., and Kleinman, L. (1986). *Phys. Rev.* **B34**, 5280.
Callaway, J. (1974). "Quantum Theory of the Solid State." Academic, London.
Capasso, F., and Margaritondo, G. (Eds.) (1987). "Heterojunction Band Discontinuities: Physics and Device Applications." North Holland, Amsterdam.
Cardona, M., and Christensen, N. E. (1987). *Phys. Rev.* **B35**, 6182.
Chambers, S. A., and Loebs, V. A. (1989). *Phys. Rev. Lett.* **63**, 640.
Chang, Y. C. (1985). *J. Appl. Phys.* **58**, 499.
Chang, Y. C., and Schulman, N. (1985). *Phys. Rev.* **B31**, 2069.
Chang, Y. C., Schulman, J. N., Bastard, G., Guldner, Y., and Voos, M. (1985). *Phys. Rev.* **B31**, 2557.
Chen, W., and Mills, D. L. (1987). *Phys. Rev. Lett.* **58**, 160.
Christensen, N. E. (1988). *Phys. Rev.* **B37**, 4528.
Christensen, N. E. (1988). *Phys. Rev.* **B38**, 12687.
Ciraci, S., and Batra, I. P. (1987). *Phys. Rev. Lett.* **58**, 2114.
Cohen, M. L. (1980). *Adv. Electron. and Electron Physics* **51**, 1.
Danan, G., Etienne, B., Mollot, F., Planel, R., Jean-Luis, A. M., Alexander, F., Jusserand, B., Le Roux, G., Marzin, J. Y., Savary, H., and Sermage, B. (1987). *Phys. Rev.* **B35**, 6207.
Dawson, P., Wilson, B. A., Tu, C. W., and Miller, R. C. (1986). *Appl. Phys. Lett.* **48**, 541.
Dingle, R. (1975). *Festkorperprobleme* **15**, 21.
Dingle, R. (Ed.) (1988). Applications of Multiquantum Wells, Selective Doping and Superlattices, In "Semiconductors and Semimetals," Vol. 24. Academic, New York.
Eberl, K., Krötz, G., Zachai, R., and Abstreiter, G. (1987). *J. de Physique* **48** (Suppl. 11), C5-329.
Feldman, L. C., Bevk, J., Davidson, B. A., Gossman, H. J., and Mannaerts, J. P. (1987). *Phys. Rev. Lett.* **5**, 664.
Faurie, J. P. (1986). *IEEE J. Quant. Electronics* **QE-22**, 1656 (and references therein).
Finkman, E., Sturge, M. D., and Tamargo, M. C. (1986). *Appl. Phys. Lett.* **49**, 1299.
Flores, F., and Tejedor, C. (1979). *J. Phys. C* **12**, 731.
Flynn, C. P. (1986). *Phys. Rev. Lett.* **57**, 599.
Folland, N. O. (1983). *Phys. Rev.* **B28**, 6068.
Frensley, W. R., and Kroemer, H. (1977). *Phys. Rev.* **B16**, 2642.
Froyen, S., Wood, D. M., and Zunger, A. (1987). *Phys. Rev.* **B34**, 4574.
Froyen, S., Wood, D. M., and Zunger, A. (1988). *Phys. Rev.* **B37**, 6893.
Fukui, T., and Saito, H. (1984). *Jpn. J. Appl. Phys.* **23**, L521.
Fukui, T., and Saito, H. (1985). *Jpn. J. Appl. Phys.* **24**, L774.
Gell, M. A., Wong, K. B., Ninno, D., and Jaros, M. (1986). *J. Phys. C* **19**, 3821.
Gell, M. A., Ninno, D., Jaros, M., and Herbert, D. C. (1986). *Phys. Rev.* **B34**, 2416.
Gell, M. A., Ninno, D., Jaros, M., Wolford, D. J., Kuech, T. F. and Bradley, J. A. (1987). *Phys. Rev.* **B35**, 1196.

Gell, M. A. (1988). *Phys. Rev.* **B38**, 7535.
Gnutzmann, U., and Clausecker, K. (1974). *Appl. Phys.* **3**, 9.
Godby, R. W., Schlüter, M., and Sham, L. J. (1986). *Phys. Rev. Lett.* **56**, 2415.
Guldner, Y., Bastard, G., Vieren, J. P., Voos, M., Faurie, J. P., and Million, A. (1983). *Phys. Rev. Lett.* **51**, 907.
Hagon, J. P., Jaros, M., and Herbert, D. C. (1989). *Phys. Rev.* **B40**, 6420.
Hass, K. C., Velicky, B., and Ehrenreich, H. (1984). *Phys. Rev.* **B29**, 3697.
Herman, M. A., and Pessa, M. (1985). *J. Appl. Phys.* **57**, 2671.
Hoffman, C. A., Mayer, J. R., Youngdale, E. R., Lindle, J. R., Bartoli, F. J., Harris, K. A., Cook, J. W., Jr., and Schetzina, J. F. (1988). *Phys. Rev.* **B37**, 6933.
Hybertsen, M. S., and Louie, S. G. (1985). *Phys. Rev. Lett.* **55**, 1418.
Hybertsen, M. S., and Louie, S. G. (1986). *Phys. Rev.* **B34**, 5390.
Hybertsen, M. S., and Schlüter, M. (1987). *Phys. Rev.* **B36**, 9683.
Hybertsen, M. S., Schlüter, M., People, R., Jackson, S. A., Lang, D. V., Pearsall, T. P., Bean, J. C., Vandenberg, J. M., and Bevk, J. (1988). *Phys. Rev.* **B37**, 10195.
Ihm, J. (1987). *Appl. Phys. Lett.* **50**, 1068.
Ishibashi, A., Mori, Y., Itabashi, M., and Watanabe, N. (1985). *J. Appl. Phys.* **58**, 2691.
Ishibashi, A., Mori, Y., Itabashi, M., and Watanabe, N. (1986). *J. Appl. Phys.* **59**, 2503.
Ivanov, I., and Pollmann, J. (1979). *Solid State Commun.* **32**, 869.
Jaros, M. (1982). "Deep Levels in Semiconductors." A. Hilger, Bristol.
Jaros, M. (1985). *Reports on Progress in Physics* **48**, 1091.
Jaros, M. (1988). *Phys. Rev.* **B37**, 7112.
Jaros, M. (1989). "Physics and Applications of Semiconductor Microstructures." Oxford University Press, Oxford.
Jaros, M., and Guimaraes, P. S. (1985). *J. Phys. C* **18**, L117.
Jaros, M., Wong, K. B., Gell, M. A., and Wolford, D. J. (1985). *Vac. Sci. Technol.* **B3**, 1051.
Jaros, M., Zoryk, A., and Ninno, D. (1987). *Phys. Rev.* **B35**, 8277.
Jen, H. R., Cherng, M. J., and Stringfellow, G. B. (1986). *Appl. Phys. Lett.* **48**, 1603.
Johnson, N. F., Ehrenreich, H., Haas, K., and McGill, T. C. (1988a). *Phys. Rev. Lett.* **59**, 2352.
Johnson, N. F., Hui, P. M., and Ehrenreich, H. (1988b). *Phys. Rev. Lett.* **61**, 1993.
Kasper, E., Kibbel, H., Jorke, H., Brugger, H., Friess, E., and Abstreiter, G. (1988). *Phys. Rev.* **B38**, 4580.
Kelly, M. J., and Weisbuch, C. (Eds.) (1986). "The Physics and Fabrication of Microstructures and Microdevices." Springer, Heidelberg.
Kirton, M. J., Brand, S., and Jaros, M. (1983). *Physica* **116B**, 79.
Kirton, M. J., Banks, P. W., Lu, D. L., and Jaros, M. (1984). *J. Phys. C* **17**, 2487.
Kobayashi, A., and Das Sharma, S. (1988). *Phys. Rev.* **B37**, 1988.
Kohn, W. (1957). *Solid State Phys.* **5**, 257.
Kondow, M., Kakibayashi, H., Tanaka, T., and Minatawa, S. (1989). *Phys. Rev. Lett.* **63**, 884.
Kowalczyk, S. P., Cheung, J. T., Kraut, E. A., and Grant, R. W. (1986). *Phys. Rev. Lett.* **58**, 1605.
Krishnamurthy, S., Sher, A., and Chen, A.-B. (1986). *Phys. Rev.* **B33**, 1026.
Kuan, T. S., Kuech, T. F., Wang, W. I., and Wilkie, E. L. (1985). *Phys. Rev. Lett.* **54**, 201.
Kunc, K., and Martin, R. M. (1981). *Phys. Rev.* **24**, 3445.
Lambrecht, W. R. L., and Segall, B. (1988). *Phys. Rev. Lett.* **61**, 1764.
Lin, L. Y. R., and Sham, L. J. (1985). *Bull. Am. Phys. Soc.* **30**, 266
Macfarlane, G. G., and Roberts, V. (1955). *Phys. Rev.* **97**, 1714.
Macfarlane, G. G., and Roberts, V. (1955). *Phys. Rev.* **98**, 1865.
Martins, L., and Zunger, A. (1986). *Phys. Rev. Lett.* **56**, 1400.
Massida, S., Min, B. I., and Freeman, A. J. (1987). *Phys. Rev.* **B35**, 9871.
Meynadier, M.-H., Nahory, R. E., Worlock, J. M., Tamargo, M. L., and de Miguel, J. L. (1988). *Phys. Rev. Lett.* **60**, 1338.

Moezawa, K., Mizutani, T., and Yanakawa, F. (1986). *Jap. J. Appl. Phys.* **25**, L557.
Moore, K. J., Dawson, P., and Foxon, C. T. (1988). *Phys. Rev.* **B38**, 3368.
Morrison, I., and Jaros, M. (1988). *Phys. Rev.* **B37**, 916.
Morrison, I., Jaros, M., and Beavis, A. W. (1989). *App. Phys. Lett.* **55**, 1609.
Morrison, I., Jaros, M., and Wong, K. B. (1987). *Phys. Rev.* **B35**, 9693.
Morrow, R. A. (1987a). *Phys. Rev.* **B35**, 8074.
Morrow, R. A. (1987b). *Phys. Rev.* **B36**, 4836.
Morrow, R. A., and Brownstein, K. R. (1984). *Phys. Rev.* **B30**, 678.
Müller, E., Nissen, H. U., Ospelt, M., and von Känel, H. (1989). *Phys. Rev. Lett.* **63**, 1819.
Munoz, A., Sanchez-Dehesa, J., and Flores, F. (1987). *Phys. Rev.* **B35**, 6468.
Nagle, J., Carriga, M., Stolz, W., Isu, T., and Ploog, K. (1987). *Int. Conf. on Modulated Semic. Structures* (MSS-III), p. 419, Montpellier, France.
Ninno, D., Wong, K. B., Gell, M. A., and Jaros, M. (1985). *Phys. Rev.* **B32**, 2700.
Ninno, D., Gell, M. A., and Jaros, M. (1986). *J. Phys. C.* **19**, 3845.
Northrop, G. A., Iyer, S. S., Wolford, D. J., and Delage, S. L. (1988). *Physics of Semiconductors*, Warsaw, p. 1041.
Ourmazd, A., and Bean, J. C. (1985). *Phys. Rev. Lett.* **55**, 765.
Patel, C. K. N., Slusher, R. E., and Fleury, P. A. (1966). *Phys. Rev. Lett.* **17**, 1011.
Pearsall, T. P. (Ed.) (1982). "GaInAsP Alloy Semiconductors." J. Wiley, New York.
Pearsall, T. P., Bevk, J., Feldman, L. C., Ourmazd, A., Bonar, J. M., and Mannaerts, J. P. (1987). *Phys. Rev. Lett.* **58**, 729.
Pearsall, T. P., Bevk, J., Bean, J. C., Bonar, J., Mannaerts, J. P., and Ourmazd, A. (1989). *Phys. Rev.* **339**, 3741.
People, R. (1985). *Phys. Rev.* **B32**, 1405.
People, R., and Jackson, S. A. (1987). *Phys. Rev.* **B36**, 1310.
Perfetti, P., Quaresima, C., Coluzza, C., Fortunato, C., and Margaritondo, G. (1986). *Phys. Rev. Lett.* **57**, 2065.
Perkowitz, S., Sudharsanan, R., Harros, K. A., Cook, J. W., Jr., Schetzina, J. F., and Schulman, J. N. (1987). *Phys. Rev.* **B36**, 9290.
Perkowitz, S., Sudharsanan, R., and Yom, S. S. (1988). *J. Vac. Sci. Technol.* **B6**, 3157.
Phillips, J. C. (1973). "Bonds and Bands in Semiconductors." Academic Press, New York.
Pickett, W. E., Louie, S. G., and Cohen, M. L. (1978). *Phys. Rev.* **B17**, 815.
Pulsford, N. J., Nicholas, R. J., Dawson, P., Moore, K. J., Duggan, G., and Foxon, C. T. B. (1989). *Phys. Rev. Lett.* **63**, 2284.
Robbins, D. J., Kubiak, R. A. A., and Parker, E. H. C. (1985). *J. Vac. Sci. Technol.* **B3**, 588.
Schmitt-Rink, S., Chemla, D. S., and Miller, D. A. B. (1985). *Phys. Rev.* **B32**, 6601.
Schulman, J. N., and McGill, T. C. (1977). *Phys. Rev. Lett.* **39**, 1680.
Schulman, J. N., and McGill, T. C. (1979). *Appl. Phys. Lett.* **34**, 669.
Schulman, J. N., and Chang, Y. C. (1986). *Phys. Rev.* **B33**, 2594.
Smith, D. L., McGill, T. C., and Schulman, J. N. (1983). *Appl. Phys. Lett.* **43**, 180.
Solomon, P. M., Wright, S. L., and Lanza, C. (1986). *Superlattices and Microstructures* **2**, 521.
Steiner, T. W., Wolford, D. J., Kuech, T. F., and Jaros, M. (1988). *Superlattices and Microstructures* **4**, 227.
Tejedor, C., Flores, F., and Louis, E. (1977). *J. Phys. C* **10**, 2163.
Tersoff, J. (1984). *Phys. Rev.* **B30**, 4874.
Tersoff, J. (1985). *Phys. Rev.* **B32**, 6968.
Tersoff, J. (1986). *Phys. Rev. Lett.* **51**, 907.
Tersoff, J. (1989). *Phys. Rev.* **B40**, 10615.
Ting, D. Z.-Y., and Chang, Y. C. (1986). *J. Vac. Sci. Technol.* **B4**, 1002.
Ting, D. Z. Y., and Chang, Y. C. (1987). *Phys. Rev.* **B36**, 4359.

Trzeciakowski, W. (1988). *Phys. Rev.* **B38**, 4322.
Turton, R., Jaros, M., and Morrison, I. (1988). *Phys. Rev.* **B38**, 8397.
Van de Walle, C. G., and Martin, R. M. (1985). *J. Vac. Sci. Technol.* **B3**, 1256.
Van de Walle, C. G., and Martin, R. M. (1986). *J. Vac. Sci. Technol.* **B4**, 1055.
Van de Walle, C. G., and Martin, R. M. (1986). *Phys. Rev.* **B34**, 5621.
Van de Walle, C. G., and Martin, R. M. (1987). *Phys. Rev.* **B35**, 8154.
Van de Walle, C. G., and Martin, R. M. (1988). *Phys. Rev.* **B37**, 4801.
Vogl, P. (1981). *Ferstkorperprobleme* **20**, 191.
Wei, S.-H., and Zunger, A. (1988). *Phys. Rev. Lett.* **61**, 1505.
Wolford, D. J., Kuech, T. F., Bradley, J. A., Gell, M. A., Ninno, D., and Jaros, M. (1986). *J. Vac. Sci. Technol.* **B4**, 1043.
Wong, K. B., and Jaros, M. (1988). *Appl. Phys. Lett.* **53**, 657.
Wong, K. B., Jaros, M., Gell, M. A., Ninno, D. (1986). *J. Phys.* **C19**, 53.
Wong, K. B., Jaros, M., and Hagon, J. P. (1987). *Phys. Rev.* **B35**, 2463.
Wong, K. B., Jaros, M., Morrison, I., and Hagon, J. P. (1988). *Phys. Rev. Lett.* **60**, 2221.
Wong, K. B., and Jaros, M. (1989). Unpublished.
Wood, D. M., Wei, S.-H., and Zunger, A. (1987). *Phys. Rev. Lett.* **58**, 1123.
Wood, D. M., Wei, S.-H., and Zunger, A. (1988). *Phys. Rev.* **B37**, 1342.
Zhu, Q. G., and Kroemer, H. (1983). *Phys. Rev.* **B27**, 3519.
Zohta, Y. (1988). *Jpn. J. Appl. Phys.* **27**, L906.
Zoryk, A., and Jaros, M. (1988). Unpublished.

Index

A

$Al_xGa_{1-x}As/GaAs$, 72, 218
$Al_{1-x}In_xAs/GaAs$, 62–63
AlSb, 27, 36
Areal strain energy, 123, 126
Auger effect, 213

B

Band gap
    dielectric model, 195
    direct, 8, 19–20, 24, 27–28, 134–143
    indirect, 36, 140–141, 181
    Krönig–Penney, 182
    quasi-direct, 169
    strain-dependent, 8
    zero-bandgap materials near $\mathbf{k} = 0$, 32
Band offset
    estimation, 146–147
    experimental determination, 87–93
    Ge/Si, 227–229
    $Hg_{1-x}Cd_xTe/Hg_{1-y}Cd_yTe$, 241–244
    optical determination, 87–90, 209–211
    strain-modified, 143–144
    table, 196
    theoretical models, 146–152, 194–197
Band structure
    band line-up, 195
        table, 196
    band non-parabolicity, 218–219
    calculation of strain effects, 26–32
    envelope function, 64–72
    interface states, 245
    magneto-optical measurements, 97–101
    photo-excitation spectroscopy, 113–114
    quantum well, 180–183
    strained valence band, 18–28, 72–82
    strain effects, 19–20
    uniaxial strain splittings, 134–140
        table, 154
    zone folding, 163–168, 223
Bi-layer strain energy, 120
Bir and Pikus, 26, 51
Box, particle-in-a, 179
Brooks deformation potential model, 35, 51
Bulk-crystal Hamiltonian, 187

C

CdTe, 27, 47, 245
Clustering, 7, 104
Confinement, *see also* Band offset
    envelope function model, 68–71, 182–185, 242
    materials systems
        HgTe/CdTe, 245
        InAs/GaAs, 105
        Si/Ge, 222, 232
    particle in-a-box, 179
    quantum wire, 214
Critical thickness
    correspondence principle, 133
    defects, 183, 201, 237
    dislocations
        equilibrium, 60–63, 83, 95, 109
        force of strain, 125–128
        measurements, 130–131
        misfit, 124–125
        self-energy, 2–4, 57, 126
        stability, 60–63, 83, 95, 127–129, 201
        threading, 124–132
    interfacial energy, 57–60, 123
    model, 122–134
        Frank–van de Merwe, 123–124
        Matthews–Blakeslee, 124–127
    multilayer structures, 132
    single-layer structures, 124–126

strain energy
  areal, 123, 126
  bi-layer, 120
stress relief, 61

D

Defects, 183, 201, 237
Deformation potentials
  Brooks, 35, 51
  interband, 21, 25-7, 36-8, 40
  intraband, 21, 28, 35-36
  Kane, 26, 35, 51
  Kleiner and Roth, 26, 51, 154
  materials systems
    AlSb, 27, 36
    CdTe, 27, 47
    GaAs, 27, 40, 47
    GaP, 27, 36, 47
    GaSb, 27, 40, 47
    Ge, 27, 36, 40, 47
    InAs, 27, 47
    InP, 27, 47
    InSb, 27, 40, 47
    Si, 27, 36, 40, 47
    ZnS, 28
    ZnSe, 28, 47
    ZnTe, 28
  orbital-uniaxial, 21, 25-28, 35-36, 40
  PIkus and Bir, 26, 51
  strain-dependent spin-orbit splitting, 21, 26
  table, 27, 28
  theoretical, 26, 134-142
  valence band, 21, 25-28, 35
  vibrational, 43-48
Diamond structure materials, 43
Dielectric model of band gap, 195
Direct band gap, 8, 19-20, 24, 27-28, 134-143
Dislocations
  equilibrium, 60-63, 83, 95, 109
  force of strain, 125-128
  measurements, 130-131
  misfit, 124-125
  self-energy, 2-4, 57, 126
  stability, 60-63, 83, 95, 127-129, 201
  threading, 124-132

E

Effective mass
  electrons, 154
  holes, 154, 158
  models
    $\mathbf{k} \cdot \mathbf{p}$ approximation, 155, 163
    Kane model with strain, 161-163
    envelope function, 182-185, 242
    non-parabolicity, 65, 69, 73-74, 100, 160-161, 217-220
    reversal, 65, 74, 100
    theory, 164-168, 180-187
    transition probability, 164-169, 182
Epitaxial growth
  clustering, 7, 104
  enthalpy of formation, 199-200
  equilibrium, 7
  InAs/GaAs, 107-108
  kinetics, 4-5, 7
  modulation doping, 9-10, 143-144
  planar, 4, 120-132
  Stranski-Krastanov mode, 4, 122
  three-dimensional, 101-104
Envelope function model
  band gap, 64-72
  quantum confinement, 68-71, 182-185, 242
Equilibrium dislocation concentration, 60-63, 83, 95, 109
Excitation
  photo-spectroscopy, 85, 94, 112-114
  virtual, 202, 243
Exciton, effect of strain on, 33

F

Frank-van der Merwe growth, 4, 122-124
Frensley-Kroemer model, 148

G

GaAs, 20, 40, 47
GaAs/$Al_xGa_{1-x}As$, 72, 218
GaAs/$Al_{1-x}In_xAs$, 62-63
GaAs/InAs, 105, 107-108
GaP, 27, 36, 47
GaSb, 27, 40, 47

INDEX

Ge, 27, 36, 40, 47
Ge/Si, 222, 227–229, 232
Grünheisen parameter
  vibrational modes, 43
  table, 47

H

Hamiltonian
  bulk crystal, 187
  k·p, 21, 23, 29
  spin-exchange, 21, 23, 29
  strain
    exciton, 33
    hydrostatic pressure, 20–24, 35, 63–64, 72–73
    orbital bands
      $k = 0$, 21–24
      $k = 0$, 35–36
    spin-orbit splitting, 21, 38
    tensor, 63, 135–136
    uniaxial, 21–24, 33–36, 63–64, 72–73, 136–139
    vibrational modes, 43–44
Hartree potential, 151–153
Heat of formation, 199
Herring and Vogt model, 35, 51, 138
Heterojunction, band line-up
  determination in III-V materials, 87–101
  estimation, 146–147
  Ge/Si, 227–229
  $Hg_{1-x}Cd_xTe/Hg_{1-y}Cd_yTe$, 241–244
  level crossing, 204–210
  optical measurements, 87–90, 209–211
  pseudomorphic, 124–126, 131–134, 143, 153–155, 159
  physics of band line-up, 195–198
  strain-modified, 143–144
  table, 196
  theoretical models, 146–152, 194–197
Heterojunction bi-polar transistor, 8
$Hg_{1-x}Cd_xTe/Hg_{1-y}Cd_yTe$, 241–244
HgTe/CdTe, 245
High-resolution X-ray diffraction, 11
Holes, strain-induced mass, 28, 32, 65, 74, 100, 156, 158
Hydrostatic pressure
  deformation potentials, 27, 36, 40
  elastic strain energy, 139
  Hamiltonian, 20, 35

I

InAs, 27, 47
InAs/GaAs, 102, 105, 108–114
Indirect transition, 36
$In_xGa_{1-x}As/GaAs$, 82–91, 126
$In_xGa_{1-x}As/In_yAl_{1-y}As$, 93, 101
$In_xGa_{1-x}As/In_yGa_{1-y}As$, 101
$In_xGa_{1-x}As/InP$, 143
InP, 27, 47, 154
In-plane strain, 135
InSb, 27, 40, 47
In segregation, 104
Intra-band transition, 21, 28, 35–36
Inter-band transition, 8, 19–20, 24, 27–28, 134–143, 168 (table), 169, 182, 208, 211, 216, 236
Interface
  charge transfer, 192
  dipole, 197
  energy model, 123
  state, 245
Inverse mass parameters, 154 (table)
Isoelectronic center, 183, 213

J

Junction, *see* Heterojunction

K

Kane model, 26, 51, 161
Kinetic effects, 4–5, 60–63, 101, 104, 107–108
Kleiner–Roth model, 26, 51, 134
k·p approximation, 21, 23, 29, 155, 163
Krönig–Penney model, 182

L

Lattice parameter
  table, 154
Level crossing
  electric-field induced, 210
  pressure-induced, 208
  X-states interface plane, 210
Linear muffin-tin orbital (LMTO) model, 149–150
Liquid-phase epitaxy, 1–3
Local density approximation (LDA), 190, 198–201

M

Magneto-optical measurements, 97–101
Mass
    electrons, 154
    holes, 154, 158
    inverse mass parameters, 154 (table)
    models
        envelope function, 182–185, 242
        Kane model with strain, 161–163
        $\mathbf{k \cdot p}$ approximation, 155–163
    non-parabolicity, 65, 69, 73–74, 100, 160–161, 217–220
    parameters
        conduction band near $\mathbf{k} = 0$, no strain, 30
        conduction band near $\mathbf{k} = 0$, with strain, 32
        valence bands near $\mathbf{k} = 0$, no strain, 28
        valence bands near $\mathbf{k} = 0$, with strain, 32
    reversal, 65, 74, 100
    strain effects, 155–163
    theory, 164–168, 180–187
    transition probability, 164–169, 182
Materials systems
    $Al_xGa_{1-x}As/GaAs$, 72, 218
    GaAs/InAs, 102
    GaSb/AlSb, 95
    Ge/Si, 6, 224
    HgTe/CdTe, 240–245
    $In_xAl_{1-x}As/GaAs$, 94
    $In_xGa_{1-x}As/GaAs$, 82
    $In_xGa_{1-x}As/In_xAl_{1-x}As$, 93
    $In_xGa_{1-x}As/In_yGa_{1-y}As$, 101
Matrix element, *see* Optical properties
Matthews–Blakesell model, 124–127
Metastability
    dynamical equation, 129
    excess stress, 129
    plastic flow, 61, 89, 127
    self-equilibrium, 62, 83, 95
    spontaneous nucleation, 127
    thermodynamic, 60–63, 198–201
Micro-crack diffusion, 128
Misfit dislocations, 2, 130–131
Modes
    Grüneisen parameter
        table, 47
    vibrational modes, 43
    optical phonon
        effect of strain, 43–48
        momentum mixing, 183, 207, 214, 230, 232, 241
        Raman effect, 63, 95, 109
Modulation-doping
    band offset, 143–144
    field-effect transistor, 9–10
Momentum
    conservation, 167
    mixing, 183, 207, 214, 230, 232, 241
Multilayer structures, 6–13, 108–111, 122, 132

N

Neutrality level, 149
Non-linear effects, 201–203
Non-parabolicity, 65, 69, 73–74, 100, 160–161, 217–220

O

Offset, band
    optical determination of, 87–90, 209–211
Optical phonons
    effect of strain, 43–48
    momentum mixing, 183, 207, 214, 230, 232, 241
    Raman effect, 63, 95, 109
Optical properties
    Auger effect, 213
    excitation spectroscopy, 85, 94, 112
    magneto-optical transmission, 83, 93, 95, 106
    non-linear effects, 201–203
    photoluminescence, 83, 85, 88, 93–95, 106, 112
    piezo-electric effects, 169–170
    plasma frequency, 195
    radiative lifetime, 212
    Raman scattering, 43–46, 63, 95, 109
    recombination, 212
    spatially indirect transitions, 212
    stress effects, 89
    transition probability
        effective mass approximation, 164–169, 182
        momentum conservation, 167
        photoluminescence, 211

pressure-induced, 208
Si/Ge, 216
strained layer, 134, 168
ultra-thin layer, 216, 234–238
transmission spectroscopy, 83, 93, 95, 106
virtual excitation, 202, 243
Orbital bands Hamiltonian
$\mathbf{k} = 0$, 21–24
$\mathbf{k} = 0$, 35–36
Orbital–uniaxial deformation potential, 21, 25–28, 35–36, 40

P

Particle-in-a-box, 179
Peierls stress, 128
Phonons
effect of strain, 43–48
momentum mixing, 183, 207, 214, 230, 232, 241
Raman effect, 63, 95, 109
Photoconductivity in Si/Ge, 237
Photo-excitation spectroscopy, 85, 113–114
Photoluminescence, 83, 85, 88, 93–95, 106, 112, 211
Piezo-electric effects, 169–170
Pikus and Bir, 26, 51
Planar nucleation, 4, 120–132
Plasma frequency, 195
Plastic flow, 61, 89, 127
Pressure-induced level crossing, 208
Probability of optical transitions
effective mass approximation, 164–169, 182
momentum conservation, 167
photoluminescence, 211
pressure-induced, 208
Si/Ge, 216
strained layer, 134, 168
ultra-thin layer, 216, 234–238
Pseudomorphic heterojunctions
GaAs/GaAs$_x$P$_{1-x}$, 134
GaAs/In$_x$Ga$_{1-x}$As, 126
Ge$_x$Si$_{1-x}$/Si, 124, 126, 131
InP/GaAs, 154
InP/In$_x$Ga$_{1-x}$As, 132, 143, 153–155, 159
Pseudopotential calcualtions
bandstructure, 191, 194–198
deformation potentials, 26

Q

Quantum confinement
envelope function model
band gap, 64–72
quantum confinement, 68–71, 182–185, 242
materials systems
HgTe/CdTe, 245
InAs/GaAs, 105
Si/Ge, 222, 232
particle-in-a-box, 179
quantum wire, 214
Quantum well bandstructure, 180–183
Quasi-direct bandgap (strain induced), 169

R

Radiative lifetime, 212
Raman scattering, 43–46, 63, 95, 109
Recombination, 212
Resonance
central valley related, 205
X-related, 205
quantum wire, 216
Reversal of effective masses, 65, 74, 100

S

Segregation, Indium, 104
Self-energy of dislocations, 2–4, 57, 126
Self-equilibrium, 62, 83, 95
Si/Ge, 6, 140–141, 146–147, 216, 222, 227–229, 232
Spatially indirect transitions, 212
Spin-exchange Hamiltonian, 21, 23, 29
Spin-orbit splitting, 20, 23–24, 28, 32, 38
table, 154
Spontaneous nucleation of dislocations, 127
Strain effects
energy
areal, 123, 126
bi-layer, 120
hydrostatic, 139
minimum, 120
exciton, 33
hydrostatic pressure, 20–24, 35, 63–64, 72–73
orbital bands
$\mathbf{k} = 0$, 21–24
$\mathbf{k} \neq 0$, 35–36

spin-orbit splitting, 21, 38
tensor, 63, 135–136
unixial, 21–24, 33–36, 63–64, 72–73, 136–139
vibrational modes, 3–4
Strained bandstructire
   band gap, 8
   strain-modified band offset, 143–144
   uniaxial strain splittings, 134–140
      table, 154
   valence band, 18–28, 72–82
Strain energy
   bi-layer, 120
   single structures, 124–126
   stability, 60–63, 83, 95, 127–129, 132, 198, 201
Stranski–Krastanov mode, 4, 122
Superlattices
   band structure, 66, 72
   effective mass theory, 164–168, 180–187
   materials systems
      $Al_xGa_{1-x}As/GaAs$, 72, 218
      GaAs/InAs, 102
      GaSb/AlSb, 95
      Ge/Si, 6, 224
      HgTe/CdTe, 240–245
      $In_xAl_{1-x}As/GaAs$, 94
      $In_xGa_{1-x}As/GaAs$, 82
      $In_xGa_{1-x}As/In_xAl_{1-x}As$, 93
      $In_xGa_{1-x}As/In_yGa_{1-y}As$, 101
   optical transitions, 134, 168
   ordered, 14
   structural stability, 132, 198–201
   ultra-thin, 228–232
   zone-folding, 163–166, 223

T

Tables
   band offset, 196
   deformation potentials, 27, 28
   inverse mass parameters, 154
   mode Grünheisen parameter, 47
   spin-orbit splitting, 154
   uniaxial strain splittings, 154
Tensor
   asymmetric, 226
   conventional strain components, 63, 135
   substrate orientation, 136
   symmetric, 221
Tetragonal splitting, 229

Thermodynamic metastability, 60–63, 198–201
Thermoionization, 220
Threading dislocations, 124–132
Three-dimensional growth mode, 101–104
Transistor
   bi-polar, 8
   field-effect, 9–10
Transition probability, optical
   effective mass approximation, 164–169, 182
   momentum conservation, 167
   photoluminescence, 211
   pressure-induced, 208
   Si/Ge, 216
   strained layer, 134, 168
   ultra-thin layer, 216, 163–167, 234–238
Transmission spectroscopy, 83, 93, 95, 106

U

Ultra-thin layer, 216, 163–167, 234–238
Uniaxial–orbital deformation potential, 21, 25–28, 35–36, 40
Uniaxial strain splittings, 21–24, 33–36, 63–64, 72–73, 134–140
   table, 154

V

Valence band
   configurations, 67
   deformation potentials, 21, 25–28, 35
   effective mass, 28, 32, 65, 74, 100, 156, 158
   electronic structure
      bulk materials, 64
      dispersion relations, 72, 97
   heterojunction offset, 195
      table, 196
   mixing, 18–28, 72–82
   spin-orbit splitting, 21, 26, 32
   strained valence band, 8, 18–28, 72–82
   superlattices, 66, 72
   table, 154
   uniaxial strain splittings, 65, 74, 82, 113, 134–140
      table, 154
van de Merwe–Frank model, 123–124
Virtual excitations, 202, 243
Vogt and Herring model, 35, 51, 138

X

X-ray diffraction
  high resolution, 11
  materials systems
    $Al_{1-x}In_xAs/GaAs$, 62–63
    GaSb/AlSb, 95
    Ge/Si, 12
    InAs/GaAs, 109
    $In_xGa_{1-x}As/GaAs$, 84
  stress relaxation, 89
  topography, 62, 84

X-states level crossing, 210
X-valley resonance, 205

Z

Zero-bandgap materials near $\mathbf{k} = 0$, 32
ZnS, deformation potentials
  electronic levels, 28
ZnSe, deformation potentials
  electronic levels, 28
  optical phonons, 47
Zone folding, 163–166, 223

Contents of Previous Volumes

Volume 1   Physics of III–V Compounds

C. Hilsum, Some Key Features of III–V Compounds
Franco Bassani, Methods of Band Calculations Applicable to III–V Compounds
E. O. Kane, The k·p Method
V. L. Bonch-Bruevich, Effect of Heavy Doping on the Semiconductor Band Structure
Donald Long, Energy Band Structures of Mixed Crystals of III–V Compounds
Laura M. Roth and Petros N. Argyres, Magnetic Quantum Effects
S. M. Puri and T. H. Geballe, Thermomagnetic Effects in the Quantum Region
W. M. Becker, Band Characteristics near Principal Minima from Magnetoresistance
E. H. Putley, Freeze-Out Effects, Hot Electron Effects, and Submillimeter Photoconductivity in InSb
H. Weiss, Magnetoresistance
Betsy Ancker-Johnson, Plasmas in Semiconductors and Semimetals

Volume 2   Physics of III–V Compounds

M. G. Holland, Thermal Conductivity
S. I. Novkova, Thermal Expansion
U. Piesbergen, Heat Capacity and Debye Temperatures
G. Giesecke, Lattice Constants
J. R. Drabble, Elastic Properties
A. U. Mac Rae and G. W. Gobeli, Low-Energy Electron Diffraction Studies
Robert Lee Mieher, Nuclear Magnetic Resonance
Bernard Goldstein, Electron Paramagnetic Resonance
T. S. Moss, Photoconduction in III–V Compounds
E. Antončik and J. Tauc, Quantum Efficiency of the Internal Photoelectric Effect in InSb
G. W. Gobeli and F. G. Allen, Photoelectric Threshold and Work Function
P. S. Pershan, Nonlinear Optics in III–V Compounds
M. Gershenzon, Radiative Recombination in the III–V Compounds
Frank Stern, Stimulated Emission in Semiconductors

Volume 3   Optical of Properties III–V Compounds

Marvin Hass, Lattice Reflection
William G. Spitzer, Multiphonon Lattice Absorption
D. L. Stierwalt and R. F. Potter, Emittance Studies
H. R. Philipp and H. Ehrenreich, Ultraviolet Optical Properties
Manuel Cardona, Optical Absorption above the Fundamental Edge
Earnest J. Johnson, Absorption near the Fundamental Edge
John O. Dimmock, Introduction to the Theory of Exciton States in Semiconductors
B. Lax and J. G. Mavroides, Interband Magnetooptical Effects

H. Y. Fan, Effects of Free Carriers on Optical Properties
Edward D. Palik and George B. Wright, Free-Carrier Magnetooptical Effects
Richard H. Bube, Photoelectronic Analysis
B. O. Seraphin and H. E. Bennett, Optical Constants

Volume 4 Physics of III–V Compounds

N. A. Goryunova, A. S. Borschevskii, and D. N. Tretiakov, Hardness
N. N. Sirota, Heats of Formation and Temperatures and Heats of Fusion of Compounds $A^{III}B^{V}$
Don L. Kendall, Diffusion
A. G. Chynoweth, Charge Multiplication Phenomena
Robert W. Keyes, The Effects of Hydrostatic Pressure on the Properties of III–V Semiconductors
L. W. Aukerman, Radiation Effects
N. A. Goryunova, F. P. Kesamanly, and D. N. Nasledov, Phenomena in Solid Solutions
R. T. Bate, Electrical Properties of Nonuniform Crystals

Volume 5 Infrared Detectors

Henry Levinstein, Characterization of Infrared Detectors
Paul W. Kruse, Indium Antimonide Photoconductive and Photoelectromagnetic Detectors
M. B. Prince, Narrowband Self-Filtering Detectors
Ivars Melngailis and T. C. Harman, Single-Crystal Lead–Tin Chalcogenides
Donald Long and Joseph L. Schmit, Mercury–Cadmium Telluride and Closely Related Alloys
E. H. Putley, The Pyroelectric Detector
Norman B. Stevens, Radiation Thermopiles
R. J. Keyes and T. M. Quist, Low Level Coherent and Incoherent Detection in the Infrared
M. C. Teich, Coherent Detection in the Infrared
F. R. Arams, E. W. Sard, B. J. Peyton, and F. P. Pace, Infrared Heterodyne Detection with Gigahertz IF Response
H. S. Sommers, Jr., Microwave-Based Photoconductive Detector
Robert Sehr and Rainer Zuleeg, Imaging and Display

Volume 6 Injection Phenomena

Murray A. Lampert and Ronald B. Schilling, Current Injection in Solids: The Regional Approximation Method
Richard Williams, Injection by Internal Photoemission
Allen M. Barnett, Current Filament Formation
R. Baron and J. W. Mayer, Double Injection in Semiconductors
W. Ruppel, The Photoconductor–Metal Contact

Volume 7 Application and Devices
Part A

John A. Copeland and Stephen Knight, Applications Utilizing Bulk Negative Resistance
F. A. Padovani, The Voltage–Current Characteristics of Metal–Semiconductor Contacts
P. L. Hower, W. W. Hooper, B. R. Cairns, R. D. Fairman, and D. A. Tremere, The GaAs Field-Effect Transistor
Marvin H. White, MOS Transistors
G. R. Antell, Gallium Arsenide Transistors
T. L. Tansley, Heterojunction Properties

Part B

T. Misawa, IMPATT Diodes
H. C. Okean, Tunnel Diodes
Robert B. Campbell and Hung-Chi Chang, Silicon Carbide Junction Devices
R. E. Enstrom, H. Kressel, and L. Krassner, High-Temperature Power Rectifiers of $GaAs_{1-x}P_x$

Volume 8 Transport and Optical Phenomena

Richard J. Stirn, Band Structure and Galvanomagnetic Effects in III–V Compounds with Indirect Band Gaps
Roland W. Ure, Jr., Thermoelectric Effects in III–V Compounds
Herbert Piller, Faraday Rotation
H. Barry Bebb and E. W. Williams, Photoluminescence I: Theory
E. W. Williams and H. Barry Bebb, Photoluminescence II: Gallium Arsenide

Volume 9 Modulation Techniques

B. O. Seraphin, Electroreflectance
R. L. Aggarwal, Modulated Interband Magnetooptics
Daniel F. Blossey and Paul Handler, Electroabsorption
Bruno Batz, Thermal and Wavelength Modulation Spectroscopy
Ivar Balslev, Piezooptical Effects
D. E. Aspnes and N. Bottka, Electric-Field Effects on the Dielectric Function of Semiconductors and Insulators

Volume 10 Transport Phenomena

R. L. Rode, Low-Field Electron Transport
J. D. Wiley, Mobility of Holes in III–V Compounds
C. M. Wolfe and G. E. Stillman, Apparent Mobility Enhancement in Inhomogeneous Crystals
Robert L. Peterson, The Magnetophonon Effect

Volume 11 Solar Cells

Harold J. Hovel, Introduction; Carrier Collection, Spectral Response, and Photocurrent; Solar Cell Electrical Characteristics; Efficiency; Thickness; Other Solar Cell Devices; Radiation Effects; Temperature and Intensity; Solar Cell Technology

Volume 12 Infrared Detectors (II)

W. L. Eiseman, J. D. Merriam, and R. F. Potter, Operational Characteristics of Infrared Photodetectors
Peter R. Bratt, Impurity Germanium and Silicon Infrared Detectors
E. H. Putley, InSb Submillimeter Photoconductive Detectors
G. E. Stillman, C. M. Wolfe, and J. O. Dimmock, Far-Infrared Photoconductivity in High Purity GaAs
G. E. Stillman and C. M. Wolfe, Avalanche Photodiodes
P. L. Richards, The Josephson Junction as a Detector of Microwave and Far-Infrared Radiation
E. H. Putley, The Pyroelectric Detector—An Update

Volume 13 Cadmium Telluride

Kenneth Zanio, Materials Preparation; Physics; Defects; Applications

Volume 14 Lasers, Junctions, Transport

N. Holonyak, Jr. and M. H. Lee, Photopumped III–V Semiconductor Lasers
Henry Kressel and Jerome K. Butler, Heterojunction Laser Diodes
A. Van der Ziel, Space-Charge-Limited Solid-State Diodes
Peter J. Price, Monte Carlo Calculation of Electron Transport in Solids

Volume 15 Contacts, Junctions, Emitters

B. L. Sharma, Ohmic Contacts to III–V Compound Semiconductors
Allen Nussbaum, The Theory of Semiconducting Junctions
John S. Escher, NEA Semiconductor Photoemitters

Volume 16 Defects, (HgCd)Se, (HgCd)Te

Henry Kressel, The Effect of Crystal Defects on Optoelectronic Devices
C. R. Whitsett, J. G. Broerman, and C. J. Summers, Crystal Growth and Properties of $Hg_{1-x}Cd_xSe$ Alloys
M. H. Weiler, Magnetooptical Properties of $Hg_{1-x}Cd_xTe$ Alloys
Paul W. Kruse and John G. Ready, Nonlinear Optical Effects in $Hg_{1-x}Cd_xTe$

Volume 17 CW Processing of Silicon and Other Semiconductors

James F. Gibbons, Beam Processing of Silicon
Arto Lietoila, Richard B. Gold, James F. Gibbons, and Lee A. Christel, Temperature Distributions and Solid Phase Reaction Rates Produced by Scanning CW Beams
Arto Lietoila and James F. Gibbons, Applications of CW Beam Processing to Ion Implanted Crystalline Silicon
N. M. Johnson, Electronic Defects in CW Transient Thermal Processed Silicon
K. F. Lee, T. J. Stultz, and James F. Gibbons, Beam Recrystallized Polycrystalline Silicon: Properties, Applications, and Techniques
T. Shibata, A. Wakita, T. W. Sigmon, and James F. Gibbons, Metal–Silicon Reactions and Silicide
Yves I. Nissim and James F. Gibbons, CW Beam Processing of Gallium Arsenide

Volume 18 Mercury Cadmium Telluride

Paul W. Kruse, The Emergence of $(Hg_{1-x}Cd_x)Te$ as a Modern Infrared Sensitive Material
H. E. Hirsch, S. C. Liang, and A. G. White, Preparation of High-Purity Cadmium, Mercury, and Tellurium
W. F. H. Micklethwaite, The Crystal Growth of Cadmium Mercury Telluride
Paul E. Petersen, Auger Recombination in Mercury Cadmium Telluride
R. M. Broudy and V. J. Mazurczyck, (HgCd)Te Photoconductive Detectors
M. B. Reine, A. K. Sood, and T. J. Tredwell, Photovoltaic Infrared Detectors
M. A. Kinch, Metal-Insulator-Semiconductor Infrared Detectors

Volume 19 Deep Levels, GaAs, Alloys, Photochemistry

G. F. Neumark and K. Kosai, Deep Levels in Wide Band-Gap III–V Semiconductors
David C. Look, The Electrical and Photoelectronic Properties of Semi-Insulating GaAs
R. F. Brebrick, Ching-Hua Su, and Pok-Kai Liao, Associated Solution Model for Ga–In–Sb and Hg–Cd–Te
Yu. Ya. Gurevich and Yu. V. Pleskov, Photoelectrochemistry of Semiconductors

Volume 20 Semi-Insulating GaAs

R. N. Thomas, H. M. Hobgood, G. W. Eldridge, D. L. Barrett, T. T. Braggins, L. B. Ta, and S. K. Wang, High-Purity LEC Growth and Direct Implantation of GaAs for Monolithic Microwave Circuits
C. A. Stolte, Ion Implantation and Materials for GaAs Integrated Circuits
C. G. Kirkpatrick, R. T. Chen, D. E. Holmes, P. M. Asbeck, K. R. Elliott, R. D. Fairman, and J. R. Oliver, LEC GaAs for Integrated Circuit Applications
J. S. Blakemore and S. Rahimi, Models for Mid-Gap Centers in Gallium Arsenide

Volume 21 Hydrogenated Amorphous Silicon
Part A

Jacques I. Pankove, Introduction
Masataka Hirose, Glow Discharge; Chemical Vapor Deposition
Yoshiyuki Uchida, dc Glow Discharge
T. D. Moustakas, Sputtering
Isao Yamada, Ionized-Cluster Beam Deposition
Bruce A. Scott, Homogeneous Chemical Vapor Deposition
Frank J. Kampas, Chemical Reactions in Plasma Deposition
Paul A. Longeway, Plasma Kinetics
Herbert A. Weakliem, Diagnostics of Silane Glow Discharges Using Probes and Mass Spectroscopy
Lester Guttman, Relation between the Atomic and the Electronic Structures
A. Chenevas-Paule, Experimental Determination of Structure
S. Minomura, Pressure Effects on the Local Atomic Structure
David Adler, Defects and Density of Localized States

Part B

Jacques I. Pankove, Introduction
G. D. Cody, The Optical Absorption Edge of a-Si:H
Nabil M. Amer and Warren B. Jackson, Optical Properties of Defect States in a-Si:H
P. J. Zanzucchi, The Vibrational Spectra of a-Si:H
Yoshihiro Hamakawa, Electroreflectance and Electroabsorption
Jeffrey S. Lannin, Raman Scattering of Amorphous Si, Ge, and Their Alloys
R. A. Street, Luminescence in a-Si:H
Richard S. Crandall, Photoconductivity
J. Tauc, Time-Resolved Spectroscopy of Electronic Relaxation Processes
P. E. Vanier, IR-Induced Quenching and Enhancement of Photoconductivity and Photoluminescence
H. Schade, Irradiation-Induced Metastable Effects
L. Ley, Photoelectron Emission Studies

Part C

Jacques I. Pankove, Introduction
J. David Cohen, Density of States from Junction Measurements in Hydrogenated Amorphous Silicon
P. C. Taylor, Magnetic Resonance Measurements in a-Si:H
K. Morigaki, Optically Detected Magnetic Resonance
J. Dresner, Carrier Mobility in a-Si:H
T. Tiedje, Information about Band-Tail States from Time-of-Flight Experiments
Arnold R. Moore, Diffusion Length in Undoped a-Si:H
W. Beyer and H. Overhof, Doping Effects in a-Si:H
H. Fritzsche, Electronic Properties of Surfaces in a-Si:H
C. R. Wronski, The Staebler–Wronski Effect
R. J. Nemanich, Schottky Barrier on a-Si:H
B. Abeles and T. Tiedje, Amorphous Semiconductor Superlattices

Part D

Jacques I. Pankove, Introduction
D. E. Carlson, Solar Cells
G. A. Swartz, Closed-Form Solution of I–V Characteristic for a-Si:H Solar Cells
Isamu Shimizu, Electrophotography
Sachio Ishioka, Image Pickup Tubes
P. G. LeComber and W. E. Spear, The Development of the a-Si:H Field-Effect Transistor and Its Possible Applications
D. G. Ast, a-Si:H FET-Addressed LCD Panel
S. Kaneko, Solid-State Image Sensor
Masakiyo Matsumura, Charge-Coupled Devices
M. A. Bosch, Optical Recording
A. D'Amico and G. Fortunato, Ambient Sensors
Hiroshi Kukimoto, Amorphous Light-Emitting Devices
Robert J. Phelan, Jr., Fast Detectors and Modulators
Jacques I. Pankove, Hybrid Structures
P. G. LeComber, A. E. Owen, W. E. Spear, J. Hajto, and W. K. Choi, Electronic Switching in Amorphous Silicon Junction Devices

Volume 22 Lightwave Communication Technology
Part A

Kazuo Nakajima, The Liquid-Phase Epitaxial Growth of InGaAsP
W. T. Tsang, Molecular Beam Epitaxy for III–V Compound Semiconductors
G. B. Stringfellow, Organometallic Vapor-Phase Epitaxial Growth of III–V Semiconductors
G. Beuchet, Halide and Chloride Transport Vapor-Phase Deposition of InGaAsP and GaAs
Manijeh Razeghi, Low-Pressure Metallo-Organic Chemical Vapor Deposition of $Ga_xIn_{1-x}As_yP_{1-y}$ Alloys
P. M. Petroff, Defects in III–V Compound Semiconductors

Part B

J. P. van der Ziel, Mode Locking of Semiconductor Lasers
Kam Y. Lau and Amnon Yariv, High-Frequency Current Modulation of Semiconductor Injection Lasers
Charles H. Henry, Spectral Properties of Semiconductor Lasers
Yasuharu Suematsu, Katsumi Kishino, Shigehisa Arai, and Fumio Koyama, Dynamic Single-Mode Semiconductor Lasers with a Distributed Reflector
W. T. Tsang, The Cleaved-Coupled-Cavity (C^3) Laser

Part C

R. J. Nelson and N. K. Dutta, Review of InGaAsP/InP Laser-Structures and Comparison of Their Performance
N. Chinone and M. Nakamura, Mode-Stabilized Semiconductor Lasers for 0.7–0.8- and 1.1–1.6-μm Regions
Yoshiji Horikoshi, Semiconductor Lasers with Wavelengths Exceeding 2 μm
B. A. Dean and M. Dixon, The functional Reliability of Semiconductor Lasers as Optical Transmitters
R. H. Saul, T. P. Lee, and C. A. Burrus, Light-Emitting Device Design
C. L. Zipfel, Light-Emitting Diode Reliability
Tien Pei Lee and Tingye Li, LED-Based Multimode Lightwave Systems
Kinichiro Ogawa, Semiconductor Noise-Mode Partition Noise

Part D

Federico Capasso, The Physics of Avalanche Photodiodes
T. P. Pearsall and M. A. Pollack, Compound Semiconductor Photodiodes
Takao Kaneda, Silicon and Germanium Avalanche Photodiodes
S. R. Forrest, Sensitivity of Avalanche Photodetector Receivers for High-Bit-Rate Long-Wavelength Optical Communication Systems
J. C. Campbell, Phototransistors for Lightwave Communications

Part E

Shyh Wang, Principles and Characteristics of Integratable Active and Passive Optical Devices
Shlomo Margalit and Amnon Yariv, Integrated Electronic and Photonic Devices
Takaaki Mukai, Yoshihisa Yamamoto, and Tatsuya Kimura, Optical Amplification by Semiconductor Lasers

Volume 23 Pulsed Laser Processing of Semiconductors

R. F. Wood, C. W. White, and R. T. Young, Laser Processing of Semiconductors: An Overview
C. W. White, Segregation, Solute Trapping, and Supersaturated Alloys
G. E. Jellison, Jr., Optical and Electrical Properties of Pulsed Laser-Annealed Silicon
R. F. Wood and G. E. Jellison, Jr., Melting Model of Pulsed Laser Processing
R. F. Wood and F. W. Young, Jr., Nonequilibrium Solidification Following Pulsed Laser Melting
D. H. Lowndes and G. E. Jellison, Jr., Time-Resolved Measurements During Pulsed Laser Irradiation of Silicon
D. M. Zehner, Surface Studies of Pulsed Laser Irradiated Semiconductors
D. H. Lowndes, Pulsed Beam Processing of Gallium Arsenide
R. B. James, Pulsed CO_2 Laser Annealing of Semiconductors
R. T. Young and R. F. Wood, Applications of Pulsed Laser Processing

Volume 24 Applications of Multiquantum Wells, Selective Doping, and Superlattices

C. Weisbuch, Fundamental Properties of III–V Semiconductor Two-Dimensional Quantized Structures: The Basis for Optical and Electronic Device Applications
H. Morkoç and H. Unlu, Factors Affecting the Performance of (Al, Ga)As/GaAs and (Al, Ga)As/InGaAs Modulation-Doped Field-Effect Transistors: Microwave and Digital Applications
N. T. Linh, Two-Dimensional Electron Gas FETs: Microwave Applications
M. Abe et al., Ultra-High-Speed HEMT Integrated Circuits
D. S. Chemla, D. A. B. Miller, and P. W. Smith, Nonlinear Optical Properties of Multiple Quantum Well Structures for Optical Signal Processing
F. Capasso, Graded-Gap and Superlattice Devices by Band-gap Engineering
W. T. Tsang, Quantum Confinement Heterostructure Semiconductor Lasers
G. C. Osbourn et al., Principles and Applications of Semiconductor Strained-Layer Superlattices

Volume 25 Diluted Magnetic Semiconductors

W. Giriat and J. K. Furdyna, Crystal Structure, Composition, and Materials Preparation of Diluted Magnetic Semiconductors
W. M. Becker, Band Structure and Optical Properties of Wide-Gap $A^{II}_{1-x}Mn_xB^{VI}$ Alloys at Zero Magnetic Field
Saul Oseroff and Pieter H. Keesom, Magnetic Properties: Macroscopic Studies
T. Giebultowicz and T. M. Holden, Neutron Scattering Studies of the Magnetic Structure and Dynamics of Diluted Magnetic Semiconductors
J. Kossut, Band Structure and Quantum Transport Phenomena in Narrow-Gap Diluted Magnetic Semiconductors
C. Riqaux, Magnetooptics in Narrow Gap Diluted Magnetic Semiconductors
J. A. Gaj, Magnetooptical Properties of Large-Gap Diluted Magnetic Semiconductors
J. Mycielski, Shallow Acceptors in Diluted Magnetic Semiconductors: Splitting, Boil-off, Giant Negative Magnetoresistance
A. K. Ramdas and S. Rodriquez, Raman Scattering in Diluted Magnetic Semiconductors
P. A. Wolff, Theory of Bound Magnetic Polarons in Semimagnetic Semiconductors

Volume 26 III–V Compound Semiconductors and Semiconductor Properties of Superionic Materials

Zou Yuanxi, III–V Compounds
H. V. Winston, A. T. Hunter, H. Kimura, and R. E. Lee, InAs-Alloyed GaAs Substrates for Direct Implantation
P. K. Bhattacharya and S. Dhar, Deep Levels in III–V Compound Semiconductors Grown by MBE
Yu. Ya. Gurevich and A. K. Ivanov-Shits, Semiconductor Properties of Superionic Materials

Volume 27 Highly Conducting Quasi-One-Dimensional Organic Crystals

E. M. Conwell, Introduction to Highly Conducting Quasi-One-Dimensional Organic Crystals
I. A. Howard, A Reference Guide to the Conducting Quasi-One-Dimensional Organic Molecular Crystals
J. P. Pouget, Structural Instabilities
E. M. Conwell, Transport Properties
C. S. Jacobsen, Optical Properties
J. C. Scott, Magnetic Properties
L. Zuppiroli, Irradiation Effects: Perfect Crystals and Real Crystals

Volume 28   Measurement of High-Speed Signals in Solid State Devices

J. Frey and D. Ioannou, Materials and Devices for High-Speed and Optoelectronic Applications
H. Schumacher and E. Strid, Electronic Wafer Probing Techniques
D. Auston, Picosecond Photoconductivity
J. Valdmanis, Electrooptic Measurement Techniques
R. Jain and J. Wiesenfeld, Direct Optical Probing of Integrated Circuits and High-Speed Devices
G. Plows, Electron Beam Probing
A. M. Weiner and R. B. Marcus, Photoemissive Probing

Volume 29   Very High Speed Integrated Circuits: Gallium Arsenide LSI

M. Kuzuhara, T. Nozaki, and H. Hashimoto, Active Layer Formation by Ion Implantation
H. Hashimoto, Focused Ion Beam Implantation
T. Nozaki and A. Higashisaka, Device Fabrication Process Technology
M. Ino and T. Takada, GaAs LSI Circuit Design
M. Hirayama, M. Ohmori, and K. Yamasaki, GaAs LSI Fabrication and Performance

Volume 30   Very High Speed Integrated Circuits: Heterostructure

H. Watanabe, T. Mizutani, and A. Usui, Fundamentals of Epitaxial Growth and Atomic Layer Epitaxy
S. Hiyamizu, Molecular Beam Epitaxy for High Quality Active Layers
T. Nakanisi, Metal Organic Vapor Phase Epitaxy for High Quality Active Layers
T. Mimura, High Electron Mobility Transistor and LSI Applications
T. Sugeta and T. Ishibashi, Hetero-Bipolar Transistor and LSI Applications
H. Matsueda, T. Tanaka, and M. Nakamura, Opto-Electronic Integrated Circuits

Volume 31   Indium Phosphide: Crystal Growth and Characterization

J. P. Farges, Growth of Dislocation-Free InP
M. J. McCollum and G. E. Stillman, High Purity InP Grown by Hybride Vapor Phase Epitaxy
T. Inada and T. Fukuda, Direct Synthesis and Growth of Indium Phosphide by the Liquid Phosphorous Encapsulated Czochralski Method
O. Oda, K. Katagiri, K. Shinohara, S. Katsura, U. Takahashi, K. Kainosho, K. Kohiro, and R. Hirano, InP Crystal Growth, Substrate Preparation and Evaluation
K. Tada, M. Tatsumi, M. Morioka, T. Araki, and T. Kawase, InP Substrates: Production and Quality Control
M. Razeghi, LP-MOCVD Growth, Characterization, and Application of InP Material
T. A. Kennedy and P. J. Lin-Chung, Stoichiometric Defects in InP